NANOTECHNOLOGY

Smart Engineering Systems: Design and Applications
Series Editor: Suman Lata Tripathi

Aimed at senior undergraduate students, graduate students, academic researchers and professionals, this series will focus on the design of smart engineering systems and their diverse applications. The series will cover important topics, including organic electronics and applications, smart engineering materials, design and development of VLSI circuits with artificial intelligence techniques, smart and intelligent solutions for energy technologies, and intelligent communications systems and sensor networks.

Internet of Things: Robotic and Drone Technology
Nitin Goyal, Sharad Sharma, Arun Kumar Rana and Suman Lata Tripathi

Nanotechnology: Device Design and Applications
Shilpi Birla, Neha Singh and Neeraj Kumar Shukla

For more information about this series, please visit: https://www.routledge.com/Smart-Engineering-Systems-Design-and-Applications/book-series/CRCSESDA

NANOTECHNOLOGY

Device Design and Applications

Edited by
Shilpi Birla, Neha Singh, and
Neeraj Kumar Shukla

CRC Press
Taylor & Francis Group
Boca Raton London New York

CRC Press is an imprint of the
Taylor & Francis Group, an **informa** business

First edition published 2022
by CRC Press
6000 Broken Sound Parkway NW, Suite 300, Boca Raton, FL 33487-2742

and by CRC Press
2 Park Square, Milton Park, Abingdon, Oxon, OX14 4RN

First edition published by CRC Press 2022

CRC Press is an imprint of Taylor & Francis Group, LLC

Library of Congress Cataloging-in-Publication Data
A catalog record for this title has been requested

ISBN: 978-1032-11523-8 (hbk)
ISBN: 978-1032-11538-2 (pbk)
ISBN: 978-1003-22035-0 (ebk)

DOI: 10.1201/9781003220350

Typeset in Times
by MPS Limited, Dehradun

Contents

Preface

Nanotechnology has majorly contributed to advancements in computing and electronics, for faster, smaller and more compact and handy systems that can handle and store huge amounts of information in the form of data. These advancing applications include: nano transistors, MRAMs, ultra-high-definition televisions and displays, flexible electronics, IoT and flash memories. The other sector that gets the maximum benefit from nanotechnology is medical and the healthcare applications. Nanotechnology has helped in broadening the medical area, knowledge and cures. Nanomedicine, one of the emerging applications of nanotechnology, has greatly helped the biomedical field by giving precise solutions for diagnosis of disease and treatment. Nanomaterials are nowadays used in pharmaceutical sciences and technology. A few other emerging areas of nanotechnology are drug delivery and biosensors. The next field that excels with the help of nanotechnology is the energy sector. Nanotechnology has given alternative energy approaches to replace the traditional energy sources to meet daily increasing energy demands. They have developed thin-film solar panels, windmill blades and quick-charging batteries as an alternative. Nanotechnology now also helps clean drinking water through rapid, low-cost detection and treatment of water impurities.

Nanotechnology is one of the emerging areas, having a promising future with many breakthroughs that have changed and further bring various new technological advances to a wide range of applications. The purpose of this book is to introduce nanotechnology on a level that explores the various applications of nanotechnology that are discussed below. It will give a flavor to the people who are in research or interested exploring future possibilities. It will also create interest in different areas where nanotechnology has created new possibilities. This books aims to explore the area of nanotechnology and its various emerging applications.

Chapter Organization

The book is organized into 18 chapters as follows:

Chapter 1 provides a quick overview of evolving memory technology with a focus on materials used in RRAM, the method for resistance swapping, RRAM output parameters, the methods used to enhance the performance of HfO2-based RRAM devices, problems associated and potential possibilities of RRAM devices.

Chapter 2 presents the latest developments in the field of nanotechnology with a focus on energy and the environment, which includes materials based on nanotechnology, their processes and applications, the new generation of highly efficient solar cells and devices for energy storage and energy saving.

Chapter 3 evaluates the efficiency, advantages, disadvantages and applications of nanocarriers in addressing micronutrient and macronutrient deficiency.

Chapter 4 focuses on understanding the process that takes place while applying nanoparticles or materials as catalysts, with some insight on potential health and environmental risks that are associated with the use of nanomaterials.

Chapter 5 discusses the impact of nanoelectronics in the semiconductor field, focusing on experience, present development and future opportunities in a clear manner.

Chapter 6 on nanoelectronics is a concise effort to outline, structure and explain various topics in the related field in a nutshell.

Chapter 7 presents gradual development of modern-day FET devices, from micron technology to nanotechnology; from MOSFET devices to FinFET and TFET devices; and finally the FET structure with 2D materials are analyzed with their advantages and challenges, as reported by several researchers.

Chapter 8 contains the design of 6T SRAM cell at 32 nm by using CMOS, FinFET and CNTFET technologies.

Chapter 9 examines nanotechnology's current developments and presents selected recent research to the most demanding challenges and exciting prospects in the foodstuff industry and farming engineering.

Chapter 10 presents the latest advancement of ZnO material for thin-film heterojunction-based structures and their recent applications in UV detectors, gas sensors, memory devices and other applications.

Chapter 11 studies nonplanar devices, like FinFET for reduced short-channel effects due to suppressed leakage current, on conventional as well as SOI wafers.

Chapter 12 sheds light on the methods used by researchers all over the world to improve the performance of wide fan-in OR logic domino circuits in the field of submicron VLSI design to overcome leakage current by modifying the evaluation network.

Chapter 13 discusses the use of nanowires of molybdates with K+ group (K2Mo3O10.4H2O) grown using chemical method possess potential as exceptionally sensitive gas-sensing materials at room temperature.

Chapter 14 briefly discusses the physical and chemical properties, synthesis and biomedical application of graphene nanomaterials.

Chapter 15 studies C-dots, which include fullerene, nanotubes, nano diamonds, carbon nanofibers, grapheme and other allotropes, for biomedical applications due to their unique and superior properties like prominent biocompatibility, high water solubility, optimal performance for energy conservation, good photo-stability, photoluminescence characteristics, ease of industrial scale up and low production cost.

Chapter 16 sheds light on use of nanotechnology for powerful innovative modifications in healthcare and biomedical applications, like diagnostic analysis and therapeutic approaches.

Chapter 17 is focused on summarizing basic trends of hybrid organolead halide perovskite solar cells and the perspectives of these materials for future solar energy harvesting applications. Various device approaches of the hybrid perovskite solar cells with different materials are concisely addressed.

Chapter 18 provides a comprehensive study and a comparison of various types of energy storage systems available to increase wind energy potential and to enhance power smoothing in a wind energy generation system.

Acknowledgements

Editors would like to thank all the authors and the reviewers without which this book would not be possible. The editors are also thankful to their respective organizations for their continuous support and encouragement.

Acknowledgements

Editors would like to thank all the authors and the reviewers without which this book would not be possible. The editors are also thankful to their respective organizations for their continuous support and encouragement.

Contributors

Sajal Agarwal
Department of Electronics Engineering
Rajeev Gandhi Institute of Petroleum Technology
Jias, India

Md. Sabir Alam
NIMS Institute of Pharmacy
NIMS University
Jaipur, Rajasthan, India

Jamilur R Ansari
Faculty of Physical Sciences
PDM University
Bahadurgarh, Haryana, India

Deepak Bhatia
Department of Electronics and Communication Engineering
Rajasthan Technical University
Kota, India

Shilpi Birla
Department of Electronics & Communication Engineering
Manipal University Jaipur
Jaipur, Rajasthan, India

Spandanagowda N. D.
Department of Electrical and Electronics Engineering
Jain (Deemed-to-be-University)
Bengaluru, Karnataka, India

Vikramaditya Dave
Department of Electrical Engineering
College of Technology and Engineering, Udaipur
Rajasthan, India

Gulshan Dhamija
University School of Basic & Applied Sciences
Guru Govind Singh Indraprastha University
Dwarka Sector 16C, New Delhi, India

S. Dwivedi
S.S. Jain Subodh P.G. (Autonomous) College
Jaipur, Rajasthan, India

S. A. Gangal
Department of Electronic Science
Savitribai Phule
Pune University
Pune, Maharashtra, India

Chandra Prakash Gupta
Department of Electronics & Communication Engineering
Manipal University Jaipur
Jaipur, Rajasthan, India

Meenal Gupta
University School of Basic & Applied Sciences
Guru Govind Singh Indraprastha University
Dwarka Sector 16C, New Delhi, India

Abhilasha Jain
Department of Electronics and Communication Engineering
Meerut Institute of Engineering and Technology
Meerut, U P, India

Abhinandan Jain
Department of Electronics and Communication Engineering
Swami Keshvanand Institute of Technology
Management and Gramothan
Jaipur, India

Praveen K. Jain
Department of Electronics and Communication Engineering
Swami Keshvanand Institute of Technology
Management and Gramothan
Jaipur, India

M.S. Sri Janani
PG Department of Microbiology
The American College
Madurai, Tamil Nadu, India

Md. Noushad Javed
Quality Assurance Lab
Department of Pharmaceutics
School of Pharmaceutical Education and Research
Jamia Hamdard University
New Delhi, India

Aditee Joshi
Department of Electronic Science
Savitribai Phule
Pune University
Pune, Maharashtra, India

Ankur Kumar
Department of Electronics and Communication Engineering
Meerut Institute of Engineering and Technology
Meerut, U P, India

Ritesh Kumar
University School of Basic & Applied Sciences
Guru Govind Singh Indraprastha University
Dwarka Sector 16C, New Delhi, India

Sujeet Kumar
Department of Electrical and Electronics Engineering
Jain (Deemed-to-be-University)
Bengaluru, Karnataka, India

Lalit Kumar Lata
Department of Electronics and Communication Engineering
Swami Keshvanand Institute of Technology
Management and Gramothan
Jaipur, India

Neha Mathur
Department of Electronics & Communication Engineering
Manipal University Jaipur
Jaipur, Rajasthan, India

Komal Mehta
Civil Engineering Department
Dr. Kiran and Pallavi Patel Global University
Vadodara, Gujarat, India

R. K. Nagaria
Department of Electronics and Communication Engineering
Motilal Nehru National Institute of Technology Allahabad
Prayagraj, U P, India

G. Boopathi Raja
Velalar College of Engineering and Technology
Erode, Tamil Nadu, India

Manisha Rastogi
Department of Biomedical Engineering
School of Biological Engineering and Sciences
Shobhit Institute of Engineering and Technology (Deemed to be University)
Meerut, U P, India

Deepika Sharma
Department of Electronics & Communication Engineering
Manipal University Jaipur
Jaipur, Rajasthan, India

Shiva Sharma
Department of Biomedical Engineering
School of Biological Engineering and Sciences
Shobhit Institute of Engineering and Technology (Deemed to be University)
Meerut, U P, India

Amit Kumar Singh
Department of Electronics & Communication Engineering
Manipal University Jaipur
Jaipur, Rajasthan, India

Neha Singh
Department of Electronics & Communication Engineering
Manipal University Jaipur
Jaipur, Rajasthan, India

Sugandha Singhal
University School of Basic & Applied Sciences
Guru Govind Singh Indraprastha University
Dwarka Sector 16C, New Delhi, India

Kunal Sinha
Department of Electronics
Asutosh College
University of Calcutta
Kolkata, West Bengal, India

Dimpy Sood
College of Technology and Engineering
Udaipur, Rajasthan, India

Ananthakumar Soosaimanickam
Crystal Growth Centre
Anna University
Chennai, Tamil Nadu, India
Institute of Materials Science
University of Valencia
Spain

Moorthy Babu Sridharan
Crystal Growth Centre
Anna University
Chennai, Tamil Nadu, India

Saravanan Krishna Sundaram
Department of Physics
Anna University
Chennai, Tamil Nadu, India

J. Immanuel Suresh
PG Department of Microbiology
The American College
Madurai, Tamil Nadu, India

Ritesh Tirole
Department of Electrical Engineering
Sir Padampat Singhania University
Udaipur, Rajasthan, India

Vikrant Varshney
Department of Electronics and Communication Engineering
Meerut Institute of Engineering and Technology
Meerut, U P, India

Editors

Dr Shilpi Birla is working as an Associate Professor in the Electronics & Communication Department at Manipal University Jaipur. She has teaching and industrial experience of more than 15 years. She did her Ph.D. in low-power VLSI design. Her research interests are low-power VLSI design, memory circuits, digital VLSI circuits, nanodevices and image processing. She has authored more than 60 research papers in journals of repute and international conferences. She has organized several workshops in HSPICE, TCAD and XILINX, summer internships in diode fabrication and Faculty Development Programs. She has worked as a session chair, conference steering committee member, editorial board member, and reviewer in international/national IEEE Journal and conferences. She has guided many M. Tech Students and guiding Ph.D. students. She is a senior member of IEEE.

Dr Neha Singh is currently working as Assistant Professor in the Department of Electronics & Communication Engineering, School of Electrical, Electronics & Communication Engineering at Manipal University Jaipur, Rajasthan, India. She has more than 17 years of experience in academics. Her areas of research interest include image processing, machine learning, VLSI design and nanodevices. She has several papers and book chapters published in journals and conferences of repute. She has served as reviewer in various international and peer-reviewed journals and conferences. She has also worked as Convener, Session Chair and organizer of various international conferences, summer internships in Diode Fabrication and Faculty Development Programs. She has guided several M. Tech Dissertations and B. Tech projects.

Dr Neeraj Kumar Shukla is currently working as an Associate Professor in the Department of Electrical Engineering at King Khalid University, Abha, Kingdom of Saudi Arabia. He has academic, research, and industry experience of more than 20 years. He did his Ph.D. in Low Power VLSI Design. He has authored more than 110 research publications in journals of repute and international conferences. Apart from this, he has 2 patents and 1 Ph.D. completed under him. He guided 20 M. Tech Dissertations, 50 B. Tech Projects and conducted 50 short-term skills development training programs and delivered 30 expert lectures. He is reviewer in several international conferences and journals. His research areas are low-power VLSI design, digital design and machine learning.

1 An Overview of Current Trends in Hafnium Oxide–Based Resistive Memory Devices

Lalit Kumar Lata, Praveen Kumar Jain, and Abhinandan Jain
Department of Electronics & Communication Engineering, Swami Keshvanand Institute of Technology, Management & Gramothan, Jaipur, India

Deepak Bhatia
Department of Electronics & Communication Engineering, Rajasthan Technical University, Kota, India

1.1 INTRODUCTION

Memory is a core component of any computing device because it stores data and programs. The main memory of the computing device directly communicates with the CPU. It holds the program and data currently required by the processor of the computer. Memory devices used for storage and backup are called auxiliary memory, and they contain large files and data. Magnetic disks and tapes are examples of auxiliary memory. Auxiliary memory stores all the other information, and this information is transferred to the computer's main memory when needed. Main memory is the central stage in a computer system, and it stores data and programs during the operation of a computer. The principal operation used in main memory is a semiconductor integrated circuit. RAM is an example of main memory.

RAM was initially used for random access memory but is now used to designate read/write memory. It may be volatile or nonvolatile. In volatile memory, all the data are destroyed when the power supply is removed, whereas in nonvolatile memory, there is no loss of data, even if the power supply is switched off. RAM has two types: SRAM (static RAM) and DRAM (dynamic RAM). These memory devices have their benefits and drawbacks. For example, the capacity and density of DRAM are high, but this memory is volatile, and power consumption is high because it needs to be refreshed every second. SRAM is fast but volatile, and large memory cells reduce its capacity. Compared to RAM, flash

DOI: 10.1201/9781003220350-1

memory is incredibly well-liked, features a very high capacity, and is nonvolatile; however, it is relatively slow. Since no existing memory meets all criteria, development is needed to continuously search for new technologies. The perfect memory should have high capability, provide a speedy response, have long retention time, use low power consumption, be nonvolatile and have higher scaling than existing technology [1–5].

RRAM has several benefits over other nonvolatile memories, i.e. simple fabrication, wonderful scalability, structural simplicity, high density of integration, rapid switchover and better compatibility with complementary metal-oxide semiconductor (CMOS) technology [6–24]. Tables 1.1 and 1.2 demonstrate how RRAM compares to the other memory devices on the market.

1.1.1 Basic Working Principle of RRAM

Figure 1.1 depicts the basic structure of a resistive memory cell.

A set voltage should be used to move the memory cell of an RRAM device from HRS (high resistance state) to LRS (low resistance state). HRS is represented by Logic 0 (OFF state), and LRS is represented by logic 1 (ON state). The mechanism by which

TABLE 1.1
Comparison of Various Memory Devices [25]

Function	Working Memory		
	RRAM	SRAM	DRAM
Non-Volatility	Yes	No	No
Program Voltage	Low	Low	Low
Reading Time (ns)	20	8	50
Writing/Erasing Time (ns)	0.3–30	1–8	8–50
Multi-Bit Storage	Yes	No	No
Endurance	∞	∞	10^[12]

TABLE 1.2
Comparison of Various Recent Nonvolatile Memory Devices [26]

Function	RRAM	FRAM	PCRAM	MRAM
Nonvolatile	Yes	Yes	Yes	Yes
Write Time (ns)	< 5	20	50	< 100
Read Time (ns)	< 10	20	< 60	< 20
Erase Time (ns)	< 5	20	120	< 100
Endurance	1E12	1E14	1E15	> 3E16
Write Operation Voltage (V)	3	0.6	3	1.8

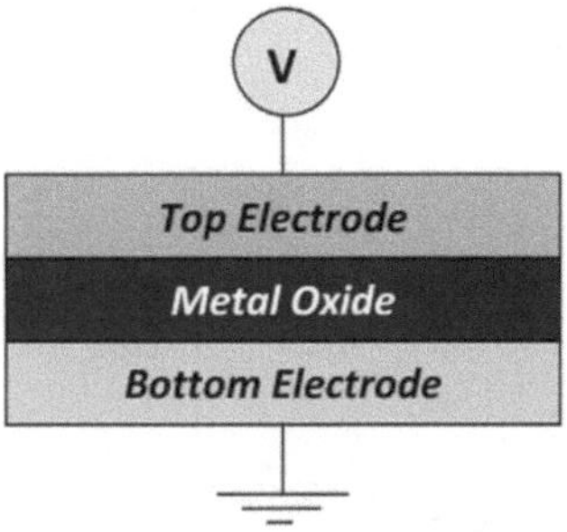

FIGURE 1.1 Schematic of RRAM structure

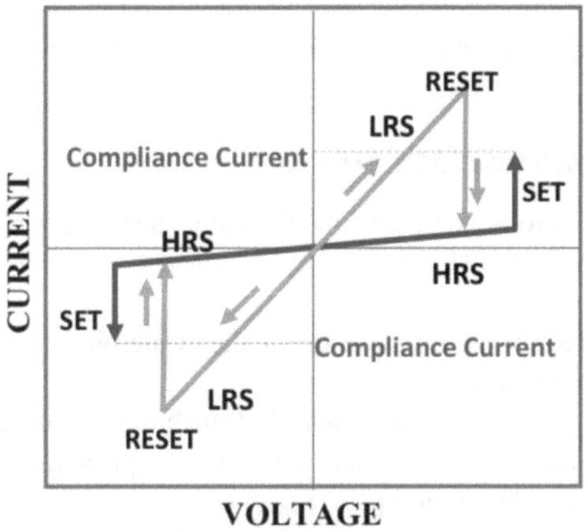

FIGURE 1.2 Graphic representation current-voltage curves for the unipolar mode of operation.

resistive action is known as electroforming. For the forming process of a fresh sample, a voltage greater than the set voltage is necessary. The electrical polarity of SET and RESET is the foundation of two modes of resistive switching action [27].

There are two kinds of operations of resistive switching behavior based on the electrical polarity of SET and RESET process:

1. Unipolar switching
2. Bipolar switching

Figures 1.2 and 1.3 show the current-voltage curve of unipolar and bipolar switching, respectively.

Compliance current is enforced, which is generally offered by the memory cell junction resistor, series resistance or semiconductor parameter analyzer, and is advised to prevent permanent dielectric failure during the forming/set process of RRAM. A minimal voltage that has little impact on the memory cell is used to read the memory cell's information and determine if the memory cell is in LRS or HRS [27,28].

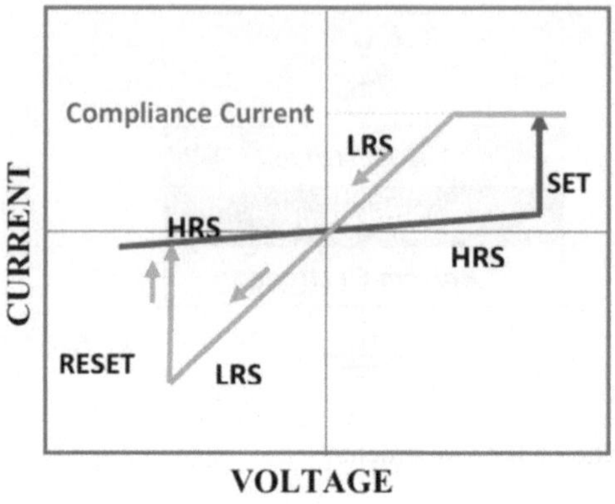

FIGURE 1.3 Graphic representations of current-voltage curves for the bipolar mode of operation.

1.1.2 Resistive Switching Mechanisms

On the basis of the various switching mechanism, there are mainly three types of RRAM devices. These RRAM devices are as follows:

1. Oxide-based RRAM, which works on the principle of conductive filament consisting of oxygen vacancies.
2. CBRRAM, which is based on a metal atom-based conductive filament.
3. Electronic mechanism type, which works on the principle of charge trapping/detrapping within the device [27,28].

1.2 CHARACTERISTIC PARAMETERS OF SWITCHING DEVICES

1.2.1 Operating Voltage and Operating Speed

If a memory device has a high operating voltage, then its power requirement will also be high, so for RRAM devices, programming and erasing voltage should be only a few volts, unlike flash memory. Operating speed represents the shortest time duration for erasing or programming a memory cell [29].

1.2.2 Endurance

The memory devices can be switched from LRS to HRS, and vice versa, many times, but each switching operation introduces degradation in memory devices. Endurance represents a number of SET or RESET cycles, after which HRS and LRS cannot be distinguished. RRAM devices should provide at least the same endurance provided by flash memory [29].

1.2.3 Retention Time

It represents the time duration for which a memory cell stays in one state after erasing or programming. It shows the capability of a memory cell to maintain its content. The retention time for a nonvolatile memory device should be high [29].

1.3 SWITCHING MECHANISM IN A RESISTIVE MEMORY CELL

Figure 1.4 depicts the fundamental switching mechanism in a resistive memory cell. Before switching RRAM devices, a voltage larger than the SET voltage is expected for the forming operation. In the presence of a strong electric field, oxygen ions travel toward the anode interface and are discharged as neutral non-lattice oxygen or react with the oxidizable anode during the device forming phase. As a result, the electrode/oxide interface serves as a reservoir for oxygen. As there is a flow of current through the conductive fields of bulk oxide, the condition is known as LRS and is considered binary "1." In the reset step, oxygen ions return to the bulk oxide and recombine with the oxygen vacancies or oxidize the metal precipitates to restore the memory cell to HRS. HRS is represented by a binary "0" [30–32].

1.4 MATERIALS FOR BINARY METAL OXIDE RRAM

It has been discovered that when an electric field is applied to some insulators, resistance changes. This resistance transition feature has recently been investigated in the development of possible NVM [33]. A variety of oxides exhibit resistance switching phenomena, but binary metal oxides have been examined in depth for possible nonvolatile memory applications. In Table 1.3, several metal oxide–based materials for RRAM are listed.

In traditional electronic devices, electrode materials are important as they serve as the transportation routes for the carriers. The RRAM electrode material has a considerable impact on the device's switching behavior. Electrodes for RRAM devices are made of a broad range of materials. Elementary substance electrodes, silicone-based electrodes, alloy electrodes, oxide–based electrodes and nitride-based electrodes are the five types of electrode materials used in RRAM devices. Table 1.4 lists the most widely used electrode materials. Figure 1.5 shows the summery of materials used for RRAM devices.

1.5 IMPROVEMENT IN DEVICE PARAMETER OF HFO_2-BASED RRAM

Several researchers have suggested various ways of improving the device parameters of HfO_2-based RRAM, including HfO_2 doping/alloying, electrode material changes, insertion of the buffer layer, oxide thickness variations and several more. With the aid of experimental data recorded by them, some common techniques used by numerous researchers around the world are described here.

1.5.1 Effect of Doping/Alloying on HfO_2-Based RRAM

The doping of the oxide layer is a key way to improve the uniformity, speed of processing and switching ratio of the device. Deng Ning et al. [66] showed that both

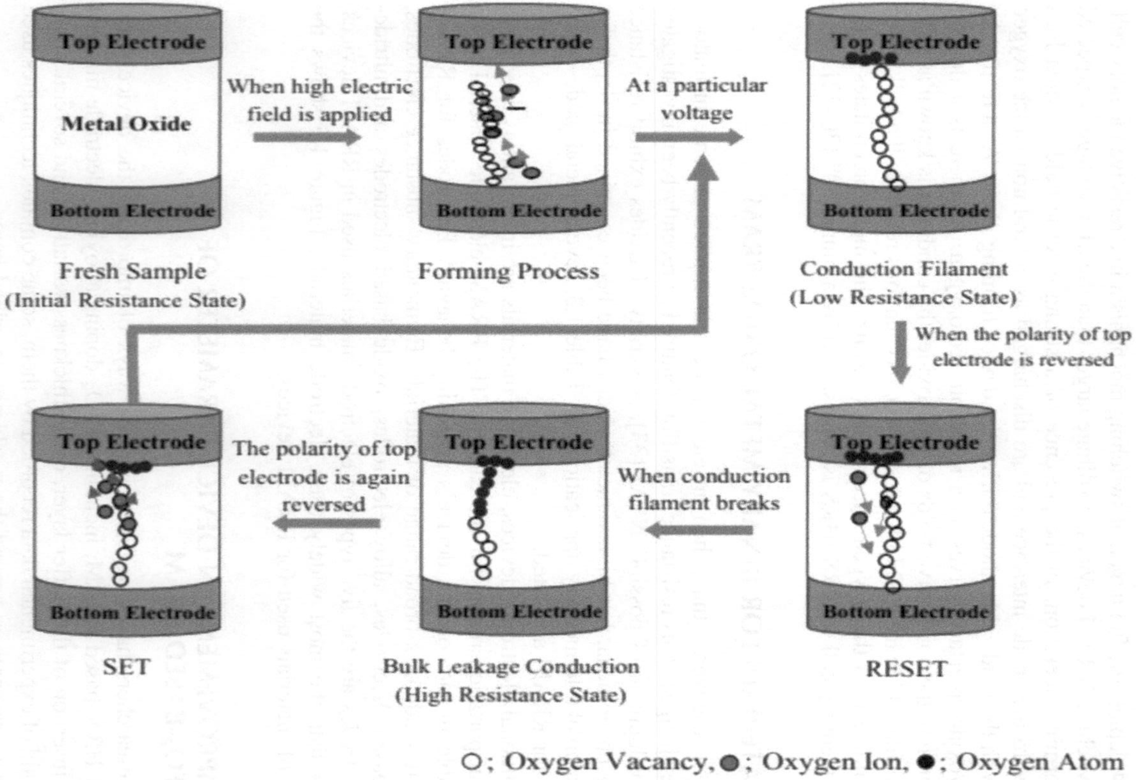

FIGURE 1.4 The representation of the switching process in the RRAM.

TABLE 1.3
Metal Oxide–Based Material for RRAM

S.No.	Metal Oxide Based Material	Reference
1.	Hafnium Oxide (HfOx)	[34,35]
2.	Titanium Oxide (TiOx)	[36,37]
3.	Tantalum Oxide (TaOx)	[38,39]
4.	Nickel Oxide (NiO)	[40,41]
5.	Zinc Oxide (ZnO)	[42,43]
6.	Zinc Titanate (Zn_2TiO_4)	[44]
7.	Manganese Oxide (MnOx)	[45,46]
8.	Magnesium Oxide (MgO)	[47]
9.	Aluminum Oxide (AlOx)	[48,49]
10.	Zirconium Dioxide (ZrO_2)	[15,50]

TABLE 1.4
Electrode Type and Material for RRAM

Electrode type	Electrode Material	Reference
Elementary Substance Electrodes	Aluminium	[48]
	Titanium	[46]
	Copper	[51]
	Graphene	[52]
	Carbon Nanotubes	[53]
	Silver	[42]
	Tungsten	[54]
	Platinum	[55]
Silicon-Based Electrodes	P- and N-type Silicon	[56]
Alloy electrodes	Copper-Titanium	[57]
	Copper-Telluride	[58]
	Platinum–Aluminium	[59].
Nitride-Based Electrodes	TiN	[60]
	TaN	[61]
Oxide-Based Electrodes	Al: ZnO	[62]
	Ga: ZnO	[63]
	ITO	[64]

undoped and doped HfO_2-based RRAM devices work on thermally oxidized Si substrates. 100 μA enforcement current was used.

The forming voltages for aluminum- and copper-doped HfO_2-based devices were decreased by 2.2 V and 1.6 V, respectively, as compared to undoped devices.

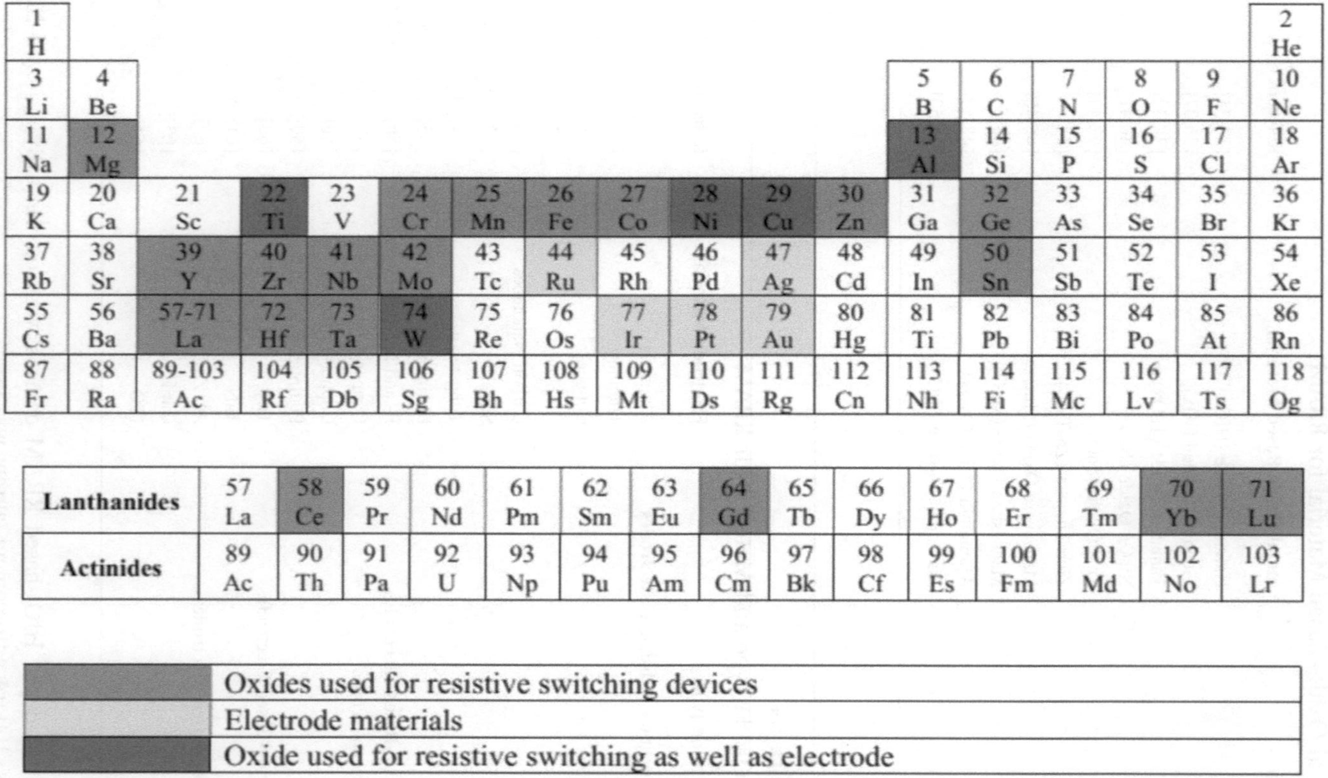

FIGURE 1.5 Summary of the Materials used for RRAM devices [65].

On the other hand, the set voltages of aluminum- and copper-doped HfO_2 devices were also reduced by 2.4 V and 3.5 V, respectively. It was observed that both aluminum and copper doping might decrease the forming and SET voltages. The doping of copper improves the uniformity of the device impressively. The above effects of dopants were observed due to various formulations of filaments at low resistance.

Ching-Shiang Peng [67] found improvement in resistive switching stability in Al doped Pt-HfO_2-Tin RRAM device due to Al-Hf-O bonding. Bipolar switching characteristics were observed for both doped and undoped devices. After doping Al, the resultant device exhibited very good switching performance for the 500 DC sweep endurance cycle and 3×10^4s retention at 0.1 V and 85°C thermal stress. The Al-doped device exhibited less variability in both the set and reset voltage.

Long et al. [68] reported the effect of doping of Mg in HfO_2-based RRAM. The proposed device structure was Ru/Mg:HfO_2/TiN/W. The effect on forming voltage was observed when the doping percentage of Mg was varied from 0% to 20%. All samples in initial resistance, emissions from Frenkel-pooles were determined to be a mechanism for conduction. There was no degradation observed in retention for doped samples, but the ON to OFF ratio was reduced.

T. Tan et al. [69] reported the effect of Au doping. The proposed device was a Cu/HfO_2: Au/Pt stack. The performance was shown to significantly improved, with improved consistency, a higher ON/OFF ratio, and a lower Vset after doping with Au. The Au–oxygen bond may be responsible for enhanced memory performance. The reset voltage (Vreset) ranged from –0.73 to –0.92V without Au doping, but after doping with Au, the Vreset ranged from –0.83 to –0.9V. In Cu/HfO_2: Au/Pt devices, the relative fluctuation of HRS and LRS were observed to be 22% and 4%, respectively, whereas in Cu/HfO_2/Pt devices, these values were observed to be 24% and 22%.

A Gd-doped HfO_2 RRAM device was demonstrated by Zhang et al. [70]. A Hf layer of 20 nm was first deposited using dc sputtering on the bottom electrode. In the O2 atmosphere, furnace annealing was performed around 20 minutes to form HfO_2 at 600°C. For the devices doped, with Gd, a 0.1% concentration of Gd ion was implanted with 80 keV energy into the HfO_2 layer; afterward, there was a five-minute furnace annealing at 700°C in a N_2 environment to activate the dopants. At last, through reactive sputtering, TiN was deposited as a top electrode material. For both undoped and doped devices, bipolar switching behavior was observed, and resistance in LRS and HRS decreased when the temperature increased. The SET voltage was observed between 0.7 V and 2.7 V for the undoped device. The standard deviation and mean value of the SET voltages were observed to be 0.32 V and 1.16V, respectively, whereas for Gd-doped device, these parameters were 0.89 V and 0.03 V, respectively. Similarly, the variations between LRS and HRS were narrowed down, and the uniformity of the reset voltage was also increased.

1.5.2 Role of Electrode Material on Switching Performance

C. Cagli et al. [71] reported the fabrication of RRAM devices using Ti, Pt and TiN as electrode materials. The presence of Ti stimulates the creation of an interface layer, a region rich in O vacancies, facilitating switching. It enables bipolar switching, lowers the formation and set voltages, and increases retention in comparison to Pt-Pt devices. The data undeniably show the significant level of interest in HfO-based dielectrics and Ti-electrodes as potential RRAM memory components.

Boubacar Traoré et al. [72] reported that when RRAM devices with inert electrodes (e.g. Pt) are compared to those with O reactive electrodes (e.g. Ti), the concentration of oxygen interstitial (Oi) ions in the oxide following CF formation is higher, resulting in device output deterioration. With Ti electrodes, the CF thermal stability and device variability are improved because the Oi concentration in the HfO_2 layer is decreased.

1.5.3 Impact of Buffer Layer on HfO_2-Based RRAM

A TiN/Ti/HfOx/TiN resistive memory device was proposed by SK.Ziaur Rahaman [73]. The mean forming voltages for RRAM devices with Ti buffer layers having five different thicknesses (3 nm, 5 nm, 7 nm, 8.5 nm, and 10 nm) were observed to be 2.6 V, 2.2 V, 2.0 V, 1.9 V and 1.8 V, respectively, and due to the Ti buffer layer's high ability for oxygen absorption, the VF (or IL) decreased (or improved) as Ti thickness increased. It was observed that putting a thick Ti layer on top of HfOx results in a higher IL, meaning that increasing the Ti layer thickness decreases the HfOx layer's dielectric strength. It was also observed that a memory device with a Ti layer thickness greater than 7 nm on the HfOx layer was unsuitable for low-power processing. When the Ti/HfOx layer thickness ratio is 1 in the Pt/Ti/HfOx/Pt structure, the resistive switching uniformity may be enhanced.

R. Nakajima et al. proposed the Ti/Hf/HfO_2/Au ReRAM device [74], and the effect of Hf layer thickness on the resistive switching properties was studied. The thickness of the Hf layer was taken at 2 nm, 6 nm, 10 nm, 12 nm, 14 nm and 20 nm. The best Hf thickness for high SET/RESET voltage reproducibility with low RESET current was discovered to be about 10 nm. Due to the extraordinarily high RESET current, the device failed during the first RESET step when the Hf layer thickness was extremely short (>2 nm). For the thickness of the Hf layer 10 nm and 12 nm, the forming voltage was minimal (4.8V). The value of SET voltage was minimum (3.4 V) for Hf layer thicknesses of 14 nm. The device Resistance ratio was maximum for Hf layer thicknesses of 10 nm. The optimum thickness of Hf for obtaining a high yield of memory operation was discovered to be about 10 nm.

1.5.4 Oxide Thickness

A Ni/HfOx/Pt RRAM was developed by D. Ito et al. [75]. For HfOx layers with thicknesses between 3.3 and 6.5 nm, the reset, set and forming voltages were observed. The thickness of the HfOx film was observed to affect these voltages. As the HfOx film thickness increases, the reset, set and forming voltages also increase. There was an obvious relationship between switching voltages and oxide thickness, indicating that conductive filament forming and breakup occurred over the entire thickness spectrum of the HfOx layer.

A review of some development in HfO_2-based RRAM devices proposed by different researcher across the globe is shown in Table 1.5.

1.6 CHALLENGES AND FUTURE OUTLOOK

In recent years, as advancement in emerging memory technology has greatly grown, several implementations of RRAM devices have demonstrated the potential of low-power and high-speed memory applications. RRAM is the greatest capable memory technology due to its simplicity of construction, compatibility with current CMOS technology, fast switching speed, and capability to downsize to the lowest possible size. RRAM is gaining popularity as a viable alternative, particularly for fast-running, mid-capacity storage applications, since flash memory struggles to shrink in size. The reliability of RRAM is one of the most important considerations that must be thoroughly investigated. A procedure should be identified to validate the detection of device failure. It is also worth noting that no single RRAM device has ever managed to combine fast switching, stable operation and low power consumption. Even though the endurance of RRAM has been achieved up to 10^{12} [84], replacing DRAM is still a challenge. RRAM has a high-enough switching speed to replace the DRAM, and the materials used in its production are nearly equivalent to those used in the DRAM. It turns out to be a challenging task to enhance the endurance properties of RRAM. The biggest problem restricting the production of RRAM to date is an accurate interpretation of the device-switching process, which researchers around the world have been debating for a long time. Due to variations in manufacturing methods, it is assumed that the uneven switching mechanism of several published RRAM devices would require more detailed study to get an improved grasp of the switching mechanism. The above problems must be effectively addressed before deploying RRAM in potential memory applications. The ability of RRAM to deal with variability issues, not just in normal operating environments but even at high temperatures, is the most important obstacle to its use in a wide variety of applications.

1.7 CONCLUSION

This chapter sheds light on current advancements in memory technology, the materials used in RRAM, the method for resistance swapping, and the performance parameters of RRAM. This chapter also covers a brief review of the methods used to boost the performance of RRAM devices based on HfO_2. Though RRAM technology is advanced, further research is needed to address issues such as high operating current, low resistance ratios and reliability concerns. Improved solutions for faster programming/erasing, reduced energy consumption and more storage capacity should be the subject of future study. According to current research and improvements, RRAM devices must be a prominent technology for emerging nonvolatile memory applications.

TABLE 1.5
A review of the Different Structure used in HfO_2-Based RRAM Devices

Structure	Top Electrode	Bottom Electrode	Switching Type	Set Voltage (V)	Reset Voltage (V)	Endurance	Ratio of High and Low Resistance	Reference
HfO_2	Cu	Pt	Bipolar	0.92	−1.51	–	-100	[76]
HfO_2	Ag	ITO	Bipolar	1.9	−1.5	–	–	[77]
HfO_2	Pt	TiN	Bipolar	1.96	−1.25	–	–	[78]
HfO_2	Al	TiN	Bipolar	−0.9	1.4	200	≈5	[79]
HfO_2	Cu	TiN	Unipolar	−0.8	1.2	120	≈2	[79]
HfO_2	Hf	TiN	Bipolar	1.2	−1.3	200	>2	[79]
HfO_2	Pt	TiN	Unipolar	1.4	−1.8	150	≈2	[79]
HfO_2	Ti	TiN	Bipolar	0.9	−1.6	200	≈4	[79]
Al:HfO_2	TiN/Ti	ITO	Bipolar	–	–	10^3	~44	[80]
HfO_2/ZrO_2	TaN	Pt	Bipolar	1.37	−1.07	10^3	10	[81]
HfO_2/TiO_x	Pt	Pt	Bipolar	1.52	− 1.12	–	100	[82]
HfO_x/Ti	TiN	TiN	Bipolar	1.15	− 0.98	–	10	[83]

REFERENCES

1. Kahng, D., & Sze, S. M. "A floating gate and its application to memory devices." *The Bell System Technical Journal*, 46, no. 6 (1967), 1288–1295.
2. Masuoka, F., Asano, M., Iwahashi, H., Komuro, T., & Tanaka, S. "A new flash E2PROM cell using triple polysilicon technology." In *1984 International Electron Devices Meeting* (pp. 464–467). IEEE, 1984, December.
3. Chen, A., Haddad, S., Wu, Y. C., Fang, T. N., Lan, Z., Avanzino, S., … & Taguchi, M. "Nonvolatile resistive switching for advanced memory applications." In *IEEE International Electron Devices Meeting, 2005. IEDM Technical Digest* (pp. 746–749). IEEE, 2005, December.
4. Ying-Tao, L., Shi-Bing, L., Hang-Bing, L., Qi, L., Qin, W., Yan, W., … & Ming, L. "Investigation of resistive switching behaviours in WO3-based RRAM devices." *Chinese Physics B*, 20, no. 1 (2011), 017305.
5. Li, Y., Long, S., Liu, Q., Wang, Q., Zhang, M., Lv, H., & Liu, M. "Nonvolatile multilevel memory effect in Cu/WO_3/Pt device structures." *physica status solidi (RRL)–Rapid Research Letters*, 4, no. 5-6 (2010), 124–126.
6. Zhuang, W. W., Pan, W., Ulrich, B. D., Lee, J. J., Stecker, L., Burmaster, A., & Ignatiev, A. "Novel colossal magnetoresistive thin film nonvolatile resistance random access memory (RRAM)." In *Digest. International Electron Devices Meeting* (pp. 193–196). IEEE, 2002, December.
7. Liu, C. Y., Wu, P. H., Wang, A., Jang, W. Y., Young, J. C., Chiu, K. Y., & Tseng, T. Y. "Bistable resistive switching of a sputter-deposited Cr-doped $SrZrO_3$/memory film." *IEEE Electron Device Letters*, 26, no. 6 (2005), 351–353.
8. Lin, C. C., Tu, B. C., Lin, C. H., & Tseng, T. Y. "Resistive switching mechanisms of V-doped $SrZrO_3$ memory films." *IEEE Electron Device Letters*, 27, no. 9 (2006), 725–727.
9. Fujii, T., Kawasaki, M., Sawa, A., Akoh, H., Kawazoe, Y., & Tokura, Y. "Hysteretic current–voltage characteristics and resistance switching at an epitaxial oxide Schottky junction $SrRuO_3$/ SrTi0.99N0.01O3." *Applied Physics Letters*, 86, no. 1 (2005), 012107.
10. Cho, B. O., Yasue, T., Yoon, H., Lee, M. S., Yeo, I. S., Chung, U. I., … & Ryu, B. I. (2006, December). "Thermally robust multi-layer nonvolatile polymer resistive memory." In *2006 International Electron Devices Meeting* (pp. 1–4). IEEE.
11. Lai, Y. S., Tu, C. H., Kwong, D. L., & Chen, J. S. "Charge-transport characteristics in bistable resistive poly (N-vinylcarbazole) films." *IEEE Electron Device Letters*, 27, no. 6 (2006), 451–453.
12. Seo, S., Lee, M. J., Kim, D. C., Ahn, S. E., Park, B. H., Kim, Y. S., … & Park, B. H. "Electrode dependence of resistance switching in polycrystalline NiO films." *Applied Physics Letters*, 87, no. 26 (2005), 263507.
13. Yang, W. Y., & Rhee, S. W. "Effect of electrode material on the resistance switching of Cu_2O film." *Applied Physics Letters*, 91, no. 23 (2007), 232907.
14. Wu, X., Zhou, P., Li, J., Chen, L. Y., Lv, H. B., Lin, Y. Y., & Tang, T. A. "Reproducible unipolar resistance switching in stoichiometric ZrO_2 films." *Applied Physics Letters*, 90, no. 18 (2007), 183507.
15. Lin, C. Y., Wu, C. Y., Wu, C. Y., Lee, T. C., Yang, F. L., Hu, C., & Tseng, T. Y. "Effect of top electrode material on resistive switching properties of ZrO_2 film memory devices." *IEEE Electron Device Letters*, 28, no. 5 (2007), 366–368.
16. Choi, B. J., Jeong, D. S., Kim, S. K., Rohde, C., Choi, S., Oh, J. H., … & Tiedke, S. "Resistive switching mechanism of TiO_2 thin films grown by atomic-layer deposition." *Journal of Applied Physics*, 98, no. 3 (2005), 033715.

17. Lee, D., Seong, D. J., jung Choi, H., Jo, I., Dong, R., Xiang, W., … & Hwang, H. "Excellent uniformity and reproducible resistance switching characteristics of doped binary metal oxides for nonvolatile resistance memory applications." In *2006 International Electron Devices Meeting* (pp. 1–4). IEEE, 2006, December.
18. Waser, R., & Aono, M. "Nanoionics-based resistive switching memories." *Nanoscience and Technology: A Collection of Reviews from Nature Journals*, 6 (2010), 158–165.
19. Tsunoda, K., Kinoshita, K., Noshiro, H., Yamazaki, Y., Iizuka, T., Ito, Y., … & Sugiyama, Y. (2007, December). "Low power and high speed switching of Ti-doped NiO ReRAM under the unipolar voltage source of less than 3 V." In *2007 IEEE International Electron Devices Meeting* (pp. 767–770). IEEE.
20. Schindler, C., Thermadam, S. C. P., Waser, R., & Kozicki, M. N. "Bipolar and unipolar resistive switching in Cu-Doped SiO_2." *IEEE Transactions on Electron Devices*, 54, no. 10 (2007), 2762–2768.
21. Guan, W., Long, S., Liu, Q., Liu, M., & Wang, W. "Nonpolar nonvolatile resistive switching in Cu doped ZrO_2." *IEEE Electron Device Letters*, 29, no. 5 (2008), 434–437.
22. Lee, H. Y., Chen, P. S., Wu, T. Y., Chen, Y. S., Wang, C. C., Tzeng, P. J., … & Tsai, M. J. "Low power and high speed bipolar switching with a thin reactive Ti buffer layer in robust HfO_2 based RRAM." In *2008 IEEE International Electron Devices Meeting* (pp. 1–4). IEEE, 2008, December.
23. Waser, R., Dittmann, R., Staikov, G., & Szot, K. "Redox-based resistive switching memories–nanoionic mechanisms, prospects, and challenges." *Advanced Materials*, 21, no. 25-26 (2009), 2632–2663.
24. Chien, W. C., Chen, Y. C., Chang, K. P., Lai, E. K., Yao, Y. D., Lin, P., & Lu, C. Y. "Multi-level operation of fully CMOS compatible WOx resistive random access memory (RRAM)." In *2009 IEEE International Memory Workshop* (pp. 1–2). IEEE, 2009, May.
25. Yang, P. K., Ho, C. H., Lien, D. H., Retamal, J. R. D., Kang, C. F., Chen, K. M., & He, J. H. "A fully transparent resistive memory for harsh environments." *Scientific Reports*, 5, no. 1 (2015), 1–9.
26. Kumar, D., Aluguri, R., Chand, U., & Tseng, T. Y. "Metal oxide resistive switching memory: materials, properties and switching mechanisms." *Ceramics International*, 43 (2017), S547–S556.
27. Wong, H. S. P., Lee, H. Y., Yu, S., Chen, Y. S., Wu, Y., Chen, P. S., & Tsai, M. J. "Metal–oxide RRAM." *Proceedings of the IEEE*, 100, no. 6 (2012), 1951–1970.
28. Li, Y., Long, S., Liu, Q., Lü, H., Liu, S., & Liu, M. "An overview of resistive random access memory devices." *Chinese Science Bulletin*, 56, no. 28 (2011), 3072–3078.
29. Guan, W., Long, S., Liu, Q., Liu, M., & Wang, W. "Nonpolar nonvolatile resistive switching in Cu doped ZrO_2." *IEEE Electron Device Letters*, 29, no. 5 (2008), 434–437.
30. Yu, S., & Wong, H. S. P. "A phenomenological model for the reset mechanism of metal oxide RRAM." *IEEE Electron Device Letters*, 31, no. 12 (2010), 1455–1457.
31. Fujimoto, M., Koyama, H., Konagai, M., Hosoi, Y., Ishihara, K., Ohnishi, S., & Awaya, N. "TiO_2 anatase nanolayer on TiN thin film exhibiting high-speed bipolar resistive switching." *Applied Physics Letters*, 89, no. 22 (2006), 223509.
32. Lee, H. D., Magyari-Köpe, B., & Nishi, Y. "Model of metallic filament formation and rupture in NiO for unipolar switching." *Physical Review B*, 81, no. 19 (2010), 193202.
33. Chang, T. C., Chang, K. C., Tsai, T. M., Chu, T. J., & Sze, S. M. "Resistance random access memory." *Materials Today*, 19, no. 5 (2016), 254–264.

34. Su, Y. T., Liu, H. W., Chen, P. H., Chang, T. C., Tsai, T. M., Chu, T. J., … & Sze, S. M. "A method to reduce forming voltage without degrading device performance in hafnium oxide-based 1T1R resistive random access memory." *IEEE Journal of the Electron Devices Society*, 6 (2018), 341–345.
35. La Torre, C., Fleck, K., Starschich, S., Linn, E., Waser, R., & Menzel, S. "Dependence of the SET switching variability on the initial state in HfOx-based ReRAM." *Physica Status Solidi (a)*, 213, no. 2 (2016), 316–319.
36. Huang, Y. J., Shen, T. H., Lee, L. H., Wen, C. Y., & Lee, S. C. "Low-power resistive random access memory by confining the formation of conducting filaments." *AIP Advances*, 6, no. 6 (2016), 065022.
37. Acharyya, D., Hazra, A., & Bhattacharyya, P. "A journey towards reliability improvement of TiO_2 based resistive random access memory: a review." *Microelectronics reliability*, 54, no. 3 (2014), 541–560.
38. Prakash, A., Deleruyelle, D., Song, J., Bocquet, M., & Hwang, H. "Resistance controllability and variability improvement in a TaOx-based resistive memory for multilevel storage application." *Applied Physics Letters*, 106, no. 23 (2015), 233104.
39. Jana, D., Dutta, M., Samanta, S., & Maikap, S. "RRAM characteristics using a new Cr/GdOx/TiN structure." *Nanoscale Research Letters*, 9, no. 1 (2014), 1–9.
40. Ma, G., Tang, X., Zhang, H., Zhong, Z., Li, J., & Su, H. "Effects of stress on resistive switching property of the NiO RRAM device." *Microelectronic Engineering*, 139 (2015), 43–47.
41. Long, S., Cagli, C., Ielmini, D., Liu, M., & Sune, J. "Reset statistics of NiO-based resistive switching memories." *IEEE Electron Device Letters*, 32, no. 11 (2011, 1570–1572.
42. Huang, Y., Shen, Z., Wu, Y., Wang, X., Zhang, S., Shi, X., & Zeng, H. "Amorphous ZnO based resistive random access memory." *RSC Advances*, 6, no. 22 (2016), 17867–17872.
43. Seo, J. W., Baik, S. J., Kang, S. J., & Lim, K. S. "Characteristics of ZnO thin film for the resistive random access memory." *MRS Online Proceedings Library*, 1250, no. 1 (2010), 1–5.
44. Chen, S. X., Chang, S. P., Hsieh, W. K., Chang, S. J., & Lin, C. C. "Highly stable ITO/Zn 2 TiO 4/Pt resistive random access memory and its application in two-bit-per-cell." *RSC Advances*, 8, no. 32 (2018), 17622–17628.
45. Zhang, S., Long, S., Guan, W., Liu, Q., Wang, Q., & Liu, M. "Resistive switching characteristics of MnOx-based ReRAM." *Journal of Physics D: Applied Physics*, 42, no. 5 (2009), 055112.
46. Yang, M. K., Park, J. W., Ko, T. K., & Lee, J. K. "Bipolar resistive switching behavior in Ti/MnO_2/Pt structure for nonvolatile memory devices." *Applied Physics Letters*, 95, no. 4 (2009), 042105.
47. Chiu, F. C., Shih, W. C., & Feng, J. J. "Conduction mechanism of resistive switching films in MgO memory devices." *Journal of Applied Physics*, 111, no. 9 (2012), 094104.
48. Wu, Y., Lee, B., & Wong, H. S. P. "AI2O3-based RRAM using atomic layer deposition (ALD) with 1μA reset current." *IEEE Electron Device Letters*, 31, no. 12 (2010), 1449–1451.
49. Lin, C. Y., Lee, D. Y., Wang, S. Y., Lin, C. C., & Tseng, T. Y. "Effect of thermal treatment on resistive switching characteristics in Pt/Ti/Al_2O_3/Pt devices." *Surface and Coatings Technology*, 203, no. 5-7 (2008), 628–631.
50. Wu, M. C., Jang, W. Y., Lin, C. H., & Tseng, T. Y. "A study on low-power, nanosecond operation and multilevel bipolar resistance switching in Ti/ZrO_2/Pt

nonvolatile memory with 1T1R architecture." *Semiconductor Science and Technology*, 27, no. 6 (2012), 065010.

51. Yang, L., Kuegeler, C., Szot, K., Ruediger, A., & Waser, R. "The influence of copper top electrodes on the resistive switching effect in TiO_2 thin films studied by conductive atomic force microscopy." *Applied Physics Letters*, 95, no. 1 (2009), 013109.
52. Son, J. Y., Shin, Y. H., Kim, H., & Jang, H. M. "NiO resistive random access memory nanocapacitor array on graphene." *ACS Nano*, 4, no. 5 (2010), 2655–2658.
53. Tsai, C. L., Xiong, F., Pop, E., & Shim, M. "Resistive random access memory enabled by carbon nanotube crossbar electrodes." *ACS Nano*, 7, no. 6 (2013), 5360–5366.
54. Prakash, A., Park, J., Song, J., Woo, J., Cha, E. J., & Hwang, H. "Demonstration of low power 3-bit multilevel cell characteristics in a TaOx-based RRAM by stack engineering." *IEEE Electron Device Letters*, 36, no. 1 (2014), 32–34.
55. Chiu, F.C., Li, P.W. and Chang, W.Y. "Reliability characteristics and conduction mechanisms in resistive switching memory devices using ZnO thin films." *Nanoscale Research Letters*, 7, no. 1 (2012), pp.1–9.
56. Tang, G. S., Zeng, F., Chen, C., Liu, H. Y., Gao, S., Li, S. Z., ... & Pan, F. "Resistive switching with self-rectifying behavior in Cu/SiOx/Si structure fabricated by plasma-oxidation." *Journal of Applied Physics*, 113, no. 24 (2013), 244502.
57. Huang, Y. C., Chou, C. H., Liao, C. Y., Tsai, W. L., & Cheng, H. C. "High-performance resistive switching characteristics of programmable metallization cell with oxidized Cu-Ti electrodes." *Applied Physics Letters*, 103, no. 14 (2013), 142905.
58. Goux, L., Opsomer, K., Degraeve, R., Müller, R., Detavernier, C., Wouters, D. J., ... & Kittl, J. A. "Influence of the Cu-Te composition and microstructure on the resistive switching of Cu-Te/Al_2O_3/Si cells." *Applied Physics Letters*, 99, no. 5 (2011), 053502.
59. Wang, J. C., Jian, D. Y., Ye, Y. R., & Chang, L. C. "Platinum–aluminum alloy electrode for retention improvement of gadolinium oxide resistive switching memory." *Applied Physics A*, 113, no. 1 (2013), 37–40.
60. Lee, M. J., Lee, D., Cho, S. H., Hur, J. H., Lee, S. M., Seo, D. H., ... & Yoo, I. K. "A plasma-treated chalcogenide switch device for stackable scalable 3D nanoscale memory." *Nature Communications*, 4, no. 1 (2013), 1–8.
61. Tang, G. S., Zeng, F., Chen, C., Gao, S., Fu, H. D., Song, C., ... & Pan, F. "Resistive switching behaviour of a tantalum oxide nanolayer fabricated by plasma oxidation." *physica Status Solidi (RRL)–Rapid Research Letters*, 7, no. 4 (2013), 282–284.
62. Cao, X., Li, X., Gao, X., Liu, X., Yang, C., Yang, R., & Jin, P. "All-ZnO-based transparent resistance random access memory device fully fabricated at room temperature." *Journal of Physics D: Applied Physics*, 44, no. 25 (2011), 255104.
63. Zheng, K., Sun, X. W., Zhao, J. L., Wang, Y., Yu, H. Y., Demir, H. V., & Teo, K. L. "An indium-free transparent resistive switching random access memory." *IEEE Electron Device Letters*, 32, no. 6 (2011), 797–799.
64. Kim, H. D., An, H. M., Hong, S. M., & Kim, T. G. "Unipolar resistive switching phenomena in fully transparent SiN-based memory cells." *Semiconductor Science and Technology*, 27, no. 12 (2012), 125020.
65. Wu, J., Cao, J., Han, W. Q., Janotti, A., & Kim, H. C. (Eds.). *Functional Metal Oxide Nanostructures* (Vol. 149). Springer Science & Business Media, 2011.
66. Ning, D., Hua, P., & Wei, W. "Effects of different dopants on switching behavior of HfO2-based resistive random access memory." *Chinese Physics B*, 23, no. 10 (2014), 107306.

67. Peng, C. S., Chang, W. Y., Lee, Y. H., Lin, M. H., Chen, F., & Tsai, M. J. "Improvement of resistive switching stability of HfO_2 films with Al doping by atomic layer deposition." *Electrochemical and Solid State Letters*, 15, no. 4 (2012), H88.
68. Long, B. M., Mandal, S., Livecchi, J., & Jha, R. "Effects of Mg-doping on HfO_2 -based ReRAM device switching characteristics." *IEEE electron device letters*, 34, no. 10 (2013), 1247–1249.
69. Tan, T., Guo, T., Chen, X., Li, X., & Liu, Z. "Impacts of Au-doping on the performance of Cu/HfO_2/Pt RRAM devices." *Applied Surface Science*, 317 (2014), 982–985.
70. Zhang, H., Liu, L., Gao, B., Qiu, Y., Liu, X., Lu, J., … & Yu, B. "Gd-doping effect on performance of HfO_2 based resistive switching memory devices using implantation approach." *Applied Physics Letters*, 98, no. 4 (2011), 042105.
71. Cagli, C., Buckley, J., Jousseaume, V., Cabout, T., Salaun, A., Grampeix, H., … & De Salvo, B. "Experimental and theoretical study of electrode effects in HfO_2 based RRAM." In *2011 International Electron Devices Meeting* (pp. 28.7.1–28.7.4). IEEE, 2011, December. doi: 10.1109/IEDM.2011.6131634.
72. Traoré, B., Blaise, P., Vianello, E., Perniola, L., De Salvo, B., & Nishi, Y. "HfO_2-based RRAM: Electrode effects, Ti/HfO_2 interface, charge injection, and oxygen (O) defects diffusion through experiment and ab initio calculations." *IEEE Transactions on Electron Devices*, 63, no. 1 (2015), 360–368.
73. Rahaman, S. Z., Lin, Y. D., Lee, H. Y., Chen, Y. S., Chen, P. S., Chen, W. S., … & Wang, P. H. "The role of Ti buffer layer thickness on the resistive switching properties of hafnium oxide-based resistive switching memories." *Langmuir*, 33, no. 19 (2017), 4654–4665.
74. Nakajima, R., Azuma, A., Yoshida, H., Shimizu, T., Ito, T., & Shingubara, S. "Hf layer thickness dependence of resistive switching characteristics of Ti/Hf/HfO_2/Au resistive random access memory device." *Japanese Journal of Applied Physics*, 57, no. 6S1 (2018), 06HD06.
75. Ito, D., Hamada, Y., Otsuka, S., Shimizu, T., & Shingubara, S. "Oxide thickness dependence of resistive switching characteristics for Ni/HfOx/Pt resistive random access memory device." *Japanese Journal of Applied Physics*, 54, no. 6S1 (2015), 06FH11.
76. Lata, L. K., Jain, P. K., Chand, U., Bhatia, D., & Shariq, M. "Resistive switching characteristics of HfO_2 based bipolar nonvolatile RRAM cell." *Materials Today: Proceedings*, 30 (2020), 217–220.
77. Ramadoss, A., Krishnamoorthy, K., & Kim, S. J. "Resistive switching behaviors of HfO_2 thin films by sol–gel spin coating for nonvolatile memory applications." *Applied Physics Express*, 5, no. 8 (2012), 085803.
78. Sun, C., Lu, S., Jin, F., Mo, W., Song, J., & Dong, K. "Control the switching mode of Pt/HfO_2/TiN RRAM devices by tuning the crystalline state of TiN electrode." *Journal of Alloys and Compounds*, 749 (2018), 481–486. 10.1016/j.jallcom.2018.03.320
79. Bertaud, T., Walczyk, D., Walczyk, C., Kubotsch, S., Sowinska, M., Schroeder, T., … & Grampeix, H. (2012). "Resistive switching of HfO_2-based Metal–Insulator–Metal diodes: Impact of the top electrode material." *Thin Solid Films*, 520, no. 14), 4551–4555.
80. Mahata, C., & Kim, S. "Modified resistive switching performance by increasing Al concentration in HfO_2 on transparent indium tin oxide electrode." *Ceramics International*, 47, no. 1 (2021), 1199–1207.
81. Ismail, M., Batool, Z., Mahmood, K., Rana, A. M., Yang, B. D., & Kim, S. "Resistive switching characteristics and mechanism of bilayer HfO_2/ZrO_2 structure deposited by

radio-frequency sputtering for nonvolatile memory." *Results in Physics*, 18 (2020), 103275.

82. Ding, X., Feng, Y., Huang, P., Liu, L., & Kang, J. "Low-power resistive switching characteristic in HfO_2/TiOx bi-Layer resistive random-access memory." *Nanoscale Research Letters*, 14, no. 1 (2019), 157.
83. Li, H., Li, K. S., Lin, C. H., Hsu, J. L., Chiu, W. C., Chen, M. C., ... & Wong, H. S. P. (2016, June). "Four-layer 3D vertical RRAM integrated with FinFET as a versatile computing unit for brain-inspired cognitive information processing." In *2016 IEEE Symposium on VLSI Technology* (pp. 1–2). IEEE.
84. Wong, H. S. P., Lee, H. Y., Yu, S., Chen, Y. S., Wu, Y., Chen, P. S., ... & Tsai, M. J. "Metal–oxide RRAM." *Proceedings of the IEEE*, 100, no. 6 (2012), 1951–1970.

2 Nanotechnology for Energy and the Environment

Professor (Dr.) Komal Mehta
Professor, Civil Engineering Department, & Deputy Director, Krishna Center of Innovation and Research, Dr. Kiran & Pallavi Patel Global University (KfPGU), Vadodara, Gujarat, India

2.1 INNOVATION AND FUNDAMENTALS

Nanotechnology is a tiny structure Cong, Q., X. Yuan, J. Qu. "A review on the removal of antibiotics by carbon nanotubes." Water SciTechnol. 68 (2013): S1679–S1687 1 nm to 100 nm in size. Drexler, K. Eric (1986). Engines of Creation: The Coming Era of Nanotechnology. Doubleday. ISBN 978-0-385-19973-5. Since 1959, the concept of nano has been known with synthesis via direct manipulation of atoms [4]. It was coined by Norio Taniguchi in 1974. This technology works at nanoscale with engineering principles. High-performance products with projected ability from bottom-up with the latest tools and techniques are the outcome of nanotechnology. One nm is 10^{-9} meter. Spacing inside molecules in between these atoms of range 0.10 to 0.15 and 2 media of DNA gives an idea of its size. As nanotechnology devices work on principles of atoms and not on Newton's Law, a lower limit is set and an upper limit is randomly set below the large structure. Material properties act differently at quantum scale compared to normal scale. With reduction in size, statistical mechanical effects, as well as quantum mechanical effects, are seen. At nanoscale, quantum effects are significant for other properties like electrical, mechanical and chemical changes in comparison to the macroscopic level. With alteration of mechanical, thermal or catalytic properties, area-to-volume ratio increases. A few noticeable differences of nanoscale are: copper can become transparent; aluminum can become combustible; and gold can become soluble. [1-7]

2.1.1 Nano from Molecular Perspective

Nowadays, synthetic chemistry has reached a level where commercial varieties of pharmaceuticals and polymers are prepared, and researchers are looking to assemble supermolecules consisting of an arrangement in a well-defined structure. With the bottom-up approach, self-assembly of molecular structure is taken care of. Compared

DOI: 10.1201/9781003220350-2

to the top-down approach, this can produce more devices at less cost. Useful nanostructures have requirements of complex arrangement, e.g. Watson–Crick base pairing and enzyme-substrate interactions. Figure 2.1 shows nanometer scale. Molecular nanotechnology is nanoscale machinery based on principles of mechanosynthesis. Compared to the difficulty level of assembling devices currently, in the opinion of Carlo Montemagno, the future is a hybrid of silicon technology and biological molecules from a nanotechnology point of view. Because of the difficulty of manipulating individual molecules, it seems impossible if we have to go for (Richard Smalley) mechano synthesis. Ho and Lee in 1999 experimented with moving CO to individual Fe on flat silver crystal to show positional molecular assembly. [33]

2.2 NANOTECHNOLOGY FOR ENERGY

Use of nanotechnology for producing cost-effective and efficient energy is common. [51]

2.2.1 Steam Generation from Sun

Experiments have shown that if sunlight is received on the surface of particles, steam can be produced, which is highly efficient energy. Developing countries may use a "solar steam device" for water purification or disinfection purposes to reduce use of electricity.

2.2.2 Producing High-Efficiency Light Bulbs

High-efficiency light bulbs, with the advantage of double efficiency and being shatter-proof, are made up of nano-engineered polymer matrices. Also, with the use of nanomaterial, high-efficiency LEDs with the use of plasmonic cavities are in progress. [10-12]

2.2.3 Generation of Increased Electricity by Wind Mills

With the use of epoxy carbon nanotubes, blades of windmills are made stronger, which can produce more electricity by windmill. [15]

2.2.4 Electricity from Heat of Waste Materials

Nanotubes can be used for generation of electricity for building thermocells while cells act at different temperatures. They could be wrapped around hot pipes to get electricity from heat, which is wasted in normal conditions. Hydrogen can be stored for fuel cell powered cars. Hydrogen can be bonded to graphene for an increase in energy that results in more storage of hydrogen and lightweight fuel tanks. The nanoparticles sodium borohydride have shown effective storage of hydrogen. [16-18]

2.2.5 Saving of Energy for Building Heating/Cooling

Nanoparticles zinc copper and silver film have positive results for absorbing sheet and heat reflecting, respectively. They can act as a supplement to current HVAC

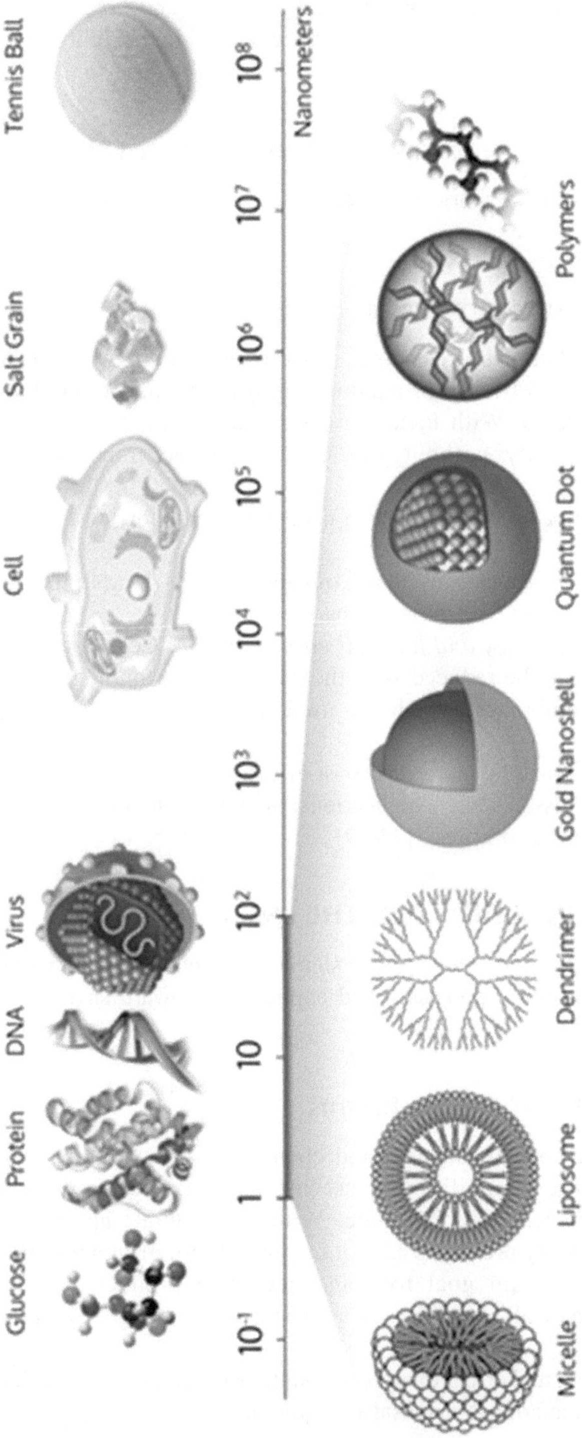

FIGURE 2.1 Nanometer scale.

(Source: http://nano.cancer.gov/learn/understanding/)

systems and reduce the requirement of energy to heat and cool buildings. For electricity from nanofibers, piezoelectric nanofibers provide flexibility to be woven into clothing. [16-18]

2.2.6 Reduction in Friction for Less Consumption of Energy

Lubricants with the use of inorganic buckyballs have been developed to reduce friction.

2.2.7 Reduction in Loss of Power

Carbon nanotubes with less resistance than wires are in use currently for electric transmission grids. Use of nanotechnology can change the distribution of electricity in the grid. With local storage capacity as a battery with power for 24 hours, electricity without loss power can be transmitted through wires.

Reduction in solar cell cost: With use of nanotechnology, low-cost solar cells are manufactured.

Efficient battery performance: With use of nanomaterials, long-lasting and faster batteries can be developed with faster recharge.

Improvement in efficiency and fuel cell cost reduction: Cost of hydrogen ion-producing catalysts can be reduced with the use of nanotechnology. It increases the efficiency of membranes, which are used for separation of hydrogen ions from gases.

Fuel production from raw materials: cost-effective fuel production from material can help save fossil fuels. Applications of nanotechnology for increasing mileage of engines are also in use. [27-29]

2.3 NANOTECHNOLOGY AND THE ENVIRONMENT

For sustainability and climate protection, a significant contribution is expected from nanotechnological products, processes and applications toward saving nonrenewable sources of energy.

2.3.1 Potential Environmental Benefits

Nanomaterials have special physical and chemical properties that make them noticeable, e.g. more durability against mechanical stress, water-resistant coatings, energy-efficient buildings, etc. With their unique application, high hopes for solving energy problems cannot be ignored. Environmentally friendly products are not the main goal for nano consumer products, but with the addition of nanomaterials, energy saving and self-cleaning products can be developed.

For analyzing the actual effect of material on the environment, assessment of the life cycle from raw material to disposal is required.

2.3.2 Nanomaterial in Active Waste Cleanup in Water

Nanofibers named titanates are used as absorbents for removal of radioactive ions from water. Research also supports the possibility of use of them in removal of cesium and iodine from water. [17-22]

2.3.2.1 Nanotechnology for Solution of Oil Spills

Though at the very initial stage, use of nanomaterial for oil spills seems a promising solution for the future. [42]

2.3.2.2 Application of Nanotechnology for Water

Use of nanomaterials in water or waste-water technology may be divided into various categories like detection, treatment and pollution prevention. Potential for transformation for field of desalination is foreseen with use of nanotechnology, e.g. use of ion concentration polarization.

A graphene-based coating for desalination membranes is more robust and scalable than current nanofiltration membrane technologies. With its use, there is the possibility of overcoming scalability issues along with developing an inexpensive membrane with the best quality

Capacitive deionization purifying method, which is cost-effective and energy efficient, is the upcoming method for purification of brackish water. It uses graphene-like nanoflakes for deionization, performing better compared to conventional activated carbon materials. [22,38]

2.3.2.3 Carbon Dioxide Capture

In place of existing costly filtration with requirements of more chemicals for CO2 storage, nanoscale fabrication of membranes could lead to change in membrane technology. [26]

2.3.2.4 Production of Hydrogen from Sunlight

Hydrogen needs to be produced, not readily available. Hydrogen fuel is a clean energy carrier, though it sometimes appears dirty due to its source. Industries are trying to claim their technology of hydrogen production as ecofriendly technology. With the use of 2.5 volts of solar energy, artificial photosynthesis breaks molecules of water into oxygen along with positively charged protons and negatively charged electrons. Due to separation and extraction, opposite charges of electrons and protons generate electric power. [20-23]

2.4 NANOTECHNOLOGY IN ENERGY

Nanotechnology has the potential to enhance efficiency of energy for all branches in the industrial sector and, with help of new and optimized production technologies, to leverage economically and contribute value addition in the energy sector. [20-23]

2.4.1 Sources of Energy

Nanotechnology can give support for developing conventional and renewable sources of energy, e.g. wear-resistant drill probes coated with nanomaterials have a longer lifespan and efficiency for developing natural gas deposits, which is cost effective. Mechanical components become lightweight with the use of nanomaterials. Solar energy through photovoltaic systems will have a major impact on the use of nanomaterials. Use of nanomaterials is intended to increase to at least 10% efficiency upon use, e.g. layer design optimization, morphology of organic semiconductor, etc. Dots and wire-type nanostructures allow solar cell efficiency of more than 60% in the long run. [20-23]

2.4.2 Conservation of Energy

Converting primary sources of energy into different forms of energy, like heat and kinetic, is more efficient. Use of nanomaterials can help avoid more carbon emissions through steam power plants and fossil-fired gas. High-efficiency power plants require high operating temperature and heat-resistant turbine materials. Nanostructured devices, such as electrodes, membranes and catalysts, can help find cost-effective applications in sectors of automobiles, buildings and electronics by converting chemical energy with the use of fuel cells. Compared to chemical energy, thermoelectric energy conversion seems more promising. [33-34]

2.4.3 Energy Distribution

To reduce loss of energy in the application of electric cables and power lines, carbon nanotubes seem promising. Research on optimization of superconductive materials for losses of current conduction to solve complex control problems and monitoring, found that sensory devices based on nanotechnology and power electronic components are promising options. [39]

2.4.4 Energy Storage

For energy storage, application of nanomaterials in batteries and supercapacitors is promising, mainly lithium-ion technology because high cell voltage along with extraordinary energy is achieving new heights. With the use of new nanomaterials that are high performance, efficiency and safety of lithium-ion batteries can be improved to a great extent. Hydrogen is also considered to be one of the promising and environmentally friendly sources for energy storage. [40]

2.4.5 Energy Usage

To save available sources of energy with optimum use and to reduce avoidable consumption, multiple options are provided by nanotechnology for sustainability, e.g. reduction in consumption of fuel with application of nanocomposite materials, wear-resistance optimization, lightweight engine components, additives, nanoparticles to achieve low rolling resistance and using a tri-biological component. Nanoporous thermal insulation in building construction can provide energetic rehabilitation to old buildings. The addition of nanomaterials in the control of light and heat is a promising path for reduction in energy consumption in buildings. [43]

2.5 NANOTECHNOLOGY STANDARDS

For any engineering or material-related work since 1960, the metric system has been accepted all over the world as its units are standard measurement to execute global-level projects. As nanotechnology is now coming to practice, U.S. researchers tried to develop conversion tables for nanometers to nano inch and nano foot and likewise other conversions. [20-22]

2.5.1 Standards

A standard is a rule, guideline or definition for technical specification or other design criteria in the form of a published document. Currently, two types of standards are in practice worldwide. First, physical measurement standards are available for mass, time and frequency according to the international system of units, and second, documentary standards are available for agreements among producers and technical specification services. Standards can increase reliability and effectiveness of services, and they are a summary of good and best practice.

2.6 NANOTECHNOLOGY FOR THE ENVIRONMENT

2.6.1 Nano-Based Materials, Processes and Applications

Nanomaterials have characteristics of high specific surface area and discontinuous properties. With their characteristics like adsorption, catalysts, increased surface area, long-lasting coatings and reagents, they can be applied as newly developed material in the process of treatment of water and waste water. [8, 13]

2.6.2 Adsorption

By definition, adsorbents are solids that adsorb gases or dissolved substances, and combinedly adsorbed molecules are called adsorbate. They have high potential to remove organic and inorganic pollutants as they are capable of efficient and after decontamination processes. Use of nano adsorbents not only helps to save adsorbent materials but also to leave a smaller footprint because a compact device can be implemented for water and waste-water treatment. Currently, researchers focus on the following types of adsorbents based on nanotechnology:

- Carbon nanotubes (CNTs)
- Metal-based polymeric zeolites [24,56,65]

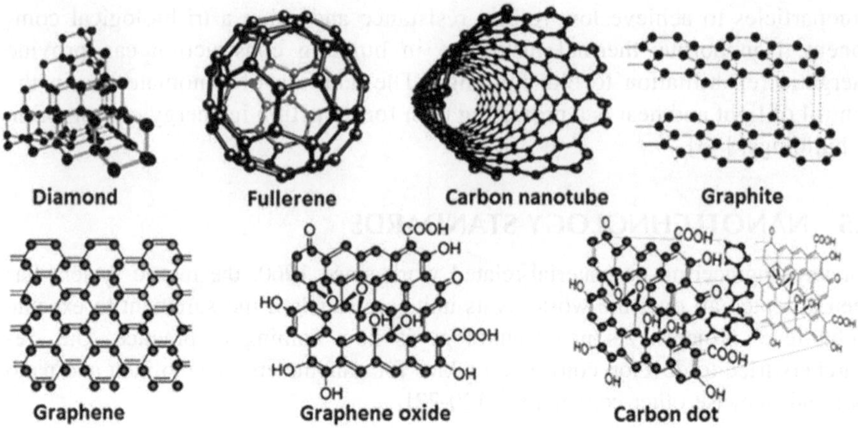

FIGURE 2.2 Carbon nanotubes.

2.6.3 CNTs

These are cylindrical in shape and nanostructured material available in different categories, such as single walled and multi-walled, as shown in Figure 2.2. They have some characteristics of increased surface area, adjustable surface chemistry and measurable adsorption site. To overcome problems of aggregation, which can reduce active surface, carbon nanotubes are stabilized in aqueous suspension. They are effective on persisting and preconcentrating and have the ability to detect contaminants. Adsorption happens because of electrostatic attraction and chemical bonding. In comparison to disinfection methods like chlorination and ozonation, no toxic byproducts are produced. With proper operating conditions, they can be regenerated.

Dendrimers (i.e. repetitively branched molecules) are polymeric nano adsorbents, which can be used for removal of organic and heavy metals. Interior hydrophobic Scheels adsorb organic compounds of dendrimer, while tailored exterior branches will adsorb heavy metals. With PH control, regeneration of adsorbent can be easily achieved. Sadeghi-Kiakhani et al. invented efficient, biodegradable, biocompatible, nontoxic bio adsorbent for removal of anionic compounds from textile waste water by preparing combined chitosan-dendrimer nanostructure, which can achieve a removal rate up to 99%. [4-7, 16, 17]

2.6.4 Zeolites

From the 1980s, zeolites with silver atoms have been known that have a porous structure capable of embedding silver ion-type nanoparticles. In solution, they are released due to the process of exchange of ions with cations. As shown in Figure 2.3, zeolites can also work as nanoparticles, as shown by Tiwari. He experimented with zeolites with the use of laser-induced fragmentation.

FIGURE 2.3 Zeolites.

2.6.5 Properties, Applications and Innovative Approaches of Nano Adsorbents

Carbon nanotubes and nanometals are highly effective nano adsorbents for arsenic-type heavy metal removal. As nanometals and zeolites can be applied with beads for absorption, they can benefit from treatment of water due to compactness and cost-effectiveness. As shown in Figure 2.4, removal of micropollutants is due to strong adsorption capacity of carbon nanotubes because of diverse interactions between carbon nanotubes and contaminants. Polymeric nano adsorbents are a promising novel product though carbon nanotubes. [33-37]

2.7 NANOMATERIALS AND NANOMETAL OXIDES

In place of activated carbon, nanoscale metal oxides are showing good results for removal of heavy metals and radionuclides as they have high specific surface area, have less distance of intraparticle diffusion and are compressible with no reduction in area. Nano iron hydroxide [α-FeO(OH)] is a strong adsorbent as it contains the property of resistance to abrasion; adsorption of arsenic waste is possible because of more specific surface area. [18,67]

2.8 NANO SILVER

The main advantage of application of nano silver is its strong and broad-spectrum antimicrobial activity without any harmful effects on humans. It is used for disinfection of water and anti-biofouling surfaces. For the disinfection and decontamination process, nano-titanium dioxide (TiO2), a cost-effective solution, is available with low human toxicity and high chemical stability. Nano titanium dioxide has the advantage of endless lifetime coatings. [21,50]

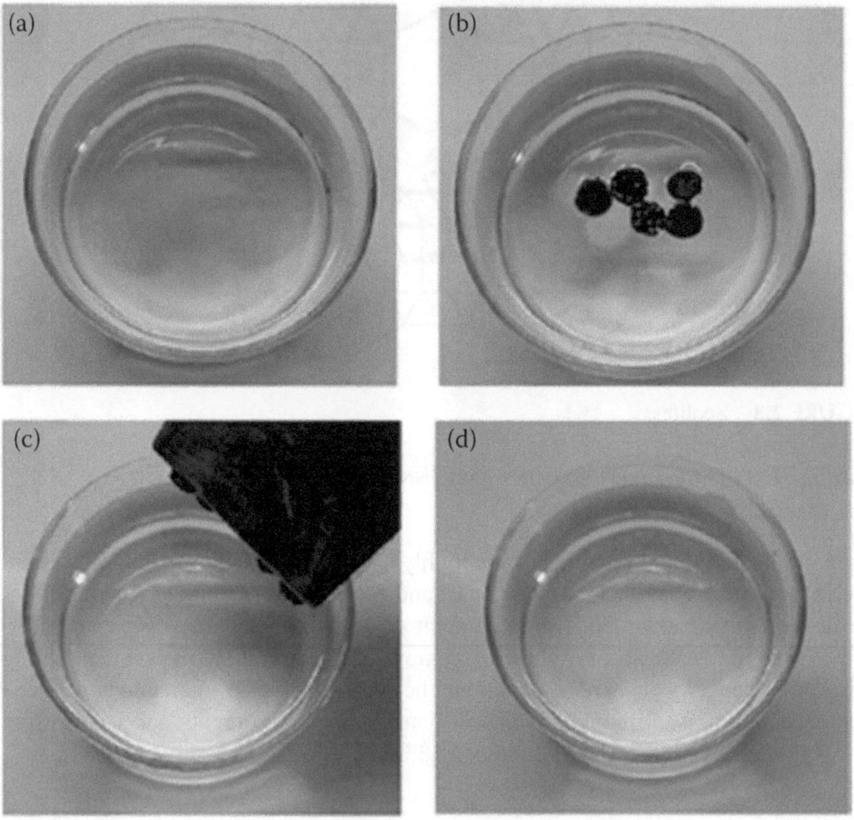

FIGURE 2.4 Heavy metal removal by nano adsorbents.

2.9 MAGNETIC NANOPARTICLES

For remediation of groundwater, especially for removal of arsenic, the application of magnetic nanoparticles (magnetite Fe3O4) to separate water pollutants is in practice. Magnetic nanoparticles can be injected into contaminated groundwater and, with magnetic field-loaded particles, can be removed. Along with remediation of groundwater, recovery of magnetic nanoparticles also works as an ideal compound that increases osmotic pressure in forward osmosis, a process in which water is drawn from low osmotic pressure to high osmotic pressure. [29-32]

2.10 NANO-ZERO VALENT IRON

In place of activated carbon, nano-zero valent iron can be the best alternative to use for groundwater remediation, which is contaminated with chlorinated hydrocarbon fluids and perchlorates. Nano-zero valent iron for in situ treatment of groundwater can be injected and shows more reactivity due to high specific surface area. Reduction rates of CCl_4 by nZVZ were more affected by solution chemistry than particle size or surface morphology. [22]

2.11 CHARACTERISTICS FOR NANOMATERIALS AND NANOMETAL OXIDES

TiO_2 is easily available at low price; it has unique characteristics of inertness and capability of chemical degradation of contaminated microorganisms. Due to said characteristics, it is reflected as a strong, long-lasting and efficient nanomaterial that can be used for small- and large-scale water and waste-water treatment processes. Until today, ultraviolet visible photocatalytic TiO_2 has been less efficient in comparison to the same type of oxidation processes, like ozonation. [51-54]

2.12 CONCLUSION AND FUTURE PROSPECTS

Biopolymers, inorganic nanoparticles, are in use for the development of bio nanocomposites with various structures, functionalities and applications that are useful for reinforcing agents in synthesis. Dispersion of nanoparticles alters physical, mechanical and structural properties. They are used by functionality providers as a substitute of biopolymers. They are biocompatible, so they are suitable for the medical field and drug delivery systems, diagnostics, cosmetics and application of biotechnology. Different micro- and nanomaterials and their composites are discussed here for applications in the mechanical, thermal and electrical fields, etc. They are easily available biocompatible-containing anti-oxidant oxidant actions. Nanofiller is approved for the healthcare sector by researchers.

Organic and inorganic compounds reacting with toxic pollutants are cleaned by clay minerals for preparation, of which interaction of lamellar materials of adsorption is reviewed. Many industries list toxic wastewater and water pollution as a major challenge for researchers to deal with. Physical treatments like adsorption; chemical treatments, like evaporation, ion exchange and membrane filtration; and biological treatments have not been found suitable for such waste. Nanotechnologies, such as clay minerals, contain characteristics of extracting radio nuclides and heavy metals from water as they are biocompatible low-cost sources. Because of interchangeable cations, pore volume and surface area, phyllosilicates are a better option compared to any other constituent for adsorption. Preboosting helps to render clays hydrophobic.

To overcome the global problems related to availability of good quality water, new technology is necessary for achieving the following goals:

- To ensure high quality of drinking water
- To eliminate micropollutants
- To increase the efficiency of industrial production processes

REFERENCES

1. Mokhtarzadeh, A., Alibakhshi, A., Hejazi, M., et al. "Bacterial-derived biopolymers: Advanced natural nanomaterials for drug delivery and tissue engineering." *TrAC – Trends Anal. Chem.* 82 (2016): 367e384. 10.1016/j.trac.2016.06.013.
2. Aredes S., Klein, B., Pawlik, M. "The removal of arsenic from water using natural iron oxide minerals." *J. Clean. Prod.* 29–30 (2012): 208–213.
3. Kapridaki, C.L., Pinho, M.J., Mosquera, et.al. "Producing photoactive, transparent and hydrophobic SiO2-crystalline TiO2 nanocomposites at ambient conditions with application as self-cleaning coatings." *Appl.Catal. B Environ.* 156e157 (2014): 416e427. 10.1016/j.apcatb.2014.03.042.
4. Cong, Q., Yuan, X., Qu, J. "A review on the removal of antibiotics by carbon nanotubes." *Water SciTechnol.* 68 (2013): S1679–S1687.
5. Buenger D., Topuz, F., Groll, J. "Hydrogels in sensing applications." *Prog. Polym. Sci.* 37 (2012): 1678e1719, 10.1016/j.progpolymsci.2012.09.001.
6. Das, R., Abd Hamid, S.B., Ali, E., et al. "Multifunctional carbon nanotubes in water treatment: The present, past and future." *Desalination* 2014, no. 354 (2014): 160–179.
7. De Volder M.F., Tawfick, S.H., Baughman, R.H., et al. "Carbon nanotubes: Present and future commercial applications." *Science* 339 (2013): 535–539.
8. Dichiara, A.B., Weinstein, S.J., Rogers, R.E. "On the choice of batch or fixed bed adsorption processes for wastewater treatment." *Ind. Eng. Chem. Res.* 2015, no. 54 (2015): 8579–8586.
9. Dotzauer, D.M., Dai, J. , Sun, L., et al. "Catalytic membranes prepared using layer-by-layer adsorption of polyelectrolyte/metal nanoparticle films in porous supports." *Nano Letters*, 6 (2006): 2268–2272.
10. Ellis, T.G. *Chemistry of wastewater. Encyclopedia of Life Support System (EOLSS), Developed under the Auspices of the UNESCO*, Eolss Publishers, Oxford, UK, 2004, http://www.eolss.net
11. Fujita, H., Miyajima, R., Sakoda, A.J.A. "Limitation of adsorptive penetration of cesium into Prussian blue crystallite." *Adsorption* 2015, no. 21 (2015): 195–204.
12. Gautam, R.K., Chattopadhyaya, M.C. "Nanomaterials in the Environment: Sources, Fate, Transport, and Ecotoxicology." In *Nanomaterials for Wastewater Remediation*; Gautam, R.K., Chattopadhyaya, M.C., Eds.; Butterworth-Heinemann, Boston, MA, 2016; pp. 311–326, 2016.
13. Gibert, O., Valderrama, C., Peterkóva, M., et al. "Evaluation of selective sorbents for the extraction of valuable metal ions (Cs, Rb, Li, U) from reverse osmosis rejected brine." *Solvent Extr. Ion Exch.* 28 (2010): 543–562.
14. Gupta, A.K., Deva, D., Sharma, A., et al. "Fe-grown carbon nanofibers for removal of arsenic (V) in wastewater." *Ind. Eng. Chem. Res.* 49 (2010): 7074–7084.
15. Hajeh M., Laurent, S., Dastafkan, K. "Nano adsorbents: classification, preparation, and applications." *ChemRev.* 113 (2013): S7728–S7768.
16. Hashim D.P., Narayanan, N.T., Romo-Herrera, J.M., et al. "Covalently bonded three-dimensional carbon nanotube solids via boron induced nanojunctions." *SciRep.* 2 (2012):363.
17. Ihsanullah. "Carbon nanotube membranes for water purification: Developments, challenges, and prospects for the future." *Sep. Purif. Technol.* 2019, no. 209 (2019): 307–337.
18. Kadam, A.A., Jang, J. "Facile synthesis of pectin-stabilized magnetic graphene oxide Prussian blue nanocomposites for selective cesium removal from aqueous solution." *Bioresour. Technol.* 2016, no. 216 (2016): 391–398. Nanomaterials 9, no. 682 2019: 19–21.

19. Margeta, Karmen, Nataša Zabukovec Logar, Mario Šiljeg, et al. "Natural zeolites in water treatment – How effective is their use?" In Elshorbagy, W. &, Chowdhury, R.K. Eds., Water Treatment. IntechOpen, 2013. doi:10.5772/50738
20. Kaewmanee, P., Manyam , J., Opaprakasit, P., et al. "Effective removal of cesium by pristine graphene oxide: Performance, characterizations and mechanisms." *RSC Adv.* 2017, no. 7 (2017): 38747–38756.
21. Kim E.S., Hwang, G., El-Din, M.G., et al. "Development of nanosilver and multi-walled carbon nanotubes thin-film nanocomposite membrane for enhanced water treatment." *J. Membr. Sci.* 394 (2012):37–48.
22. Mehta Komal, P. *Nano Technology in Water and Waste Water Treatment.* Scholar's Press, Omniscriptam Publication, Germany, 2019.
23. Lakshmi, P.L., Kodurua, J.R., Karri, R.R. "A comprehensive review of applications of magnetic graphene oxide based nanocomposites for sustainable water purification." *J. Environ. Manage.* 231, no. 1 (2019): 622–634.
24. Liu, X., Wang, M., Zhang, S., et al. "Application potential of carbon nanotubes in water treatment: A review." *J. Environ. Sci. (China)* 25 (2013): S1263–S1280.
25. Liu, J., Li, X., Rykov, A.I., et al. "Zinc-modulated Fe-Co Prussian blue analogues with well-controlled morphologies for the effecient sorption of cesium." *J. Mater. Chem. A* 2017, no. 5 (2017): 3284–3292.
26. Liu, X.T., Mu, X.Y., Wu, X.L., et al. "Toxicity of multi-walled carbon nanotubes, graphene oxide, and reduced graphene oxide to zebrafish embryos." *Biomed. Environ. Sci.* 2014, no. 27 (2014): 676–683.
27. Lu, Y., Wang, L., Cheng, J., et al. "Prussian blue: A new framework of electrode materials for sodium batteries." *Chem. Commun.* 2012, no. 48 (2012): 6544–6546.
28. Fashandi, M., Leung, S.N. "Preparation and characterization of 100% bio-based polylactic acid/palmitic acid microcapsules for thermal energy storage." *Mater. Renew. Sustain. Energy* 6 (2017): 14. doi:10.1007/s40243-017-0098-0.
29. Mahendra, C., Bera, S., Aanad Babu, C., et.al. "Separation of cesium by electro-dialysis ion exchange using AMP-PAN." *Sep. Sci. Technol.* 2013, no. 48 (2013): 2473–2478.
30. Uddin, M.K. "A review on the adsorption of heavy metals by clay minerals, with special focus on the past decade." *Chem. En. J.* 15 (2017 January): 438–462.
31. Nakajima, L., Yusof, N.N.M., Kobayashi, T. "Calixarene-composited host–guest membranes applied for heavy metal ion adsorbents." *Arab. J. Sci. Eng.* 2015, no. 40 (2015): 2881–2888.
32. Nakotte, H., Shrestha, M., Adak, S., et.al. "Magnetic properties of some transition-metal Prussian Blue Analogs with composition M3[M0(C,N)6]2_xH2O." *J. Sci. Adv. Mater. Devices* 2016, no. 1 (2016): 113–120.
33. Nasir, S., Hussein, M.Z., Zainal, Z., et al. "Carbon-based nanomaterials/allotropes: A glimpse of their synthesis, properties and some applications." *Materials* 2018, no. 11 (2018): 295.
34. Nora S., Mamadou, S.D. "Nanomaterials and water purification: Opportunities and challenges." *J Nanopart. Res.* 7 (2005): 331–342.
35. Nowack, B, Krug, H.F., HeightM. "120 years of nanosilver history: Implications for policy makers." *Environ Sci. Technol.* 45, no. 4 (2011). 1177–1183.
36. Nechyporchuk, O., Belgacem, M.N., Bras, J. "Production of cellulose nanofibrils: A review of recent advances." *Ind. Crops Prod.* 93 (2016): 2e25. doi:10.1016/j.indcrop.2016.02.016.
37. Zabihi, O., Ahmadi, M., Nikafshar, S., et al. "A technical review on epoxy-clay nanocomposites: Structure, properties, and their applications in fiber reinforced composites." *Composites Part B: Engineering* 135, no. 15 (2018): 1–24. doi:10.1016/j.compositesb.2017.09.066]

38. Prachi, P.G., Madathil, D., Brijesh Nair, A.N. "Nanotechnology in waste water treatment: A review." *Int. J. ChemTech Res. CODEN(USA)*: IJCRGG 5, no. 5 (2013): 2303–2308.fdwed
39. Protima, R., Rauwel, E. "Towards the extraction of radioactive Cesium-137 from water via graphene/CNT and nanostructured prussian blue hybrid nanocomposites: A review." *Nanomaterials* 9 (2019): 682. doi:10.3390/nano9050682 www.mdpi.com/journal/nanomaterialsNanomaterials
40. Yan, Qi-Long, Gozin, M.Zhao, F.-Q., et al. "Highly energetic compositions based on functionalised carbon nano materials, nano scale." *Royal Society of Chemistry* 8 (2016): 4799–4851.
41. Qi, W., Tian, L., An, W., et al. "Curing the toxicity of multi-walled carbon nanotubes through native small-molecule drugs." *Sci. Rep.* 2017, no. 7 (2017): 2815.
42. Qu X., Alvarez, P.J., Li, Q. "Applications of nanotechnology in water and wastewater treatment." *Water Res.* 47 (2013): 3931–3946.
43. Razali, M., Kim, J.F., Attfield, M., et al. "Sustainable wastewater treatment and recycling in membrane manufacturing." *Green Chem.* 2015, no. 17 (2015): 5196–5205.
44. Rathinasabapathy, M., Kang, S.M., Lee, I., et al. "201 Nano scale, royal society of chemisterr.,9. Highly stable Prussian blue nanoparticles containing graphene oxide-chitosan matrix for selective radioactive cesium removal." *Mater. Lett.* 241 (2019): 194–197.
45. Ruankaew, N., Yoshida, N., Watanabe, Y. "Size-dependent adsorption sites in a Prussian blue nanoparticle: A 3D-RISM study." *Chem. Phys. Lett.* 2017, no. 684 (2017): 117–125.
46. Ghaffar, S.H., Fan, M. "Lignin in straw and its applications as an adhesive", *Int. J. Adhes. Adhes.* 48 (2014): 92e101. doi:10.1016/j.ijadhadh.2013.09.001.
47. Satyawali Y., Balakrishnan, M. "Wastewater treatment in molasses-based alcohol distilleries for COD and color removal: A review." *J. Environ. Manage.* 86 (2008): 481–497.
48. Schewe J, Heinke, J., Gerten, D., et al. "Multimodal assessment of water scarcity under climate change." *Proc. Natl. Acad. Sci. USA.* 111 (2013):3245–3250.
49. Shah, M.A., Ahmed, T. *Principles of Nanoscience and Nanotechnology*. Narosa Publishing House, New Delhi, pp. 34–47, 2011.
50. Sharma, V.K., Yngard, R.A., Lin, Y. "Silver nanoparticles: Green synthesis and their antimicrobial activities." *Adv. Colloid Interface Sc.* 145 (2009): 83–96.
51. Shelley, S.A. "Nanotechnology: Turning basic science into reality." In *Nanotechnology: Environmental Implications and Solutions*, Theodore, L., Kunz, R.G., Eds., pp. 61–107. John Wiley & Sons, Inc, New York, 2005.
52. Shon H.K., Vigneswaran, S., Kandasamy, J., et al. *Characteristics of Effluent Organic Matter in Wastewater (EOLSS), Developed under the Auspices of the UNESCO*, Eolss Publishers, Oxford, UK, 2007. http://www.eolss.net
53. Seth, Jaffe. "Nano technology and site remediation: Is the promise beginning to come to frutition?" Law and the Environment. 2009, https://www.lawandenvironment.com/
54. Mekonnen, T., Mussone, P., Khalil, H., et al. "Progress in bio-based plastics and plasticizing modifications." *J. Mater. Chem. A* 1 (2013): 13379. doi:10.1039/c3ta12555f
55. Theron J., Walker, J.A., Cloete, T.E. "Nanotechnology and water treatment: Applications and emerging opportunities." *Critical Reviews in Microbiology* 34 (2008): 43–69.
56. Pham, T.D., Bui, T.T., Nguyen, V.T., et al. "Adsorption of polyelectrolyte onto nano silica synthesized from rice husk: Characteristics, mechanism and application for antibiotic removal." *Polymers* 10 (2018): 220. doi:10.3390/polym10020220

57. Torad, N.L., Hu, M., Imura, M., et al. "Large Cs adsorption capability of nanostructured prussian blue particles with high accessible surface areas." *J. Mater. Chem.* 2012, no. 22 (2012): 18261–18267.
58. Yeul, V.S., Rayalu, S.S. "Unprecedented chitin and chitosan: A chemical overview." *J. Polym. Environ.* (2013). doi:10.1007/s10924-012-0458-x.
59. Sharma, Vikas, Sharma, A. "Nanotechnology: An emerging future trend in wastewater treatment with its innovative products and processes." *Int. J. Enhanc. Res. Sci. Technol. Eng.* 1, no. 2 (2012). Issn No: 2319–7463.
60. Visa, M. "Synthesis and characterization of new zeolite materials obtained from fly ash for heavy metals removal in advanced wastewater treatment." *Powder Technol.* 2016, no. 294 (2016): 338–347.
61. Fan, W., Lai, Q., Zhang, Q., et al. "Nanocomposites of TiO2 and reduced graphene oxide as efficient photocatalysts for hydrogen evolution." *J. Phys. Chem. C* 115 (2011) 10694e10701, 10.1021/jp2008804.
62. Wegmann M., Michen, B., Graule, T. "Nanostructured surface modification of microporous ceramics for efficient virus filtration." *J. Eur. Ceram. Soc.* 28 (2008): 1603–1612.
63. Xiao, L., Erdei, L., McDonagh, A., et al. "Photocatalytic nanofibers. ICONN, 2008." International Conference on Nanoscience and Nanotechnology, Melbourne, Vic, Australia, 2008.
64. Xiaolei Qu, Alvarez, P.J.J., Qilin, L. "Applications of nanotechnology in water and wastewater treatment." *Water Res.* 4, no. 7 (2013): 3932–3934.
65. Yan H.Y., Han, Z.I., Yu, S.F., et al. "Carbon nanotube membranes with ultrahigh specific adsorption capacity for water desalination and purification." *Nat. Commun.* 13, no. 4 (2013): 2220.
66. Zhang H., Quan, X., Chen, S., et al. "Fabrication of photocatalytic membranes and evaluation of its efficiency in removal of organic pollutants from water." *Sep. Purif. Technol.* 50 (2006): 147–155.
67. Zheng, Y., Qiao, J., Yuan, J., et al. "Electrochemical removal of radioactive cesium from nuclear waste using the dendritic copper hexacyanoferrate/carbon nanotube hybrids." *Electrochim. Acta* 2017, no. 257 (2017): 172–180.

3 Nanotechnology to Address Micronutrient and Macronutrient Deficiency

Shiva Sharma, and Manisha Rastogi
Department of Biomedical Engineering, School of Biological Engineering and Sciences, Shobhit Institute of Engineering and Technology (Deemed to be University), Meerut, UP, India

Neha Singh
Department of Electronics & Communication Engineering, Manipal University Jaipur, Rajasthan, India

3.1 PHYSIOLOGICAL IMPORTANCE OF MICRONUTRIENTS AND MACRONUTRIENTS

Micro- and macronutrients serve as the central components in maintaining and regulating the biological and cellular processes behind normal physiological functions [1]. Adequate intake of micro- and macronutrients is essential to drive intermediary energy metabolism pathways, growth and overall homeostasis. The primary resource to obtain micro- and macronutrients is through dietary intake only, with no synthesizing mechanisms within the body [2]. Macronutrients (carbohydrates, fats, proteins and water) are defined as the nutrients required in large amounts by the human body, which play a vital role in structuring and metabolic activities, whereas fats and carbohydrates are known as the primary agent for bulk energy or calories. Protein and their constituent amino acids are essential for providing nitrogen, which is eventually needed for the synthesis of deoxyribonucleic acid (DNA) and related molecules [3].

Carbohydrates: Carbohydrates comprise more than 50% of the total required calories and are largely consumed in the form of dietary starch and sugars. Clinically, two forms (simple and complex carbohydrate) are consumed. While monosaccharides (glucose and fructose) and disaccharides (sucrose and lactose) together constitute simple carbohydrates, complex carbohydrates mainly include polysaccharides (glycogen, fiber and starch). Plant-based food is a good resource

DOI: 10.1201/9781003220350-3

for starch and fiber, whereas animal-based food products contain glycogen; glycogen is mostly found in liver and muscle tissues of animals [4, 5].

Fat: Fat contributes to the structure of the cellular membrane and some signaling molecule pathways. Another important function of dietary fats is in the absorption of fat-soluble vitamins A, D, E and K. Similarly, polyunsaturated fatty acids (PUFA) obtained through human nutrition have been implicated in maintaining the structural integrity of cell membranes apart from their functioning as signaling molecules [6]. Linoleic acid and alpha-linolenic acid, also commonly known as the essential fatty acids (EFAs), are also derived from the diet [7].

Protein: To maintain the normal physiology and growth of human tissue, proteins are essentially required. A major amount of amino acids is used in the synthesis of neurotransmitters, hormones and nucleic acids as well as structuring cell membranes. In plasma membrane, proteins are involved in structuring of membrane and transportation of substances outside and inside of the membrane. Proteins maintain a colloid osmotic pressure to manage the fluid in vascular spaces of the membrane. According to the Food and Nutrition Board of the Institute of Medicine, diet is considered to be the source of nine vital amino acids for all age groups, apart from the portions synthesized within human body [8]. These necessary amino acids include histidine, methionine, isoleucine, lysine, threonine, leucine, phenylalanine, valine and tryptophan [9].

Proteins and carbohydrates combine to provide approximately 17 kJ or 4 kcal, while fats provide 37 kJ (9 kcal) per gram. Protein is also considered to be second to adipose tissue in the storage of energy (Brouwer et al., 2019). However, the energy obtained from the carbohydrates and proteins depends on absorption and digestion factors; it varies considerably [10].

Micronutrients: Micronutrients (vitamins and minerals) are the nutrients that are needed in smaller amounts by organisms throughout life for a range of physiological functions (UNICEF, 2006). Micronutrients play a vital role in key physiological functions, such as synthesis of enzymes, hormones and other chemical messengers. Several micronutrients, including iodine, zinc, iron, copper, selenium and vitamins (A, E, C, D, B2 B6 and folate), have public health importance [11]. Vitamins are crucial to maintain intra- and extracellular chemical reactions, such as growth. Vision depends on vitamin A; calcium metabolism depends on vitamin D; antioxidants and skin disease are protected by vitamins C and E; vitamin B2 and B6 support the healthy activities of the nervous system; folate (vitamin B9) is essential for healthy fetal growth and development of the brain and spine.

Minerals are broadly categorized as a macro (major) or micro (trace) elements [12–16]. Macro-minerals include calcium, sodium, magnesium, chloride, phosphorus, potassium and sulfur. While maintenance of adequate fluid balance and muscular contraction is the key responsibility of sodium, chloride and potassium, calcium maintains healthy bones and teeth.

Magnesium, which is abundant in bones, is needed for protein synthesis and muscle contraction [17]). The micro-minerals, or trace elements, include at least cobalt, iron, selenium, copper, manganese, iodine, molybdenum and zinc. Micronutrients, like zinc, copper, manganese and molybdenum, are the integral components of a range of enzymes involved in multiple catabolic and metabolic reactions. Red blood cells contain the majority of iron content in the form of hemoglobin that transports required oxygen for energy metabolism; fluoride promotes bone and teeth formation; chromium regulates blood sugar (glucose) levels in association with insulin; iodine, found in thyroid hormone, is essential to regulate metabolic activities; and selenium possesses good antioxidant potential [18].

3.2 MICRONUTRIENT AND MACRONUTRIENT DEFICIENCY: AN UPDATE

Globally, micronutrient and macronutrient deficiencies (MNDs) have a great impact on public health. The key characteristics of MNDs, generally termed undernutrition, include low weight-to-height ratio (wasting), low height-to-age ratio (stunting), and low weight-to-age ratio (underweight). MNDs also serve as the major risk factor for the enhanced incidence of diet-related noncommunicable diseases [19]. These deficiencies are prevalent in low-income countries, transition countries and industrialized countries for vulnerable groups, specifically women, children and the elderly. Consequences can be seen in the form of poor mental and physical health [20].

It was estimated by the World Health Organization that approximately two billion people are suffering from malnutrition (MNDs), making it a global health concern. Around 462 million adults are underweight, while 47 million and 14.3 million children under 5 years of age have low and severe weight-to-height ratio, respectively, and 159 million are stunted [21, 22]. It is alarming that approximately 45% of mortality and morbidity among children of this age group is connected to undernutrition, primarily in low- and middle-income countries. These estimates further raise concerns in terms of the anemic population, accounting an additional 528 million, or 29%, of women of reproductive age globally [23, 24].

Deficiency of all macronutrients causes energy deficiency and is commonly known as protein-energy undernutrition (PEU), earlier named protein-energy malnutrition. Factors responsible for PEU encompass inadequate nutrient intake, co-morbid conditions and chronic drug intake that interfere with nutrient bioavailability. Worldwide, the vulnerable population for primary PEU is again children and the elderly; however, among the elderly population, depression has also been linked with PEU occurrence. Fasting or anorexia nervosa can be another reason for PEU. Its severity can be graded into mild, moderate or severe level based upon the calculation for weight as a percentage of projected weight for length or height by means of international standards (normal, 90 to 110%; mild PEU, 85 to 90%; moderate, 75 to 85%; severe, <75%). In children, chronic primary PEU has two common forms: Marasmus and Kwashiorkor [25].

Unlike PEU, health impacts of micronutrient deficiencies are not intensely noticeable; these deficiencies are also termed hidden hunger [26]. They are associated with optimal or life-threatening physiological effects. Iron and folic acid deficiency has been found to be the most prevalent nutrition problem in the world. However, in industrialized countries, deficiency of vitamin D is extremely widespread among the elderly population. Vitamin D deficiency severely hampers the quality of life of the elderly population through osteoporosis and bone fractures [20].

The requirements for macronutrients and micronutrients can be easily met through intake of a balanced diet. It is assumed this is the reason for the low prevalence of micronutrient deficiency in developed countries.

Inadequate intake of micronutrients, a weight-controlling diet and imbalanced nutrients in food, extensive exercise, and emotional and psychological stress may be the reason for malnutrition and its associated health issues. Unaccomplished nutrition demands among pregnant and lactating women, growing children, smokers and alcoholics also increase malnutrition [27–29]. Moreover, lifestyle (occupational and personal) pressure; lifestyle-associated behavior, such as unhealthy choice of food, chronic and periodical dieting, rushed meals, excessive alcohol, smoking and coffee consumption make middle-age adults at the risk for malnutrition [30]. Not only chronic but also meek micronutrient discrepancy can result in impaired immunity, impaired cognitive skills, fatigue and lack of general well-being [31, 32]. Research studies suggest that optimal intake of certain vitamins is vital to prevent cancer, osteoporosis and coronary heart disease [33–36]

3.3 STRATEGIES TO ADDRESS MICRONUTRIENT AND MACRONUTRIENT DEFICIENCY AND CHALLENGES

In order to cope up with MND, there are three major interventional strategies. Supplementation of the specific micronutrients, fortification of foods with micronutrients and biofortification have been extensively used. However, the WHO defines the latter two strategies as the primary strategies to mitigate the deficiency of essential micronutrients, to improve food quality, to enhance disease resistance and to prevent or cure a wide range of diseases [37].

Micronutrient supplementation involves the provision of the intake of either a single or a combination of micronutrients in multiple formulations, such as capsules, tablets, liquid or syrup. A functional food delivers multiple micronutrients as a component of food to provide health benefits beyond basic nutrition as it also contains natural bioactive compounds with immense health benefits.

Biofortification has been developed as a progressive agricultural approach to address micronutrient malnutrition. The process involves the enhancement of nutrient content of staple crops by fortifying the fertilizers. Therefore, it promotes enhanced absorption of nutrients by the plants. A range of factors need to be considered during biofortification. For instance, many times plants may not be able to absorb the micronutrients due to their small roots, pore size or plant disease despite the availability of micronutrients in the soil [38]. The biofortification process includes several challenges. First, appropriate selection of a vehicle food that is to be used for fortification has the greatest impact on its consumption, accessibility and availability to the targeted population. Second is the cost of the intervention, dietary habits, availability and the consumption of the vehicle food [39, 40]. Third is the higher demand of micronutrient compounds to be used in fortified products. Fourth is the optimal absorption, distribution, metabolism, excretion and bioavailability of the minerals upon consumption in the human body [41]. Fifth is the fortificant stability, longer shelf life, and unaltered color, taste, or appearance of the food [42].

The proportion of the fortified food that can be absorbed, metabolized in the gastrointestinal (GI) tract, and further distributed to other tissues and organs indicates its bioavailability [43]. Bioaccessibility, bioavailability, bioconversion, bioequivalence and bioactivity are the major parameters on which the efficiency of the fortified food depends, and they are attributed as key factor in food fortification; however, internal and external GI tract factors have a great impact on bioavailablity [43–45]. Physicochemical characteristics, food matrices, processing methods, storage and delivery method are major external factors. Internal factors, such as physicochemical and physiological conditions, including pH of the GI tract, mechanical pressure, food composition, enzymatic activities and surfactants, etc. (Figure 3.1), have a major impact on nutracutical bioavailability [4]. Solubility, biochemical transformations, interaction with GI tract fluid and release of biochemical constituents from nutraceuticals also depends on internal and external factors of the GI tract [47]. Dysfunctioning of gastric factors, such as acidity, motility, intestinal secretion, emptying speed and residence timing of food, and permeability and solubility in the gut, also affect gut microflora and thereby digestion and health benefits of food [48, 49]. Apart from low bioavailability and unfavorable water solubility, different challenges like crystallization and chemical

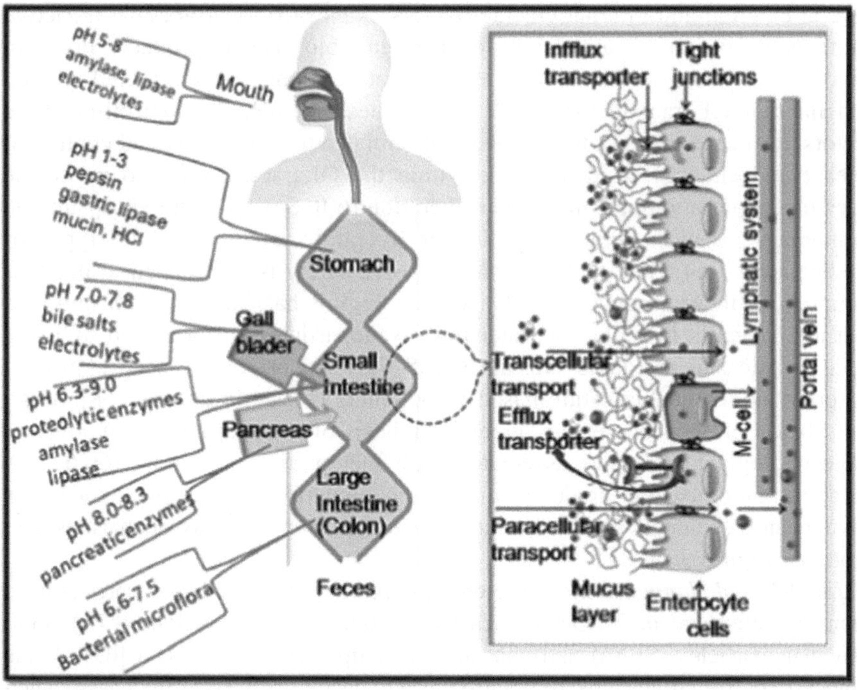

FIGURE 3.1 Stages of nutraceutical transport in the gastrointestinal tract.

(Source: [46])

instability need to be addressed while developing fortified food products with these molecules in bioactive food products [50–52]. Selection of appropriate fortification vehicles is the another challenge. The food needs to be consumed by the majority of the targeted population. Formulation of daily recommended doses that are acceptable worldwide is also the need of the hour [53]. However, the ultimate benefits of the these strategies cannot be fully utilized due to pharmacokinetic issues, including but not limited to insufficient gastric residence time, restricted bioavailability and adsorption in intestinal epithelium, low transportability and solubility within the gut, and unsteadiness of nutraceuticals in the acidic gastric environment.

3.4 NANOTECHNOLOGY TO ADDRESS PHARMACOKINETIC ISSUES

Nanotechnology techniques and principles have great potential to resolve pharmacokinetic issues related to fortification by improving the solubility, release and absorption of nutraceuticals. These are specifically addressed by introducing nanocarriers that enhance the efficiency, bioavailability, stability, solubility and localization of micronutrients to specific tissue [51,54–59]. Due to their smaller size in nm, nanocarriers have a larger interfacial surface and superficial electrical charge. Moreover, the available range of nanocarrier materials means diversified bioactive-loaded nanocarriers that have advantages in terms of protection from several external and internal GI tract factors. Nanocarriers protect the bioactive components of nutraceuticals in fortified and biofortified products that can be affected by exterior issues such as temperature, light, oxygen, humidity, etc., and internal GI tract factors, such as enzymes and pH. By controlling these factors, nanocarriers work to improve issues related to insolubility and controlled release and enhance residence time within the GI tract. They also improvise intestinal penetration or transcellular liberation [60]. Several nano-encapsulation techniques are also available to deliver lipophilic and hydrophilic types of nutraceuticals [54,55,61,61,62], which improve safety issues related to the consumption of these products [63]. Nanotechnology is basically used for nutrients that are basically insoluble in water [64]. Nanoparticles are produced by using both organic and inorganic substances [65–67], and presently encapsulation to different biodegradable nature-based biopolymers is used frequently [68]. Singh and Shukla et al. documented "green" synthesis of nanoparticles or innovative biotechnological strategies associated with nanoparticle synthesis [66,67]. A range of nontoxic, food-grade, economic, biocompatible and robust nanocarriers have been developed in the recent decade from lipids, polysaccharides and proteins (Figure 3.2).

Lipid-Based Nanocarriers: Different lipid particles functioning as carriers have been identified and investigated, such as liposomes, lipid emulsions (LEs), solid lipid nanoparticles (SLNs) and nanostructured lipid carriers (NLCs) [69]. These nanocarriers from lipids have been introduced to decrease adverse effects associated with strongly potent drugs, to enhance treatment efficacy, to express superior efficiency in gene transfer, and to use with applications related to the food and cosmetic industry [70,71].

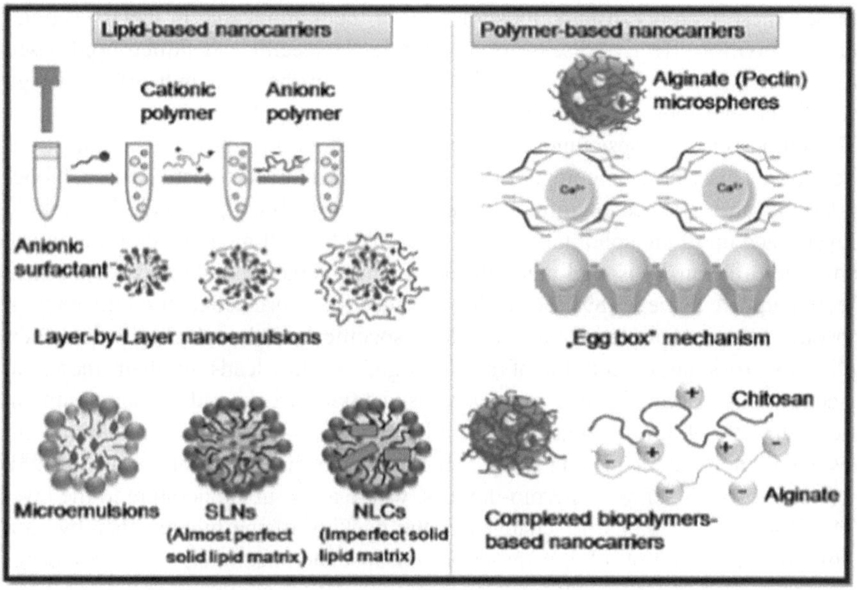

FIGURE 3.2 Different types of nanocarriers.

(Resource: Dima et al. [46].)

Micelles and Nanoemulsions: Micelles are defined as spherical vesicles made up of surfactants with the ability to self-assemble. Nanoemulsions are surfactant-assisted harmonized deferment with nano-sized drop (Pavlic et al., 2009; [72]). They facilitate monitored expulsion of cargo [73–75]. They have intrinsic biocompatibility, and due to a modifiable exterior side, they possess enhanced efficacy and potency [72,76]. Nanoemulsions are applied to encapsulate hydrophobic substances, while polymeric micelles are applied to incorporate hydrophilic as well as hydrophobic compounds. The application of micelles and nanoemulsions comprises capsulation of weakly water-soluble bioactive substances for integration into aqueous products. Such drug-encapsulated nanoemulsions are clinically accessible for ocular, intranasal, oral and topical administration (Yukuyama et al., 2017; [72]). In the food industry, different nanoemulsions are being applied as carriers that are naturally available; however, they have weakly soluble colors, tastes, additives and antioxidants [72,77].

Nanoliposomes: Liposomes are customary versions of lipid-based preparations that have been extensively examined more than the last decades [78,79] for pharmaceutical, dermal and cosmetic investigations [80]. They are the first class that has been accepted as therapeutic nanoparticles for cancer treatment, and they were identified in 1965 by Alec D. Bangham [81]. Nanometric bilayer phospholipid vesicles, or nanoliposomes, possess the potential to simultaneously encapsulate both hydrophobic and hydrophilic compounds, thereby establishing synergistic influence. Further, they can improve their storage stability, enhance bioavailability, provide protection to sensitive bioactive compounds and ensure their sustained release. These features of nanoliposomes make them applicable in dietary supplements (DISs) for effective disease prevention [82]. Recently, identified lipid-based nanocarriers, nanophytosomes, have been utilized in the food industry to design novel functional food products [83].

Nanoliposomes possess unique advantages, such as drug protection against degradation of enzymes, biocompatibility, flexibility, reduced toxicity, entire biodegradability and absence of immunogenicity [84–88]. The liposomal amphiphilic bilayer made up of phospholipid closely resembles the cell membrane of mammals, leading to efficient liposomal and cell membrane interactions, thereby resulting in effective uptake by cells [89]. To further enhance the pharmacokinetic efficiency of liposomes, these formulations are added to specific ligands, which alleviate their ability to cross target cell membranes. Together, this leads to their increased concentrations within the cell with decreased toxic effects and higher treatment efficacy. [88,90–92]. Liposome encapsulation decreases renal drug elimination with substantial increase in plasma drug half-life and overall bioavailability [78,93]. Further, liposomes serve as thermo-devices, where a rise in temperature to the level of 40°C – 41°C accounts for changes in the packing of the bilayer, thereby supporting the expulsion of a large amount of encapsulated agent, e.g. cytotoxic drugs to the tumor site treated with heat, avoiding the surrounding normal tissue [94]. Limitations associated with nanoliposomes include poor stability, low solubility, high production cost and short half-life. Furthermore, phospholipids of liposome may undergo oxidation, and encapsulated molecule leakage has been identified, particularly in molecules with low molecular weight [95,96]. They indicated low efficacy of encapsulation and fast clearance via reticulo endothelial system (RES), interactions of cells and transfer between membranes [97,98].

SLNs (Solid Lipid Nanoparticles): These are substitute carrier systems for emulsion, liposomes and polymeric nanoparticles [99,100]. SLNs may have applications for both hydrophobic and hydrophilic drugs [101,102]. Oral administration of chitosan (CS) coated SLNs loaded with curcumin (CUR) demonstrated increased physical stability and bioavailability of encapsulated curcumin when compared with their suspensions [103].

In comparison to other systems, SLNs have advantages, including simple preparation, reduced cost, ease of high-scale production using a high-pressure homogenization technique, magnificent physical stability, versatile chemistry, good release profile, drug protection against severe environmental conditions, better safety than polymeric carriers as their preparation is without organic solvents, absence of lipid carrier system toxicity, cheaper price than polymeric carrier systems, reliable biocompatibility, and biodegradability of lipids [70,104,105]. SLN disadvantages include growth of lipid particles, gelation tendency, the dynamic nature of polymorphic transition, their crystalline structure of solid lipid leading to inherent reduced incorporation rate, low drug-loading efficiency [70,105–107] and the possibility of drug expulsion during storage conditions. Another disadvantage that usually occurs with these formulations is initial burst release [108].

NLCs (Nanostructured Lipid Carriers): NLCs are modified SLNs that have been discovered as the next generation of SLNs to overcome the difficulties that have occurred in SLNs [109,110]. NLCs showcase improved stability and loading capacity pertaining to their amorphous matrix, which is made up of both solid and liquid (oil) phases [111].

They have various advantages similar to SLNs, such as biocompatibility, controlled drug release, accessibility for cost-effective large-scale production through the existing infrastructure, prevention of first-pass metabolism and protection of active drug moiety from biochemical breakdown [69]. Compared to SLNs, NLCs possess some less water in dispersion, more loading capacity for some drugs and the ability to minimize or prevent the expulsion of drugs during storage. However, no remarkable differences between biotoxicity of NLCs and SLNs have been documented [99, 112–115]. Recent findings indicated that during preparation and storage, NLCs showed less susceptibility than SLNs to gelation. NLCs facilitate nanoparticle separation from the medium as well as dosage formulations for parenteral administration [116, 117]. NLCs are solid at room temperature, which enhances their physical firmness; a main blockade was found for the emulsion-based preparation [118]. The available large-scale manufacture methods of NLCs overcome the costly technical necessity for the massive manufacture of liposomes [118]. Additionally, NLCs are biocompatible systems contributing their unique features compared to other lipid-based preparations [118, 119].

Protein-Based Nanocarriers: Proteins may be used as nanostructures to release bioactive compounds in a restrained manner in the surroundings [120]. The development of many protein-based nanostructures focuses on the method of preparation and application of environmental reactive proteins [121]. These methods include protein-based structures' collapsing or expansion ability when pH or temperature varies. Such methods may be applied to activate the release of bioactive compounds, thereby enhancing the efficiency and cost-effectiveness of functional food items [122]. The small size of protein-based nanostructures in the range of one to several hundred nanometers makes them a favorable system for encapsulation and delivery of bioactive compounds. Their small sizes may improve bioavailability of poorly absorbed active compounds [123]. Protein-based nanostructures provide an excellent strategy to improve the chemical stability and solubility of bioactive compounds [124].

Casein: Casein is designated as a biocompatible and eco-friendly GRAS (generally regarded as safe) protein [125]. Caseins are amphiphilic proteins that are copolymers of polar or nonpolar amino acid residues with the strong ability to self-assemble in the form of spherical casein micelles with 50–500 nm diameter [125,126]. Casein-based nanoparticles are used as drug delivery systems [127] because of excellent structural, physicochemical, surface-active and stabilizing characteristics, including tendency for attachment with ions and small molecules, emulsification and self-packing potential, and gelation and water-binding abilities [128]. Casein nanoparticles enhance curcumin solubility by increasing their biological activity [129].

Casein nanostructures are easy to prepare and scale-up protein nanoparticles during manufacture [127]. Proteins exhibit the potential of less opsonization via RES (reticuloendothelial system) through an aqueous steric barrier [125]. Proteins have excellent functional properties, such as gelation, emulsification, water-binding capacity and foaming [127, 130]. Protein nanocarriers possess exceptional properties, including high nutritional value, non-antigenicity, biodegradability and extraordinary binding ability with different drugs [125]. Moreover, protein nanoparticles are naturally metabolized through digestive secretions, therefore inducing different in vivo physiological effects [125, 130]. Interestingly, nanoparticle delivery systems consisting of groups of different types of nanoparticles with different functional properties that may result in excellent benefits [51], such as labile nonpolar bioactive compounds, would be more beneficial when encapsulated within protein-based nanoparticles and then added with lipid-based nanoparticles, providing a base to the digestible triglycerides and enhancing solubility of bioactive compounds in the secretions of the GI tract.

Being heterogeneous mixtures of varied sizes and having a wide range of molecular weights, proteins form heterogeneous nanoparticles displaying batch-to-batch variation [125, 131], which may inhibit the scaling-up procedure of protein nanoparticle formation. However, recombinant protein technology can overcome this problem. Another drawback of animal protein-based nanoparticles is that the hydrophilic property and quick solubilization of protein in aqueous surroundings makes them unable to attain sustained drug release. In order to provide hardening to protein nanoparticles, chemical crosslinkers, such as formaldehyde and glutaraldehyde, are used. The residual unreacted crosslinkers within the protein nanoparticles cause risk of toxic product formation via reacting with tissue during biodegradation in vivo [132]. However, this drawback could be overcome with the application of hydrophobic plant proteins, avoiding the need for crosslinking [125]. Protein-based nanostructures often show high unstablility to aggregation because of relatively high hydrophobicity present on the surface. However, the use of an emulsifier layer can overcome this problem, as shown in zein particles stabilization with the help of sodium caseinate [133]. Furthermore, undesirable burst expulsion usually occurred with protein-based nanostructures.

Polysaccharide-Based Nanocarriers: Polysaccharides have different enzymatic susceptibilities for specific degradation in the intestine. When polysaccharides are used as nanoparticle coating, they delay the nonspecific expulsion of encapsulated drug components as far as coating is displayed to its destined site. That is why coated nanoparticles can be diverged to different organs of the GI tract, where they are absorbed by enterocytes, thus facilitating enhanced oral bioavailability [134, 135]. Curcumin and oligo-hyaluronic acid-CUR polymer-loded RES nanoparticle showed enhanced stability, controlled release and free radical scavenging features when compared to single preparation. These outcomes make this system highly recommended for the preparation of favorable nanofood DISs using yogurt, juice and nutritional supplements [134, 136]. In addition, ternary complexes of protein-polysaccharide-surfactant prepared by using propylene glycol alginate, zein, lecithin or rhamnolipid through anti-solvent co-precipitation methods demonstrated enhanced photostability and bioaccessibility of curcumin. This suggests their applications in hydrophobic nutraceutical delivery to be used in pharmaceuticals, foods and nutritional supplements [134, 137].

Biopolymeric Nanocarriers: This is an important type of nanocarrier prepared with natural and biodegradable polymers, categorized as GRAS, such as polylactic acid, proteins, polyethylene glycol (PEG), polylactic-co-glycolic acid (PLGA) and polysaccharides [54]. Several parameters, including pH, solubility, ionic strength, conditions, chemical structure, environment temperature, chemical reactivity, glass transition temperature, susceptibility to enzymes, electrical properties and surface properties, are taken into consideration to select the polymer . Conjugated/complexed biopolymeric nanocarriers are extensively used in nutraceutical delivery systems. These are complex bipolymeric nanocarriers designed with protein, polysaccharides and biopolymers [46]. They are used for a wide variety of applications by virtue of their high encapsulation efficiency; improved bioavailability; ability to resist external factors, such as light, pressure, oxygen and temperature; and ability to modify charge [46].

3.5 IMPACT OF NANOCARRIERS/NANOPARTICLES OVER MACRONUTRIENT AND MICRONUTRIENT DEFICIENCY

Fortification of food components with macro- or micronutrients tremendously assists in balancing the total nutrient profile, which otherwise remains disturbed due to insufficient nutrient intake and related deficiencies [53]. When naturally occurring compounds, including curcumin in turmeric, vitamins from fruits, and ω-3-fatty acids in fish oil, are encapsulated in a suitable nanocarrier, they will be delivered at the required target site and subsequently utilized based on their nutritional values [138]. Food or DISs often function as medicines; however, they belong to a special category of foods, such as minerals, vitamins, essential fatty acids, amino acids, probiotics, natural products, etc., as active ingredients. DISs manage proper functioning of the human body by providing required compounds not gained sufficiently in regular food intake. DISs are available in the shape of pills, tablets, syrup or capsules (European Commission Food Supplements, 2018; [139]). They are not food additives that are used to change color, flavor or longevity [140]. The increased bioavailability of bioactive compounds could be accomplished via improving their GI solubility or protection from acidic GI secretions, or through their controlled release within the GI tract. Further, physical state, particle size and surface properties of nanomaterials applied in a food enhancer have important distinctive characteristics that affect the nutritional worth of food [141].

Iron: Iron deficiency is a majorly occurring nutritional disease that results in global public health problems. Although a number of commercial supplements are available, their poor tolerability results in noncompliance as well as ineffective cure of iron-deficiency anemia. Also, they are not properly absorbed as they are poorly soluble in water [142]. Interestingly, iron nanoformulations exhibit enhanced bioavailability, limited adverse effects, good product stability and no change in color and taste of fortified foods [143]. Further, iron nanoparticle safety has been revealed by in vitro and in vivo studies [144]. Ferritin containing an iron-oxide nanocore with an external protein shell has shown good absorbance [134, 145]. Recently, small-sized tartarate modified nano-dispersed ferrihydrite was found to provide efficient delivery of Fe (III) into the GI tract in murine models with free radical scavenging properties, therefore suggesting its potential to act as an iron-based nanoformulation for anemic populations with no side effect [146]. Studies conducted in humans showed that iron hydroxide adipate tartrate (IHAT) nanoparticles can potentially eliminate iron deficiency and subsequent improvements in hemoglobin levels without adversely affecting the GI tract [147, 148]. Absorption of iron from nano FePO4 can further be enhanced by adding zinc or magnesium oxides to the formulations [149]. Srinivasu et al. formulated nano-ferric-pyrophosphate as a potential diet supplement for a cure for anemia with oral availability of ferric-pyrophosphate nanoparticles to be 103.02% with respect to the ferrous sulfate (reference salt). The authors did not observe any adverse effects of the tested DISs [150].

Calcium: Bioavailability of calcium from traditional preparation is ~10% to 15% [151] ([134, 152]). However, recent in vivo experiments confirmed that nano-sized calcium DISs synthesized from nano CaCO3, citrate or oyster shell could improve absorption of calcium up to 89% and overall bioavailability up to 42% as well as the amount of phosphorous in bone [153–155]. A study in rats demonstrated that Ca alginate nanoparticles loaded with collagen peptide chelated Ca significantly enhanced absorption of calcium and remarkably increased mineral density and Ca concentration in femur bones. Therefore, they may serve as new supplements of calcium to be used for Ca deficiency prevention [134, 156]. Further, nanocomposite prepared from whey protein hydrolysate and calcium exhibited excellent absorbance and stability in both acidic and basic surroundings, hence showing advantage for calcium absorption in the human GI tract. In vitro studies also suggested that their application enhances bioavailability of Ca and hence can be considered as potential DISs for the improvement of bone health in humans [134, 157]. Microcrystals of calcium-hydroxyapatite have been implicated in fracture healing, repair and osteoporosis inhibition upon oral administration [158]. Furthermore, nanohydroxyapatite composite showed beneficial effects on bone regeneration and antistaphylococcal activity [134, 159]. In addition, chicken eggshell powder can be used as source of Ca for human nutrition (6% eggshell supplementation was found to be the best) [160, 161]. High-calcium yogurt was fortified with 10-nm crystals of nano-powdered eggshell (NPES) that increases calcium content by about 15% in yogurt used as food to cope with osteoporosis [162].

Selenium: Nanoform of selenium shows increased bioavailability and reduced toxicity in contrast to inorganic and organic forms [163–167]. Further, the efficiency of selenium nanoparticles (SeNPs) is highly dependent upon size that is smaller [168]. The particle size also influences the cellular absorption of nanoparticles; for instance, in vitro absorption of 0.1 μm particles was found to be higher by 2.5 to 6 folds in contrast to 1 μm and 10 μm particles, respectively [163, 169]. Owing to this fact, the choice of appropriate particle size, morphology and encapsulation material should be considered when DISs are prepared [163, 170]. Nano-selenium has advantage of selenium application in zero oxidation state (Se0) that represents excellent bioavailability and low toxicity compared to other states of oxidation, such as Se+IV, Se+VI [165, 168]. Although this system is highly unstable with easy transformation into nonfunctional form, stable nanoformulations can be achieved through encapsulation in favorable nanovehicles like chitosan (CS) [163, 171].

Magnesium: Food systems fortified with magnesium lead to improvement in health of humans [172]. Magnesium is successfully encapsulated in aqueous spheres prepared from w/w emulsions using multiple lipids (olein, rapeseed oil, miglyol and olive oil). Poly-glycerol-poly-ricinoleate (PGPR) and sodium-caseinate were used as lipophilic and hydrophilic emulsifiers, correspondingly. The oil applied in preparation affected release of magnesium from emulsions, with higher delivery rate for oils having lower viscocity and presence of higher saturated fatty acids [173, 174]. Moreover, the effect of magnesium addition in milk over protein coagulation showed different outcomes when added in free form or in encapsulated form. While the addition of free magnesium resulted in quick protein coagulation in milk upon heating, the addition of encapsulated magnesium led to no coagulation at all [173, 174].

Vitamins: Vitamin D3 is successfully encapsulated in nanocarriers prepared from corn and potato containing high amylase and starch, respectively. Nanocarriers prepared from these materials displayed granular structure with particle size in the array of 32.04 nm to 99.2 nm, and encapsulation efficiency in the variety of 22.34–94.8%. Ultrasonic cure causes increased hydrocarbon chain length that facilitates vitamin D3 binding with potato starch with enhanced thermal stability [175]. Similarly, stable vitamin E nanocapsules have been reported to be prepared from octenyl-succinic-anhydride altered starches and are used in beverage preparation [176, 177]. Vitamin D2 loaded within spray-dried and freeze-dried casein micelles when mixed with low-fat yogurt remained until 90% of their active form in comparison to the free vitamins, which remained active form up to 67% only [178]. Vitamin D-loaded re-assembled casein micelles (r-CMs) demonstrated a powerful protective effect against gastric breakdown with enhanced bioavailability up to fourfold when compared with free vitamin D [179]. Vitamin D–potato protein nanocomplexes provided strong protection against the loss of vitamin D at the time of pasteurization or under different storage conditions, thereby suggesting the use of potato protein for supplementation of hydrophobic nutraceuticals in food and liquid consumable foodstuffs with valuable effect on human health [180]. Controlled release of hydrophilic vitamin B12 in simulated GI condition was found when encapsulated in protein–lipid compound nanoparticles with three-layered organization (α-tocopherol layer, phospholipid layer and barley protein layer) with hydrated interior. Further, an in vivo study indicated that when compared with free vitamin B12, nanoparticles loaded with vitamin B12 resulted in enhanced serum vitamin B12 concentration and decreased methylmalonic acid concentration with no toxic effects. Therefore, these nanoparticles could be applied to increase absorption of vitamin B12 when administered orally [181]. Further, vitamin A palmitate-loaded NLC-enriched food items have also been implicated in several health benefits [182]. Vitamin E-containing extended chain triglyceride based-nanoemulsion showed high-quality preservation at room temperature [176]. Similarly, enhanced bioavailability of β-carotene from DISs was implicated from long chain triglycerides nanoemulsions [183]. Vitamin D encapsulated in medium chain triglycerides nanoemulsions were relatively stable at ambient temperatures, and the thermal stability was further enhanced upon the addition of sodium-dodecyl sulfate (co-surfactant) [184]. The edible (coconut) oil nanoemulsions have been observed to be stable and biocompatible with enhanced bioavailability of encapsulated α-tocopherol [185].

It has been found that casein micelles protected ω-3- polyunsaturated fatty acid, docosa-hexaenoic acid (DHA) and vitamin D2 against oxidation and UV-light-induced breakdown, respectively [186].

3.6 CONCLUSION

Nanotechnology has provided newer avenues to address the challenges associated with the development of nutritional strategies to curb MND. The primary potential area where nanotechnology applications could be a great help is to resolve pharmacokinetic issues associated with nutritional interventions. The other aspect where nanotechnology will provide promising outcomes is targeted delivery of nutrients to obtain improved health outcomes and prevent chronic disorders induced by nutrient deficiency. Applications of nanotechnology in food and nutrition research are wide, with much potential yet to unfold.

REFERENCES

1. Mousa, Aya, Amreen Naqash, and Siew Lim. "Macronutrient and micronutrient intake during pregnancy: An overview of recent evidence." *Nutrients* 11, no. 2 (2019): 443.
2. Boushey, Carol J., Ann M. Coulston, Cheryl L. Rock, and Elaine Monsen, eds. *Nutrition in the Prevention and Treatment of Disease*. Elsevier, 2001.
3. Ross, A. Catharine, Benjamin Caballero, Robert J. Cousins, Katherine L. Tucker, and Thomas R. Ziegler. *Modern Nutrition in Health and Disease*. Ed. 11. Lippincott Williams & Wilkins, 2012.
4. Morris, Alyssa L., and Shamim S. Mohiuddin. "Biochemistry, nutrients." *StatPearls [Internet]* (2020).
5. Mudgil, Deepak, and Sheweta Barak "Classification, technological properties, and sustainable sources." In *Dietary Fiber: Properties, Recovery, and Applications* Galanakis Charis , pp. 27–58. Academic Press, 2019.
6. Sweeney, Brian, Prem Puri, and Denis J. Reen. "Modulation of immune cell function by polyunsaturated fatty acids." *Pediatric Surgery International* 21, no. 5 (2005): 335–340.
7. Harris, William S. "Achieving optimal n–3 fatty acid status: The vegetarian's challenge … or not." *The American Journal of Clinical Nutrition* 100, no. suppl_1 (2014): 449S–452S.
8. Pencharz, Paul B., and Ronald O. Ball. "Different approaches to define individual amino acid requirements." *Annual Review of Nutrition* 23, no. 1 (2003): 101–116.
9. Reeds, Peter J. "Dispensable and indispensable amino acids for humans." *The Journal of Nutrition* 130, no. 7 (2000): 1835S–1840S.
10. Jefferson, Angie, and Katie Adolphus. "The effects of intact cereal grain fibers, including wheat bran on the gut microbiota composition of healthy adults: A systematic review." *Frontiers in nutrition* 6 (2019): 33.
11. Ekweagwu, E., A. E. Agwu, and E. Madukwe. "The role of micronutrients in child health: A review of the literature." *African Journal of Biotechnology* 7, no. 21 (2008).
12. Linkon, K. M. M. R., Mohammed A. Satter, S. A. Jabin, Nusrat Abedin, M. F. Islam, Laisa Ahmed Lisa, and Dipak Kumar Paul. "Mineral and heavy metal contents of some vegetable available in local market of Dhaka city in Bangladesh." *IOSR Journal of Environmental Science, Toxicology and Food Technology* 9 (2015): 2319–2399.

13. Malhotra, N., and A. Mithal. "Osteoporosis in Indians." *Indian Journal of Medical Research* 127, no. 3 (2008).
14. Martins, Joana T., Ana I. Bourbon, Ana C. Pinheiro, Luiz H. Fasolin, and António A. Vicente. "Protein-based structures for food applications: From macro to nanoscale." *Frontiers in Sustainable Food Systems* 2 (2018): 77.
15. Soetan, K. O., C. O. Olaiya, and O. E. Oyewole. "The importance of mineral elements for humans, domestic animals and plants – A review." *African Journal of Food Science* 4, no. 5 (2010): 200–222.
16. Soni, Shashank, Veerma Ram, and Anurag Verma. "Updates on approaches to increase the residence time of drug in the stomach for site specific delivery: Brief review." *International Current Pharmaceutical Journal* 6, no. 11 (2018): 81–91.
17. Zamberlin, Šimun, Neven Antunac, Jasmina Havranek, and Dubravka Samaržija. "Mineral elements in milk and dairy products." *Mljekarstvo: časopis za unaprjeđenje proizvodnje i prerade mlijeka* 62, no. 2 (2012): 111–125.
18. Abdel-Aziz, Shadia M., Mohamed S. Abdel-Aziz, and Neelam Garg. "Health benefits of trace elements in human diseases." In *Microbes in Food and Health* Garg Neelam, Mohammad Abdel-Aziz Shadia, Aeron Abhinav, pp. 117–142. Springer, Cham, 2016.
19. Berti, Peter R., Cynthia Fallu, and Yesmina Cruz Agudo. "A systematic review of the nutritional adequacy of the diet in the Central Andes." *Revista Panamericana de Salud Pública* 36 (2014): 314–323.
20. Tulchinsky, Theodore H. "Micronutrient deficiency conditions: Global health issues." *Public Health Reviews* 32, no. 1 (2010): 243–255.
21. Jacobsen, Kathryn H. *Introduction to Global Health.* Jones & Bartlett Publishers, 2014.
22. Jafari, Seid Mahdi, Sabike Vakili, and Danial Dehnad. "Production of a functional yogurt powder fortified with nanoliposomal vitamin D through spray drying." *Food and Bioprocess Technology* 12, no. 7 (2019): 1220–1231.
23. Scrinis, Gyorgy. "Reframing malnutrition in all its forms: A critique of the tripartite classification of malnutrition." *Global Food Security* 26 (2020): 100396.
24. Senapati, Sudipta, Arun Kumar Mahanta, Sunil Kumar, and Pralay Maiti. "Controlled drug delivery vehicles for cancer treatment and their performance." *Signal Transduction and Targeted Therapy* 3, no. 1 (2018): 1–19.
25. Keller, Heather, Marian A.E. de van der Schueren, G.L.I.M. Consortium, Gordon L. Jensen, Rocco Barazzoni, Charlene Compher, M. Isabel TD Correia et al. "Global leadership initiative on malnutrition (GLIM): Guidance on validation of the operational criteria for the diagnosis of protein-energy malnutrition in adults." *Journal of Parenteral and Enteral Nutrition* 44, no. 6 (2020): 992–1003.
26. Ritchie, Hannah, and Max Roser. "Micronutrient deficiency." *Our World in data* (2017). https://ourworldindata.org/micronutrient-deficiency
27. Andersson, Maria, Bruno de Benoist, Ian Darnton-Hill, and François Delange. "Iodine deficiency in Europe: A continuing public health problem, Geneva: World Health Organization." *Ref Type: Report* (2007).
28. Centers for Disease Control and Prevention (CDC). "Safer and healthier foods." *MMWR. Morbidity and Mortality Weekly Report* 48, no. 40 (1999): 905–913.
29. Scrimshaw, Nevin S. "Fifty-five-year personal experience with human nutrition worldwide." *Annual Review of Nutrition* 27 (2007): 1–18.
30. Huskisson, E., S. Maggini, and M. Ruf. "The influence of micronutrients on cognitive function and performance." *Journal of International Medical Research* 35, no. 1 (2007): 1–19.

31. Monsen, Elaine R. "Dietary reference intakes for the antioxidant nutrients: Vitamin C, vitamin E, selenium, and carotenoids." *Journal of the American Dietetic Association* 100, no. 6 (2000): 637–640.
32. Pitkin, Roy M., L. Allen, L. B. Bailey, and M. Bernfield. "Dietary reference intakes for thiamin, riboflavin, niacin, vitamin B6, folate, vitamin B12, pantothenic acid, biotin and choline." National Academy Press, 2000.
33. Fairfield, Kathleen M., and Robert H. Fletcher. "Vitamins for chronic disease prevention in adults: Scientific review." *JAMA* 287, no. 23 (2002): 3116–3126.
34. Feng, Tao, Ke Wang, Fangfang Liu, Ran Ye, Xiao Zhu, Haining Zhuang, and Zhimin Xu. "Structural characterization and bioavailability of ternary nanoparticles consisting of amylose, α-linoleic acid and β-lactoglobulin complexed with naringin." *International Journal of Biological Macromolecules* 99 (2017): 365–374.
35. Fletcher, Robert H., and Kathleen M. Fairfield. "Vitamins for chronic disease prevention in adults: Clinical applications." *JAMA* 287, no. 23 (2002): 3127–3129.
36. Fouache, Jacques, and Daniel Lincot. "Study of atomic layer epitaxy of zinc oxide by in-situ quartz crystal microgravimetry." *Applied Surface Science* 153, no. 4 (2000): 223–234.
37. Dable-Tupas, Genevieve, Maria Catherine B. Otero, and Leslie Bernolo. "Functional foods and health benefits." In *Functional Foods and Nutraceuticals* Aluko Rotimi E., pp. 1–11. Springer, Cham, 2020.
38. Kaur, Sumeet, and Madhusweta Das. "Functional foods: An overview." *Food Science and Biotechnology* 20, no. 4 (2011): 861.
39. Dary, Omar, and Richard Hurrell. "Guidelines on food fortification with micronutrients." *World Health Organization, Food and Agricultural Organization of the United Nations: Geneva, Switzerland* (2006): 1–376.
40. World Health Organization. "Guidelines on food fortification with micronutrients." (2006).
41. Rapaka, Rao S., and Paul M. Coates. "Dietary supplements and related products: A brief summary." *Life Sciences* 78, no. 18 (2006): 2026–2032.
42. Huma, Nuzhat, Salim-Ur-Rehman, Faqir Muhammad Anjum, M. Anjum Murtaza, and Munir A. Sheikh. "Food fortification strategy – Preventing iron deficiency anemia: A review." *Critical Reviews in Food Science and Nutrition* 47, no. 3 (2007): 259–265.
43. Parada, J., and J. M. Aguilera. "Food microstructure affects the bioavailability of several nutrients." *Journal of Food Science* 72, no. 2 (2007): R21–R32.
44. Aguilera, José Miguel. "The food matrix: implications in processing, nutrition and health." *Critical Reviews in Food Science and Nutrition* 59, no. 22 (2019): 3612–3629.
45. Thomopoulos, Rallou, Cédric Baudrit, Nadia Boukhelifa, Rachel Boutrou, Patrice Buche, Elisabeth Guichard, Valérie Guillard et al. "Multi-criteria reverse engineering for food: Genesis and ongoing advances." *Food Engineering Reviews* 11, no. 1 (2019): 44–60.
46. Dima, Cristian, Elham Assadpour, Stefan Dima, and Seid Mahdi Jafari. "Bioavailability of Nutraceuticals: Role of the Food Matrix, Processing Conditions, the Gastrointestinal Tract, and Nanodelivery Systems." *Comprehensive Reviews in Food Science and Food Safety* 19, no. 3 (2020): 954–994.
47. Jafari, Seid Mahdi, and David Julian McClements. "Nanotechnology approaches for increasing nutrient bioavailability." *Advances in Food and Nutrition Research* 81 (2017): 1–30.
48. Nedovic, Viktor, Ana Kalusevic, Verica Manojlovic, Steva Levic, and Branko Bugarski. "An overview of encapsulation technologies for food applications." *Procedia Food Science* 1 (2011): 1806–1815.

49. Pressman, Peter, Roger A. Clemens, and A. Wallace Hayes. "Bioavailability of micronutrients obtained from supplements and food: A survey and case study of the polyphenols." *Toxicology Research and Application* 1 (2017): DOI:10.1177/2397847317696366
50. Augustin, Mary Ann, and Luz Sanguansri. "Challenges and solutions to incorporation of nutraceuticals in foods." *Annual Review of Food Science and Technology* 6 (2015): 463–477.
51. McClements, David Julian. "Recent progress in hydrogel delivery systems for improving nutraceutical bioavailability." *Food Hydrocolloids* 68 (2017): 238–245.
52. McClements, David Julian. "The future of food colloids: Next-generation nanoparticle delivery systems." *Current Opinion in Colloid & Interface Science* 28 (2017): 7–14.
53. Dwyer, Johanna T., Kathryn L. Wiemer, Omar Dary, Carl L. Keen, Janet C. King, Kevin B. Miller, Martin A. Philbert et al. "Fortification and health: Challenges and opportunities." *Advances in Nutrition* 6, no. 1 (2015): 124–131.
54. Assadpour, Elham, and Seid Mahdi Jafari. "An overview of biopolymer nanostructures for encapsulation of food ingredients." *Biopolymer Nanostructures for Food Encapsulation Purposes* (2019): 1–35.
55. Assadpour, Elham, and Seid Mahdi Jafari. "Advances in spray-drying encapsulation of food bioactive ingredients: From microcapsules to nanocapsules." *Annual Review of Food Science and Technology* 10 (2019): 103–131.
56. Rostamabadi, Hadis, Seid Reza Falsafi, and Seid Mahdi Jafari. "Nanostructures of starch for encapsulation of food ingredients." In *Biopolymer Nanostructures for Food Encapsulation Purposes* Jafari Seid Mahdi, pp. 419–462. Academic Press, 2019.
57. Rostamabadi, Hadis, Seid Reza Falsafi, and Seid Mahdi Jafari. "Nano-helices of amylose for encapsulation of food ingredients." In *Biopolymer Nanostructures for Food Encapsulation Purposes*, pp. 463–491. Academic Press, 2019.
58. Rostamabadi, Hadis, Seid Reza Falsafi, and Seid Mahdi Jafari. "Nanoencapsulation of carotenoids within lipid-based nanocarriers." *Journal of Controlled Release* 298 (2019): 38–67.
59. Rostamabadi, Hadis, Seid Reza Falsafi, and Seid Mahdi Jafari. "Starch-based nanocarriers as cutting-edge natural cargos for nutraceutical delivery." *Trends in Food Science & Technology* 88 (2019): 397–415.
60. Shin, Gye Hwa, Jun Tae Kim, and Hyun Jin Park. "Recent developments in nanoformulations of lipophilic functional foods." *Trends in Food Science & Technology* 46, no. 1 (2015): 144–157.
61. Rafiee, Zahra, Mohammad Nejatian, Marjan Daeihamed, and Seid Mahdi Jafari. "Application of curcumin-loaded nanocarriers for food, drug and cosmetic purposes." *Trends in Food Science & Technology* 88 (2019): 445–458.
62. Yousefi, Mohammad, and Seid Mahdi Jafari. "Recent advances in application of different hydrocolloids in dairy products to improve their techno-functional properties." *Trends in Food Science & Technology* 88 (2019): 468–483.
63. Katouzian, Iman, and Seid Mahdi Jafari. "Protein nanotubes as state-of-the-art nanocarriers: Synthesis methods, simulation and applications." *Journal of Controlled Release* 303 (2019): 302–318.
64. Augustin, Mary Ann, and Yacine Hemar. "Nano-and micro-structured assemblies for encapsulation of food ingredients." *Chemical Society Reviews* 38, no. 4 (2009): 902–912.
65. Bhushan, Bharat, Dan Luo, Scott R. Schricker, Wolfgang Sigmund, and Stefan Zauscher, eds. *Handbook of Nanomaterials Properties*. Springer Science & Business Media, 2014.

66. Shukla, Ashutosh Kumar, and Siavash Iravani, eds. *Green Synthesis, Characterization and Applications of Nanoparticles*. Elsevier, 2018.
67. Singh, Om V., ed. *Bio-nanoparticles: Biosynthesis and Sustainable Biotechnological Implications*. John Wiley & Sons, 2015.
68. Jampilek, J. O. S. E. F., and K. Kralova. "Application of nanobioformulations for controlled release and targeted biodistribution of drugs." In *Nanobiomaterials: Applications in Drug Delivery*; Sharma, AK, Keservani, RK, Kesharwani, RK Eds. (2018): 131–208.
69. Natarajan, J., V. Karri, and D. Anindita. "Nanostructured Lipid Carrier (NLC): A promising drug delivery system." *Global Journal of Nanomedicine* 1, no. 5 (2017): 120–125.
70. Pardeshi, Chandrakantsing, Pravin Rajput, Veena Belgamwar, Avinash Tekade, Ganesh Patil, Kapil Chaudhary, and Abhijeet Sonje. "Solid lipid based nanocarriers: An overview." *Acta Pharmaceutica* 62, no. 4 (2012): 433–472.
71. Wen, Zhen, Bo Liu, Zongkun Zheng, Xinkui You, Yitao Pu, and Qiong Li. "Preparation of liposomes entrapping essential oil from Atractylodes macrocephala Koidz by modified RESS technique." *Chemical Engineering Research and Design* 88, no. 8 (2010): 1102–1107.
72. Vega-Vásquez, P., N.S. Mosier and J. Irudayaraj. 2020. "Nanoscale drug delivery systems: From medicine to agriculture." *Frontiers in Bioengineering and Biotechnology* 8: 79.
73. Godfroy, Isaac. "Polymeric micelles – The future of oral drug delivery." *Journal of Biomaterial Applications* 3 (2009): 216–232.
74. Joo, Kye-Il, Liang Xiao, Shuanglong Liu, Yarong Liu, Chi-Lin Lee, Peter S. Conti, Michael K. Wong, Zibo Li, and Pin Wang. "Crosslinked multilamellar liposomes for controlled delivery of anticancer drugs." *Biomaterials* 34, no. 12 (2013): 3098–3109.
75. Joseph, Nitin, Abhinav Kumar, Harjas Singh, Mohammed Shaheen, Kriti Das, and Apurva Shrivastava. "Nutritional supplement and functional food use among medical students in India." *Journal of Dietary Supplements* 15, no. 6 (2018): 951–964.
76. Vabbilisetty, Pratima, and Xue-Long Sun. "Liposome surface functionalization based on different anchoring lipids via Staudinger ligation." *Organic & Biomolecular Chemistry* 12, no. 8 (2014): 1237–1244.
77. Donsì, Francesco. "Applications of nanoemulsions in foods." In *Nanoemulsions* Jefari Seid, D. Julian, McClements, pp. 349–377. Academic Press, 2018.
78. Beltrán-Gracia, Esteban, Adolfo López-Camacho, Inocencio Higuera-Ciapara, Jesús B. Velázquez-Fernández, and Alba A. Vallejo-Cardona. "Nanomedicine review: Clinical developments in liposomal applications." *Cancer Nanotechnology* 10, no. 1 (2019): 1–40.
79. Mukherjee, S., S. Ray, and R. S. Thakur. "Solid lipid nanoparticles: A modern formulation approach in drug delivery system." *Indian Journal of Pharmaceutical Sciences* 71, no. 4 (2009): 349.
80. Naseri, Neda, Hadi Valizadeh, and Parvin Zakeri-Milani. "Solid lipid nanoparticles and nanostructured lipid carriers: Structure, preparation and application." *Advanced Pharmaceutical Bulletin* 5, no. 3 (2015): 305.
81. Allen, Theresa M., and Pieter R. Cullis. "Liposomal drug delivery systems: From concept to clinical applications." *Advanced Drug Delivery Reviews* 65, no. 1 (2013): 36–48.
82. Khorasani, S., M. Danaei, and M. R. Mozafari. "Nanoliposome technology for the food and nutraceutical industries." *Trends in Food Science & Technology* 79 (2018): 106–115.

83. Ghanbarzadeh, Babak, Afshin Babazadeh, and Hamed Hamishehkar. "Nanophytosome as a potential food-grade delivery system." *Food Bioscience* 15 (2016): 126–135.
84. Bozzuto, Giuseppina, and Agnese Molinari. "Liposomes as nanomedical devices." *International Journal of Nanomedicine* 10 (2015): 975.
85. Ekambaram, P., A. Abdul Hasan Sathali, and K. Priyanka. "Solid lipid nanoparticles: A review." *Scientific Reviews and Chemical Communications* 2, no. 1 (2012): 80–112.
86. Kamble, M.S., Vaidya, K.K., Bhosale, A.V., Chaudhari, P.D. 2012. "Solid lipid nanoparticles and nanostructured lipid carriers – An overview". *International Journal of Pharmaceutical, Chemical and Biological Sciences* 2, no. 4 (2012): 681–691.
87. Kumar, Ajay, Shital Badde, Ravindra Kamble, and Varsha B. Pokharkar. "Development and characterization of liposomal drug delivery system for nimesulide." *International Journal of Pharmacy and Pharmaceutical Sciences* 2, no. 4 (2010): 87–89.
88. Zamani, Parvin, Amir Abbas Momtazi-Borojeni, Maryam Ebrahimi Nik, Reza Kazemi Oskuee, and Amirhossein Sahebkar. "Nanoliposomes as the adjuvant delivery systems in cancer immunotherapy." *Journal of Cellular Physiology* 233, no. 7 (2018): 5189–5199.
89. Gonda, Amber, Nanxia Zhao, Jay V. Shah, Hannah R. Calvelli, Harini Kantamneni, Nicola L. Francis, and Vidya Ganapathy. "Engineering tumor-targeting nanoparticles as vehicles for precision nanomedicine." *Med One* 4 (2019): e190021.
90. Fouladi, Farnaz, Kristine J. Steffen, and Sanku Mallik. "Enzyme-responsive liposomes for the delivery of anticancer drugs." *Bioconjugate Chemistry* 28, no. 4 (2017): 857–868.
91. Gao, Yong, Chanuka Wijewardhana, and Jamie FS Mann. "Virus-like particle, liposome, and polymeric particle-based vaccines against HIV-1." *Frontiers in Immunology* 9 (2018): 345.
92. Lombardo, Domenico, Pietro Calandra, Davide Barreca, Salvatore Magazù, and Mikhail A. Kiselev. "Soft interaction in liposome nanocarriers for therapeutic drug delivery." *Nanomaterials* 6, no. 7 (2016): 125.
93. Bulbake, Upendra, Sindhu Doppalapudi, Nagavendra Kommineni, and Wahid Khan. "Liposomal formulations in clinical use: An updated review." *Pharmaceutics* 9, no. 2 (2017): 12.
94. Nardecchia, Stefania, Paola Sánchez-Moreno, Juan de Vicente, Juan A. Marchal, and Houria Boulaiz. "Clinical trials of thermosensitive nanomaterials: An overview." *Nanomaterials* 9, no. 2 (2019): 191.
95. Akbarzadeh, Abolfazl, Rogaie Rezaei-Sadabady, Soodabeh Davaran, Sang Woo Joo, Nosratollah Zarghami, Younes Hanifehpour, Mohammad Samiei, Mohammad Kouhi, and Kazem Nejati-Koshki. "Liposome: classification, preparation, and applications." *Nanoscale Research Letters* 8, no. 1 (2013): 1–9.
96. Ramalho, Maria J., Manuel AN Coelho, and Maria C. Pereira. "Nanoparticles for delivery of vitamin D: Challenges and opportunities." In *A Critical Evaluation of Vitamin D—Clinical Overview*; Gowder, S., Ed: 11. IntechOpen, 2017.
97. Dwivedi, Chandraprakash, Roshni Sahu, Sandip Prasad Tiwari, Trilochan Satapathy, and Amit Roy. "Role of liposome in novel drug delivery system." *Journal of Drug Delivery and Therapeutics* 4, no. 2 (2014): 116–129.
98. Dwyer, Johanna T., Paul M. Coates, and Michael J. Smith. "Dietary supplements: Regulatory challenges and research resources." *Nutrients* 10, no. 1 (2018): 41.

99. Pardeike, Jana, Aiman Hommoss, and Rainer H. Müller. "Lipid nanoparticles (SLN, NLC) in cosmetic and pharmaceutical dermal products." *International Journal of Pharmaceutics* 366, no. 1-2 (2009): 170–184.
100. Patidar, Ajay, Devendra Singh Thakur, Peeyush Kumar, and Jhageshwar Verma. "A review on novel lipid based nanocarriers." *International Journal of Pharmacy and Pharmaceutical Sciences* 2, no. 4 (2010): 30–35.
101. Das, Surajit, and Anumita Chaudhury. "Recent advances in lipid nanoparticle formulations with solid matrix for oral drug delivery." *AAPS PharmSciTech* 12, no. 1 (2011): 62–76.
102. Kim, Byung-Do, Kun Na, and Hoo-Kyun Choi. "Preparation and characterization of solid lipid nanoparticles (SLN) made of cacao butter and curdlan." *European Journal of Pharmaceutical Sciences* 24, no. 2–3 (2005): 199–205.
103. Ramalingam, Prakash, Sang Woo Yoo, and Young Tag Ko. "Nanodelivery systems based on mucoadhesive polymer coated solid lipid nanoparticles to improve the oral intake of food curcumin." *Food Research International* 84 (2016): 113–119.
104. Feng, Lan, and Russell J. Mumper. "A critical review of lipid-based nanoparticles for taxane delivery." *Cancer Letters* 334, no. 2 (2013): 157–175.
105. Yoon, Goo, Jin Woo Park, and In-Soo Yoon. "Solid lipid nanoparticles (SLNs) and nanostructured lipid carriers (NLCs): recent advances in drug delivery." *Journal of Pharmaceutical Investigation* 43, no. 5 (2013): 353–362.
106. Mendes, A. I., A. C. Silva, J. A. M. Catita, F. Cerqueira, C. Gabriel, and Carla Martins Lopes. "Miconazole-loaded nanostructured lipid carriers (NLC) for local delivery to the oral mucosa: Improving antifungal activity." *Colloids and Surfaces B: Biointerfaces* 111 (2013): 755–763.
107. Wang, Shu, Rui Su, Shufang Nie, Ming Sun, Jia Zhang, Dayong Wu, and Naima Moustaid-Moussa. "Application of nanotechnology in improving bioavailability and bioactivity of diet-derived phytochemicals." *The Journal of Nutritional Biochemistry* 25, no. 4 (2014): 363–376.
108. Makwana, Vivek, Rashmi Jain, Komal Patel, Manish Nivsarkar, and Amita Joshi. "Solid lipid nanoparticles (SLN) of Efavirenz as lymph targeting drug delivery system: Elucidation of mechanism of uptake using chylomicron flow blocking approach." *International Journal of Pharmaceutics* 495, no. 1 (2015): 439–446.
109. Guimarães, Kleber L., and Maria Inês Ré. "Lipid nanoparticles as carriers for cosmetic ingredients: The first (SLN) and the second generation (NLC)." In *Nanocosmetics and nanomedicines*, pp. 101–122. Springer, Berlin, Heidelberg, 2011.
110. Müller, Rainer H., Sven Staufenbiel, and C. M. Keck. "Lipid Nanoparticles (SLN, NLC) for innovative consumer care & household products." *Household and Personal Care Today* 9 (2014): 18–24.
111. Beloqui, Ana, María Ángeles Solinís, Anne des Rieux, Véronique Préat, and Alicia Rodríguez-Gascón. "Dextran–protamine coated nanostructured lipid carriers as mucus-penetrating nanoparticles for lipophilic drugs." *International Journal of Pharmaceutics* 468, no. 1–2 (2014): 105–111.
112. Araujo, J., E. Gonzalez-Mira, M. A. Egea, M. L. Garcia, and E. B. Souto. "Optimization and physicochemical characterization of a triamcinolone acetonide-loaded NLC for ocular antiangiogenic applications." *International Journal of Pharmaceutics* 393, no. 1–2 (2010): 168–176.
113. Arora, Shweta, Javed Ali, Alka Ahuja, Roop K. Khar, and Sanjula Baboota. "Floating drug delivery systems: a review." *AAPS PharmSciTech* 6, no. 3 (2005): E372–E390.

114. Doktorovova, Slavomira, Eliana B. Souto, and Amélia M. Silva. "Nanotoxicology applied to solid lipid nanoparticles and nanostructured lipid carriers – A systematic review of in vitro data." *European Journal of Pharmaceutics and Biopharmaceutics* 87, no. 1 (2014): 1–18.
115. Tamjidi, Fardin, Mohammad Shahedi, Jaleh Varshosaz, and Ali Nasirpour. "Nanostructured lipid carriers (NLC): A potential delivery system for bioactive food molecules." *Innovative Food Science & Emerging Technologies* 19 (2013): 29–43.
116. Beloqui, Ana, María Ángeles Solinís, Alicia Rodríguez-Gascón, António J. Almeida, and Véronique Préat. "Nanostructured lipid carriers: Promising drug delivery systems for future clinics." *Nanomedicine: Nanotechnology, Biology and Medicine* 12, no. 1 (2016): 143–161.
117. Shidhaye, S. S., Reshma Vaidya, Sagar Sutar, Arati Patwardhan, and V. J. Kadam. "Solid lipid nanoparticles and nanostructured lipid carriers-innovative generations of solid lipid carriers." *Current Drug Delivery* 5, no. 4 (2008): 324–331.
118. Haider, Mohamed, Shifaa M. Abdin, Leena Kamal, and Gorka Orive. "Nanostructured lipid carriers for delivery of chemotherapeutics: A review." *Pharmaceutics* 12, no. 3 (2020): 288.
119. Seong Keun, Cheol Seong Hwang, Sang-Hee Ko Park, and Sun Jin Yun. "Comparison between ZnO films grown by atomic layer deposition using H2O or O3 as oxidant." *Thin Solid Films* 478, no. 1–2 (2005): 103–108.
120. Sun, Hongcheng, Quan Luo, Chunxi Hou, and Junqiu Liu. "Nanostructures based on protein self-assembly: From hierarchical construction to bioinspired materials." *Nano Today* 14 (2017): 16–41.
121. McClements, David Julian, Eric Andrew Decker, Yeonhwa Park, and Jochen Weiss. "Structural design principles for delivery of bioactive components in nutraceuticals and functional foods." *Critical Reviews in Food Science and Nutrition* 49, no. 6 (2009): 577–606.
122. Gupta, Pratima, and Kush Kumar Nayak. "Characteristics of protein-based biopolymer and its application." *Polymer Engineering & Science* 55, no. 3 (2015): 485–498.
123. Zimet, Patricia, and Yoav D. Livney. "Beta-lactoglobulin and its nanocomplexes with pectin as vehicles for ω-3 polyunsaturated fatty acids." *Food Hydrocolloids* 23, no. 4 (2009): 1120–1126.
124. Sponton, Osvaldo E., Adrián A. Perez, Javier V. Ramel, and Liliana G. Santiago. "Protein nanovehicles produced from egg white. Part 1: Effect of pH and heat treatment time on particle size and binding capacity." *Food Hydrocolloids* 73 (2017): 67–73.
125. Elzoghby, Ahmed O., Wael M. Samy, and Nazik A. Elgindy. "Protein-based nanocarriers as promising drug and gene delivery systems." *Journal of Controlled Release* 161, no. 1 (2012): 38–49.
126. Horne, David S. "Casein micelle structure: models and muddles." *Current Opinion in Colloid & Interface Science* 11, no. 2–3 (2006): 148–153.
127. Elzoghby, Ahmed O., Wael S. Abo El-Fotoh, and Nazik A. Elgindy. "Casein-based formulations as promising controlled release drug delivery systems." *Journal of Controlled Release* 153, no. 3 (2011): 206–216.
128. Livney, Yoav D. "Milk proteins as vehicles for bioactives." *Current Opinion in Colloid & Interface Science* 15, no. 1–2 (2010): 73–83.
129. Pan, Kang, Qixin Zhong, and Seung Joon Baek. "Enhanced dispersibility and bioactivity of curcumin by encapsulation in casein nanocapsules." *Journal of Agricultural and Food Chemistry* 61, no. 25 (2013): 6036–6043.

130. Chen, Lingyun, Gabriel E. Remondetto, and Muriel Subirade. “Food protein-based materials as nutraceutical delivery systems.” *Trends in Food Science & Technology* 17, no. 5 (2006): 272–283.
131. Malafaya, Patrícia B., Gabriela A. Silva, and Rui L. Reis. “Natural–origin polymers as carriers and scaffolds for biomolecules and cell delivery in tissue engineering applications.” *Advanced Drug Delivery Reviews* 59, no. 4–5 (2007): 207–233.
132. Chiono, Valeria, Ettore Pulieri, Giovanni Vozzi, Gianluca Ciardelli, Arti Ahluwalia, and Paolo Giusti. “Genipin-crosslinked chitosan/gelatin blends for biomedical applications.” *Journal of Materials Science: Materials in Medicine* 19, no. 2 (2008): 889–898.
133. Patel, Ashok R., Elisabeth CM Bouwens, and Krassimir P. Velikov. “Sodium caseinate stabilized zein colloidal particles.” *Journal of Agricultural and Food Chemistry* 58, no. 23 (2010): 12497–12503.
134. Jampilek, Josef, Jiri Kos, and Katarina Kralova. “Potential of nanomaterial applications in dietary supplements and foods for special medical purposes.” *Nanomaterials* 9, no. 2 (2019): 296.
135. Sampathkumar, Kaarunya, and Say Chye Joachim Loo. “Targeted gastrointestinal delivery of nutraceuticals with polysaccharide-based coatings.” *Macromolecular Bioscience* 18, no. 4 (2018): 1700363.
136. Guo, Chunjing, Jungang Yin, and Daquan Chen. “Co-encapsulation of curcumin and resveratrol into novel nutraceutical hyalurosomes nano-food delivery system based on oligo-hyaluronic acid-curcumin polymer.” *Carbohydrate Polymers* 181 (2018): 1033–1037.
137. Dai, Lei, Yang Wei, Cuixia Sun, Like Mao, David Julian McClements, and Yanxiang Gao. “Development of protein-polysaccharide-surfactant ternary complex particles as delivery vehicles for curcumin.” *Food Hydrocolloids* 85 (2018): 75–85.
138. Rao, Pooja J., and Madhav M. Naidu. “Nanoencapsulation of bioactive compounds for nutraceutical food.” In *Nanoscience in Food and Agriculture 2*, pp. 129–156. Springer, Cham, 2016.
139. National Institutes of Health. 2018. Dietary Supplements: background Information. (https://ods.od.nih.gov/factsheets/DietarySupplements-HealthProfessional/)
140. Kuhnert, Peter. “Foods, 3. Food additives.” In *Ullmann’s Encyclopedia of Industrial Chemistry*, pp. 1–52., VCH, 2000.
141. Oehlke, Kathleen, Marta Adamiuk, Diana Behsnilian, Volker Gräf, Esther Mayer-Miebach, Elke Walz, and Ralf Greiner. “Potential bioavailability enhancement of bioactive compounds using food-grade engineered nanomaterials: A review of the existing evidence.” *Food & Function* 5, no. 7 (2014): 1341–1359.
142. Zanella, Daniele, Elena Bossi, Rosalba Gornati, Carlos Bastos, Nuno Faria, and Giovanni Bernardini. “Iron oxide nanoparticles can cross plasma membranes.” *Scientific Reports* 7, no. 1 (2017): 1–10.
143. Hosny, Khaled Mohamed, Zainy Mohammed Banjar, Amani H. Hariri, and Ali Habiballah Hassan. “Solid lipid nanoparticles loaded with iron to overcome barriers for treatment of iron deficiency anemia.” *Drug Design, Development and Therapy* 9 (2015): 313.
144. Gornati, Rosalba, Elisa Pedretti, Federica Rossi, Francesca Cappellini, Michela Zanella, Iolanda Olivato, Enrico Sabbioni, and Giovanni Bernardini. “Zerovalent Fe, Co and Ni nanoparticle toxicity evaluated on SKOV-3 and U87 cell lines.” *Journal of Applied Toxicology* 36, no. 3 (2016): 385–393.
145. Lönnerdal, Bo, Annika Bryant, Xiaofeng Liu, and Elizabeth C. Theil. “Iron absorption from soybean ferritin in nonanemic women.” *The American Journal of Clinical Nutrition* 83, no. 1 (2006): 103–107.

146. Powell, Jonathan J., Sylvaine FA Bruggraber, Nuno Faria, Lynsey K. Poots, Nicole Hondow, Timothy J. Pennycook, Gladys O. Latunde-Dada, Robert J. Simpson, Andy P. Brown, and Dora IA Pereira. "A nano-disperse ferritin-core mimetic that efficiently corrects anemia without luminal iron redox activity." *Nanomedicine: Nanotechnology, Biology and Medicine* 10, no. 7 (2014): 1529–1538.
147. Pereira, Dora IA, Sylvaine FA Bruggraber, Nuno Faria, Lynsey K. Poots, Mani A. Tagmount, Mohamad F. Aslam, David M. Frazer, Chris D. Vulpe, Gregory J. Anderson, and Jonathan J. Powell. "Nanoparticulate iron (III) oxo-hydroxide delivers safe iron that is well absorbed and utilised in humans." *Nanomedicine: Nanotechnology, Biology and Medicine* 10, no. 8 (2014): 1877–1886.
148. Pereira, Dora IA, Nuredin I. Mohammed, Ogochukwu Ofordile, Famalang Camara, Bakary Baldeh, Thomas Mendy, Chilel Sanyang et al. "A novel nano-iron supplement to safely combat iron deficiency and anaemia in young children: The IHAT-GUT double-blind, randomised, placebo-controlled trial protocol." *Gates Open Research* 2 (2018): 48.
149. Hilty, Florentine M., Myrtha Arnold, Monika Hilbe, Alexandra Teleki, Jesper TN Knijnenburg, Felix Ehrensperger, Richard F. Hurrell, Sotiris E. Pratsinis, Wolfgang Langhans, and Michael B. Zimmermann. "Iron from nanocompounds containing iron and zinc is highly bioavailable in rats without tissue accumulation." *Nature Nanotechnology* 5, no. 5 (2010): 374–380.
150. Srinivasu, Bindu Y., Gopa Mitra, Monita Muralidharan, Deepsikha Srivastava, Jennifer Pinto, Prashanth Thankachan, Sudha Suresh et al. "Beneficiary effect of nanosizing ferric pyrophosphate as food fortificant in iron deficiency anemia: Evaluation of bioavailability, toxicity and plasma biomarker." *RSC Advances* 5, no. 76 (2015): 61678–61687.
151. Park, H. S., B. J. Jeon, J. Ahn, and H. S. Kwak. "Effects of nanocalcium supplemented milk on bone calcium metabolism in ovariectomized rats." *Asian-Australasian Journal of Animal Sciences* 20, no. 8 (2007): 1266–1271.
152. Khashayar, Patricia, Abbasali Keshtkar, Mehdi Ebrahimi, and Bagher Larijani. "Nano calcium supplements: Friends or foes?" *Journal of Bone Biology and Osteoporosis* 1 (2015): 32–33.
153. Choi, Hyeon-Son, JeungHi Han, Seungsik Chung, Yang Hee Hong, and Hyung Joo Suh. "Nano-calcium ameliorates ovariectomy-induced bone loss in female rats." *Food Science of Animal Resources* 33, no. 4 (2013): 515–521.
154. Erfanian, Arezoo, Hamed Mirhosseini, Mohd Yazid Abd Manap, Babak Rasti, and Mohd Hair Bejo. "Influence of nano-size reduction on absorption and bioavailability of calcium from fortified milk powder in rats." *Food research international* 66 (2014): 1–11.
155. Erfanian, Arezoo, Babak Rasti, and Yazid Manap. "Comparing the calcium bioavailability from two types of nano-sized enriched milk using in-vivo assay." *Food Chemistry* 214 (2017): 606–613.
156. Guo, Honghui, Zhuan Hong, and Ruizao Yi. "Core-shell collagen peptide chelated calcium/calcium alginate nanoparticles from fish scales for calcium supplementation." *Journal of Food Science* 80, no. 7 (2015): N1595–N1601.
157. Cai Xixi, Zhao Lina, Wang Shaoyun, and Rao Pingfan. "Fabrication and characterization of the nano-composite of whey protein hydrolysate chelated with calcium." *Food & Function* 6, no. 3 (2015): 816–823.
158. Noor, Zairin. "Nanohydroxyapatite application to osteoporosis management." *Journal of Osteoporosis* 2013 (2013): 679025.
159. Severin, A. V., S. E. Mazina, and I. V. Melikhov. "Physicochemical aspects of the antiseptic action of nanohydroxyapatite." *Biophysics* 54, no. 6 (2009): 701–705.

160. Chakraborty, Aastha P., and Madhavi Gaonkar. "Eggshell as calcium supplement tablet." *International Journal of Animal Biotechnology and Applications* 2, no. 1 (2016). DOI: 10.37628/ijaba.v2i1.55
161. Ray, Subhajit, Amit Kumar Barman, Pradip Kumar Roy, and Bipin Kumar Singh. "Chicken eggshell powder as dietary calcium source in chocolate cakes." *The Pharma Innovation* 6, no. 9 (2017): 1.
162. El-Shibiny, Safinaze, Mona Abd El-Kader Mohamed Abd El-Gawad, Fayza Mohamed Assem, and Samah Mosbah El-Sayed. "The use of nano-sized eggshell powder for calcium fortification of cow's and buffalo's milk yogurts." *Acta Scientiarum Polonorum Technologia Alimentaria* 17, no. 1 (2018): 37–49.
163. Hosnedlova, Bozena, Marta Kepinska, Sylvie Skalickova, Carlos Fernandez, Branislav Ruttkay-Nedecky, Qiuming Peng, Mojmir Baron et al. "Nano-selenium and its nanomedicine applications: A critical review." *International Journal of Nanomedicine* 13 (2018): 2107.
164. Shi, Liguang, Wenjuan Xun, Wenbin Yue, Chunxiang Zhang, Youshe Ren, Lei Shi, Qian Wang, Rujie Yang, and Fulin Lei. "Effect of sodium selenite, Se-yeast and nano-elemental selenium on growth performance, Se concentration and antioxidant status in growing male goats." *Small Ruminant Research* 96, no. 1 (2011): 49–52.
165. Wang, Huali, Jinsong Zhang, and Hanqing Yu. "Elemental selenium at nano size possesses lower toxicity without compromising the fundamental effect on selenoenzymes: Comparison with selenomethionine in mice." *Free Radical Biology and Medicine* 42, no. 10 (2007): 1524–1533.
166. Wang, Shengpeng, Tongkai Chen, Ruie Chen, Yangyang Hu, Meiwan Chen, and Yitao Wang. "Emodin loaded solid lipid nanoparticles: preparation, characterization and antitumor activity studies." *International Journal of Pharmaceutics* 430, no. 1-2 (2012): 238–246.
167. Zhang, Jinsong, Xufang Wang, and Tongwen Xu. "Elemental selenium at nano size (Nano-Se) as a potential chemopreventive agent with reduced risk of selenium toxicity: Comparison with se-methylselenocysteine in mice." *Toxicological Sciences* 101, no. 1 (2008): 22–31.
168. Torres, S. K., V. L. Campos, C. G. León, S. M. Rodríguez-Llamazares, S. M. Rojas, M. Gonzalez, C. Smith, and M. A. Mondaca. "Biosynthesis of selenium nanoparticles by Pantoea agglomerans and their antioxidant activity." *Journal of Nanoparticle Research* 14, no. 11 (2012): 1–9.
169. Desai, Manisha P., Vinod Labhasetwar, Elke Walter, Robert J. Levy, and Gordon L. Amidon. "The mechanism of uptake of biodegradable microparticles in Caco-2 cells is size dependent." *Pharmaceutical Research* 14, no. 11 (1997): 1568–1573.
170. De Jong, Wim H., and Paul JA Borm. "Drug delivery and nanoparticles: Applications and hazards." *International Journal of Nanomedicine* 3, no. 2 (2008): 133.
171. Zhai, Xiaona, Chunyue Zhang, Guanghua Zhao, Serge Stoll, Fazheng Ren, and Xiaojing Leng. "Antioxidant capacities of the selenium nanoparticles stabilized by chitosan." *Journal of Nanobiotechnology* 15, no. 1 (2017): 1–12.
172. Gharibzahedi, Seyed Mohammad Taghi, and Seid Mahdi Jafari. "The importance of minerals in human nutrition: Bioavailability, food fortification, processing effects and nanoencapsulation." *Trends in Food Science & Technology* 62 (2017): 119–132.
173. Bonnet, M., Maud Cansell, A. Berkaoui, Marie-Hélène Ropers, Marc Anton, and F. Leal-Calderon. "Release rate profiles of magnesium from multiple W/O/W emulsions." *Food Hydrocolloids* 23, no. 1 (2009): 92–101.
174. Zarrabi, Ali, Mandana Alipoor Amro Abadi, Sepideh Khorasani, M. Mohammadabadi, Aniseh Jamshidi, Sarabanou Torkaman, Elham Taghavi, M. R.

Mozafari, and Babak Rasti. "Nanoliposomes and tocosomes as multifunctional nanocarriers for the encapsulation of nutraceutical and dietary molecules." *Molecules* 25, no. 3 (2020): 638.

175. Hasanvand, Elham, Milad Fathi, and Alireza Bassiri. "Production and characterization of vitamin D 3 loaded starch nanoparticles: Effect of amylose to amylopectin ratio and sonication parameters." *Journal of Food Science and Technology* 55, no. 4 (2018): 1314–1324.
176. Hategekimana, Joseph, Kingsley George Masamba, Jianguo Ma, and Fang Zhong. "Encapsulation of vitamin E: Effect of physicochemical properties of wall material on retention and stability." *Carbohydrate Polymers* 124 (2015): 172–179.
177. Hategekimana, Joseph, Moses VM Chamba, Charles F. Shoemaker, Hamid Majeed, and Fang Zhong. "Vitamin E nanoemulsions by emulsion phase inversion: Effect of environmental stress and long-term storage on stability and degradation in different carrier oil types." *Colloids and Surfaces A: Physicochemical and Engineering Aspects* 483 (2015): 70–80.
178. Moeller, Henrike, Dierk Martin, Katrin Schrader, Wolfgang Hoffmann, and Peter Chr Lorenzen. "Spray-or freeze-drying of casein micelles loaded with Vitamin D2: Studies on storage stability and in vitro digestibility." *LWT* 97 (2018): 87–93.
179. Cohen, Yifat, Moran Levi, Uri Lesmes, Marielle Margier, Emmanuelle Reboul, and Yoav D. Livney. "Re-assembled casein micelles improve in vitro bioavailability of vitamin D in a Caco-2 cell model." *Food & function* 8, no. 6 (2017): 2133–2141.
180. David, Shlomit, and Yoav D. Livney. "Potato protein based nanovehicles for health promoting hydrophobic bioactives in clear beverages." *Food Hydrocolloids* 57 (2016): 229–235.
181. Liu, Guangyu, Weijuan Huang, Oksana Babii, Xiaoyu Gong, Zhigang Tian, Jingqi Yang, Yixiang Wang et al. "Novel protein–lipid composite nanoparticles with an inner aqueous compartment as delivery systems of hydrophilic nutraceutical compounds." *Nanoscale* 10, no. 22 (2018): 10629–10640.
182. Kong, Rong, Qiang Xia, and Guang Yu Liu. "Preparation and characterization of vitamin A palmitate-loaded nanostructured lipid carriers as delivery systems for food products." In *Advanced Materials Research* 236 (2011): 1818–1823.
183. Salvia-Trujillo, Laura, and David Julian McClements. "Improvement of β-carotene bioaccessibility from dietary supplements using excipient nanoemulsions." *Journal of Agricultural and Food Chemistry* 64, no. 22 (2016): 4639–4647.
184. Guttoff, Marrisa, Amir Hossein Saberi, and David Julian McClements. "Formation of vitamin D nanoemulsion-based delivery systems by spontaneous emulsification: Factors affecting particle size and stability." *Food Chemistry* 171 (2015): 117–122.
185. Saxena, Varun, Abshar Hasan, Swati Sharma, and Lalit M. Pandey. "Edible oil nanoemulsion: An organic nanoantibiotic as a potential biomolecule delivery vehicle." *International Journal of Polymeric Materials and Polymeric Biomaterials* 67, no. 7 (2018): 410–419.
186. Zimet, Patricia, Dina Rosenberg, and Yoav D. Livney. "Re-assembled casein micelles and casein nanoparticles as nano-vehicles for ω-3 polyunsaturated fatty acids." *Food Hydrocolloids* 25, no. 5 (2011): 1270–1276.

4 Nanomaterials

Professor (Dr.) Komal Mehta
Professor Civil Engineering Department & Deputy Director, Krishna Center of Innovation and Research, Dr. Kiran & Pallavi Patel Global University (KPGU), Vadodara, Gujarat, India

4.1 NANOMATERIAL APPLICATIONS AND LIMITATIONS

Various nanomaterials are applied in different sectors, like food, paint, automotive industry, coatings, etc., for improved life and efficiency. To evaluate the health risks of nanomaterials, various properties and studies are carried out, such as LCA, technology implications, toxicity, pathways and mixing of nanomaterials with water bodies.

4.2 BIOMATERIALS

Biopolymers and inorganic solids of 1 nm to 100 nm range are bio nanocomposites with different structural properties, functionalities and applications. Such properties can be achieved by mixing various combinations of nanocomposites. Interface is an important parameter to decide properties of composites. Biocompatibility, antimicrobial activity and biodegradability have increased applications. Due to biocompatibility, their performance is high, and due to light weight, they are also considered as a replacement for non-biodegradable products. They are also used in the packaging industry to a large extent. [1–7]

4.2.1 Nanoparticles

In dimensional nanoparticles (spherical, cubic and shapeless), size is nanoscale, like 100 nm.

4.2.2 Nanofibers

Nanofibers have two dimensions, e.g. single or multiple sheets of graphene, carbon nanotubes. Other examples are nanorods and whiskers. Nanofiber characteristics of stiffness, strength and resilience make them the best nanocomposites. [8]

4.2.3 Nanoplatelets

For formulations of biocompactible nanocomposites, phyllosilicates, silicic acid (magadiite), double-layered hydroxides [M6Al2 (OH)16CO3nH2O; M = Mg, Zn],

DOI: 10.1201/9781003220350-4

zirconium phosphates [Zr (HPO42H2O], and di-chalcogenides [(PbS)1.18(TiS2)2, MoS2] or plate-like clay minerals are used. [9–10]

4.2.4 Bio Nanocomposites: Synthesis and Applications

Research results are supported by derivatives of starchextracted from particular polymers for specific applications and cellulose, polylactic acid (PLA), polycaprolactone (PCL), poly (butylene succinate) (PBS) and polyhydroxy butyrate (PHB): MMT, Ag, SiO2, TiO2 and ZnO. [14]

4.2.5 Eco-Friendly Materials

Environmentally friendly substitutes for composite material is a current concern for researchers.

4.3 BIOPOLYMER

"Bio" means degradable without any harm to the environment. Polymers contain biomolecules to prepare large-size molecules. Biopolymers are prepared from natural sources, such as microbes, crops, sucrose and fastest. They are complicated compared to artificial material. According to type of repeating units, their groups can be made by source and application of biopolymer combined with biopolymers extracted from particular polymers for specific applications. [23–25]

4.3.1 Biocompactible Polymeric Material

These materials are made from renewable sources and can be degraded. They can be substitutes for polymeric materials.

4.4 CELLULOSE

Cellulose is a biopolymer that is natural and can be obtained from sustainable resources, e.g. cotton, wool and hemp. Sugar is a polysaccharide. An organic cellulose is (C6H10O5)n. [26]

4.4.1 Nanocellulose

Nanocellulose is biodegradable at the nanometer scale. First-time synthesis of cellulose is done by homogenization in the presence of high temperature and pressure. In mechanical disintegration, fibers are transferred into nano- or microfibers. Methods are homogenization, micro fluidization and grinding. Cry crushing includes mechanical and chemical treatment. [27–28]

4.4.2 Chitin

Chitin is a polysaccharide, found mostly in the shells of crabs (in the waste of shells 33% id chintin polymer), and it is basic, meaning it is easily available in nature.

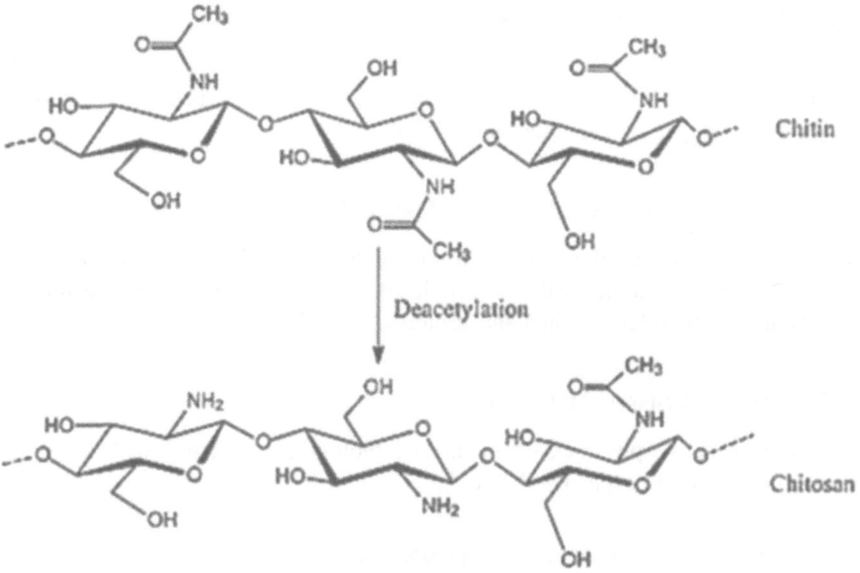

FIGURE 4.1 Chitin.

Source: Adapted from Mohammad Zuber, January 13 [30].

The difference from cellulose is it contains acetamide group, b-1, 4-N-acetyl-D-glucosamine monomer units, as presented in Figure 4.1. [33]

4.4.2.1 Lignin

Lignin contains antioxidant, antimicrobial properties with its own density and good stiffness, so it is a possible substitute for existing fuel. After cellulose, lignin is considered to be the best option to substitute for fossil fuel sources as it is a biopolymer. [32]

4.4.2.2 Vulcanized Natural Rubber Particles

Vulcanized natural rubber particles are mainly used in the field of medicine, in tires, and in the production of soft products with thin walls as it is a biopolymer that is stretchable and prepared from renewable sources. [34]

4.5 BIOCOMPATIBLE NANOMATERIALS

4.5.1 TiO_2

It is photoactive nanomaterial and an oxidizing agent with photocatalysts, water-splitting reaction, sensors, solar cells, etc. It is also a green material with reduction of toxic agents.

4.5.2 ZnO

It is a semiconductor that is a widely used component in the medical field, solar cells, ultraviolet absorption, etc., presenting a broadband gap and high binding energy, which corresponds to 3.37 eV. [38–40]

4.5.3 Coupling of Nanomaterials

Limitations of materials can be overcome by coupling of them, e.g. ZnO/TiO2, CuO/TiO2, SnO2/TiO2, CdS/TiO2, and ZnS/TiO2, which enhances photocatalytic activity due to larger contact interface. [45–47]

4.5.4 Nano Encapsulation and Its Green Properties

Nano encapsulation is used mostly in the food industry as it has promising behavior in the nutrition and public health sector. It also safeguards bioactive constituents like vitamins, antioxidants, lipids and proteins. [45–47]

4.5.5 Nature-Inspired Hydrogels

They have attractive physical, chemical and biological properties with high absorption and retention capacity of liquid.

4.5.6 Versatility of Green Nanocomposites

Green nanocomposites include methods of stirring and hydrolysis, and their synthesis can be done with sources of plants or animals. [3]

4.6 EFFECT OF NANOCLAY

Due to high performance and environmentally friendly characteristics, geopolymers are synthesized with the use of solid alumino silicate with alkaline solutions. It is possible to get building materials by conversion of alumino silicate. Still, geopolymers suffer from brittle failure mode with compressive strength 45MPa and flexural strength between 1.7 and 16.8 MPa. Alumina nanosilicates not only act as a filler but also reduce porosity and absorption. The addition of fly ash into nanotubes increases the properties (mechanical and electrical). Nanoclay has better properties compared to other nanocomposites. [47–50]

4.6.1 Materials

According to the tables with said chemical composition, fly ash was used to prepare nanocomposites. Table 4.1 presents chemical composition of fly ash, and Table 4.2 presents the physical properties of nanoclay. Natural montmorillonite clay with alkaline activators was used. [17–20]

TABLE 4.1
% of Fly ssh for Nanocomposites

SiO_2	Al_2O_3	CaO	Fe_2O3	K_2O	MgO	Na_2O	P_2O_5	SO_3	TiO_2	MnO	BaO	LOI
63.13	24.88	2.58	3.07	2.01	0.61	0.71	0.17	0.18	0.96	0.05	0.07	1.45

TABLE 4.2
Nano clay – Physical Properties

Property	Value
Color	Off white
Density (g/cm3)	1.98
d-Spacing (0 0 1) (nm)	1.85
Aspect ratio	200–1000
Surface area (m2/g)	750
Typical dry particle sizes, 90% volume	< 13 _m
50% volume	< 6 _m
10% volume	< 2 _m

TABLE 4.3
Sample Preparation

Sample	Fly-ash (g)	NaOH Solution (g)	Na2SiO3 Solution (g)	Nano-clay (g)
GP	1000	214.5	535.5	0
GPNC-1	1000	214.5	535.5	10
GPNC-2	1000	214.5	535.5	20
GPNC-3	1000	214.5	535.5	30

4.6.2 Preparation of Geopolymer Nanocomposites

Sodium silicate to hydroxide was prepared with a fixed ration of 0.75 and 2.5 alkaline solution, respectively, added to fly ash. Nanoclay and alkaline solution was dry-mixed, and then high-speed mixed while adding to fly ash. Table 4.3 shows that the resultant was vibrated for 2 minutes, and for 24 hours it was cured in an oven at a temperature of 80° C for polymer preparation. [38–40]

4.6.2.1 Physical Properties

To determine quality for geopolymer nanocomposites, density, porosity and water absorption tests were carried out.

4.6.2.2 Mechanical Properties

Flexural strength and hardness were determined for a set of samples.

4.6.2.3 Characterization

The samples were converted into powder. Scanning was done through D8 Advance Diffractometer (Bruker-AXS, Germany). Patterns of scanning were observed using CuK lines. At room temperature, FTIR analysis was performed in 4000–500 cm–1 range. Solid samples were placed for judging thermal behavior through thermogravimetric analysis. [43–49]

4.6.3 Outcome

Density, porosity and water absorption: the geopolymer exhibits behavior of higher density and lower porosity. Nanoclay can increase density and reduce porosity if added in optimum doses. Otherwise, the reverse effect will be observed. It shows similar patterns of decreasing porosity of cement paste with the addition of oregano clay composites. [43–49]

4.7 X-RAY DIFFRACTION (XRD)

With 1, 2, 3 % wt., respectively, nanoclay XRD patterns are indicated in Figure 4.2. Phases were indexed in diffraction patterns. Content of amorphous phase in proportion to strength of geopolymer is exhibited by the test.

4.7.1 Mechanical Properties

Flexural tests indicate bending responses and give an idea of efficiency performance of composites. Geopolymer addition and increase in content during the amorphous phase produces high geopolymer content. Compressive strength is inversely proportional to porosity. [18]

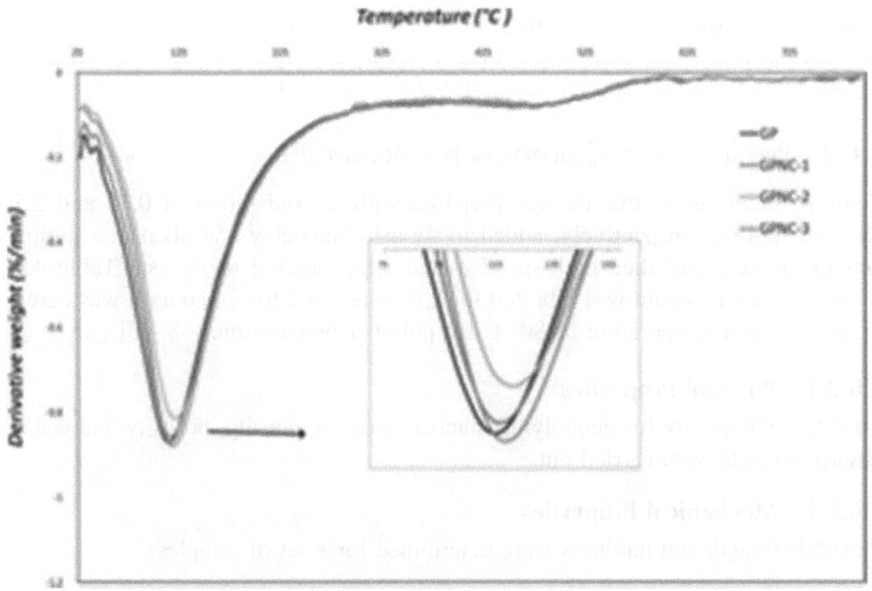

FIGURE 4.2 DTG curves for geopolymer composites.

4.7.2 Thermal Stability

Thermal stability was found by weight loss % through thermal gravimetric analysis, indicating organic or inorganic materials. No major changes in curves presenting nanocomposites were noticed. Figure 4.2 shows curves for nanocomposites based on geopolymer. Curves of pure geopolymers are steeper than nanocomposite nanopolymers. [18–20]

4.8 HEAVY METAL REMOVAL BY SYNTHETIC PHYLLOSILICATE

Phyllosilicate-like structures are proving promising for nanotechnology synthesis. Solvent PH, nature, catalysts, temperature and concentration are major factors affecting the Solgel method.

Lagadic et al. (2001) [58] prepared phyllosilicates with thiol groups with mercapto propyl trime thoxysilane (named Mg-MTMS). Jaber et al. (2005) [59] presented a saponite-like network with $Na_x[(RSi)_{(4-x)}Al_xMg_3O_{(8+x)}(OH)_2$ at room temperature. Later, Badshah et al. (2011) [60] showed organotalc with high retention for metals, such as Cu, Cd, Pb (4.02, 1.87 and 7.09 $mmol \cdot g^{-1}$). Park et al. (2019) [61] presented random elective nonreversible adsorption (type 2:1 clays). ^{137}Cs can be adsorbed reversibly on planner and edge sites. Cs^+ adsorption is shown with two different scenarios. Maximum efficiency of adsorption of phyllosilicates for metal ions is discussed in Table 4.4. Maximum adsorptive performance of different LDHs is shown in Table 4.5. Efforts to experiment with phyllosilicates in ion exchange found that parameters such as temperature and radioactivity can contribute to degradation of the structure. [55–60]

4.8.1 Summary, Discussions and Future Trends

Phyllosilicates can be seen as having high-quality adsorption capacity to be used as a solution for waste-water treatment of radioactive elements. Adsorption of toxic substances is achieved in layers between clay. Ongoing research is studying the use of clay minerals as nanocomposites. Chemical modification by sol gel process is in practice as they are biocompactible and are improving adsorption. The best heavy metal and radionuclide adsorption capacities are presented in Table 4.5, which shows high adsorption capacities or good adsorbents like carbon nanotubes, MoS_2 nanosheets, zeolites, etc. for long-term use. Commercial viability needs to be tested.

4.9 RISK ASSOCIATED WITH USE OF NANOMATERIAL

In the coming era, nanotechnology will bring many changes, which will improve quality of life. Although only advantages may be seen, when nanomaterial is used on a large scale, potential health risks and damage to the environment should be considered. Risk assessment and review of regulations are discussed here. Researchers are in the process of finding and establishing references for testing nanomaterials.

TABLE 4.4
Maximum Efficiency of Adsorption for Various Metal Cations

Heavy Metal	Adsorbent	Maximum Adsorption
Cd (II)	Smectite	8.64
Co (II)	Chemically treated bentonite	2.34
Cr (III/VI)	Polyaniline/Montmorillonite composite	5.94
Cu (II)	Bentonite	0.85
Hg (II)	Montmorillonite	1.92
Mn (II)	Kaolinite	2.72
Ni (II)	Kaolinite	2.40
Pb (II)	Illite	1.15
Zn (II)	Kaolinite	3.82

TABLE 4.5
Maximum Adsorptive Performance (Comparatively) of Different LDHs Toward a Variety of Metal Cations

	Capacity ($mmol·g^{-1}$)	
Heavy Metal	**Adsorbent (LDH)**	**Maximum Adsorption**
Cd (II)	Graphite oxide aerogels/MgAl	0.85
Cr (III/VI)	Fe^{2+}/MgAl	12.5
Cu (II)	MoS_4^{2-}/MgAl	2.85
Co (II)	Polysulfide/MgAl	1.41
Hg (II)	Polysulfide/MgAl	4.05
Ni (II)	Polysulfide/MgAl	1.81
Pb (II)	$CaFe_2O_4$/polyophenylenediamine/MgAl	4.83
Zn (II)	Polysulfide/MgAl	2.22

4.9.1 Benefits of Nanotechnologies

Nanotechnologies are helpful for many day-to-day issues and can improve quality of life. They are also capable of energy storage and efficiency, better diagnosis and treatment of diseases, etc. [15]

4.9.2 Potential Health Concerns

The type of nanomaterial cannot be measured with any equipment, and because of that, measurement of exposure to nanomaterials is challenging. Nano silver can be present in products in the food industry. Application of nanomaterials in the food industry can be beneficial or harmful. (EFSA 2008). It is always recommended to be careful of the health effects of nanomaterials in terms of inhalation, ingestion or absorption until there is specific evidence of their effect. [47]

4.9.3 Detection and Analysis

To produce facts for risk identification, assessment and management, more instruments with guidelines for use are required to meet the requirement of obtaining a certain amount of data. Application of nanomaterials is possible for detection and analysis. [46–49]

4.9.4 Nonmaterial Preparations for Biological Testing

To administer nanoparticles in biological systems, suspension or dry powder is used. The best nanoparticle dispersions are studied, including albumin, protein and phospholipids. These coatings will change the properties of nanomaterial, and with change, its biological reactions will also be changed. [9]

4.9.5 The Interface between Nanomaterials and Biological Systems

With interaction of nanomaterials with fluid, irrespective of single or agglomerate, they may have a coating of protein that will make a difference in biomolecules, and the end result of biological changes can also differ.

4.9.6 Translocation of Nanomaterials

Dependency will be physical or chemical properties, such as size of particles, surface charge and hydrophobicity. In CNTs, aspect ratios were of more concern than safety of nanotubes.

4.9.7 General Principles

The tendency of particles to agglomerate mixes well with dissolved, colloidal and particulate matter in surroundings. When nanoparticles come in contact of the atmosphere, dissolution, speciation, biological or chemical transformation, mineralization, settling, etc. takes place.

4.9.8 What Is Risk Assessment and Management?

Risk aspects of the application of nanomaterials on the health and safety of people is an area of research that needs consideration before use. A risk assessment is required to take care of necessary prevention steps for people's health while using nanomaterials for work.

4.10 CONCLUSION AND FUTURE PROSPECTS

Mechanical, thermal and microstructural property characterization showed that the addition of nanoclay increases flexural and compressive strength if nanoclay is added in optimum doses. Moreover, the content of nanoclay has more porosity and fewer dispersions. XRD and FTIR indicated that the addition of nanoclay increases the geopolymerization to geopolymer paste, and SEM indicated denser matrix. Clay minerals contain characteristics of extracting radio nuclides and heavy metals from water, as they are a biocompactible low-cost source. Because of interchangeable cations, pore volume and surface area, phyllosilicates are a better option compared to any other constituent for adsorption. Preboosting helps to render clays hydrophobic. Nanomaterials are available to be adapted to customer-specific applications like adsorbents, metals, membranes and photocatalysts based on properties of nanotechnology. The majority of these materials are compatible with existing treatment technologies and can be integrated with conventional treatments. The risk associated with the application of nanotechnology needs evaluation, but with use of nanotechnology in energy and the environment, a great revolution and cost-effective, eco-friendly solutions to existing technical methods are expected in the near future.

REFERENCES

1. Alemdar, A. and Sain, M., "Bio composites from wheat straw nanofibers: Morphology, thermal and mechanical properties." *Composites Science and Technology*, 68, no. 2 (2008), 557–565.
2. Alemdar, A. and Sain, M., "Isolation and characterization of nanofibers from agricultural residues – Wheat straw and soy hulls." *Bio Resource Technology*, 99, no. 6 (2008), 1664–1671.
3. Kaushik, A., Singh, M., and Verma, G., "Green nanocomposites based on thermoplastic starch and steam exploded cellulose nanofibrils from wheat straw." *Carbohydrate Polymers*, 82, no. 2 (2010), 337–345.
4. Uddin, A.J., Araki, J., Fujie, M., Sembo, S., and Gotoh, Y., "Interfacial interaction and mechanical properties of chitin whisker-poly(vinyl alcohol) gel-spun nanocomposite fibers." *Polymer International*, 61 (2012), 1010e1015. doi: 10.1002/pi.4174.
5. Srivastava, Anshuman, Jana, Karun Kumar, Maiti, Pralay, Kumar, Devendra, and Parkash, Om, "Poly(vinylidene fluoride)/ CaCu3Ti4O12 and La doped CaCu3Ti4O12 composites with improved dielectric and mechanical properties." *Material Research Bulletin*, 70 (2015), 735–742.
6. Srivastava, Anshuman, Parkash, Om, Kumar, Devendra, and Maiti, Pralay, "Structural and dielectric properties of lanthanum doped CaCu3Ti4O12 for capacitor application." *American Journal of Materials Synthesis and Processing*, 2, no. 6 (2017), 90–93
7. Srivastava, Anshuman, Maiti, Pralay, Kumar, Devendra, and Parkash, Om, "Mechanical and dielectric properties of CaCu3Ti4O12 and La doped CaCu3Ti4O12 poly(vinylidene fluoride) composites." *Composites Science and Technology*, 93 (2014), 83–89.
8. Wang, B. and Sain, M., "Dispersion of soybean stock-based nanofiber in a plastic matrix." *Polymer International*, 56, no. 4 (2007), 538–546.

9. Ulery, B.D., Nair, L.S., and Laurencin, C.T., "Biomedical applications of biodegradable polymers." *Journal of Polymer Science Part B: Polymer Physics*, 49 (2011), 832e864. doi:10.1002/polb.22259.
10. Barakat, M.A., "New trends in removing heavy metals from industrial wastewater." *Arabian Journal of Chemistry*, 2011, no. 4 (2011), 361–377.
11. Burakov, A.E., Galunin, E.V., Burakova, I.V. et al. "Adsorption of heavy metals on conventional and nanostructured materials for wastewater treatment purposes: A review." *Ecotoxicology and Environment Safety*, 2018, no. 148 (2018), 702–712.
12. Dupont, L., and Guillon, E., "Removal of hexavalent chromium with a legionellosis substrate extracted from wheat bran." *Environmental Science and Technology*, 37 (2003), 4235–4241.
13. Teixeira, E.M., Corrêa, A.C., de Oliveira, C.R. et al., "Cellulose nanofibers from white and naturally colored cotton fibers." *Cellulose*, 17, no. 3 (2010), 595–606.
14. FACT SHEET, Emerging Contaminants – Nanomaterials, *Solid Waste and Emergency Response (5106P),EPA 505 – F - 09-011*, United States Environmental Protection Agency, September 2009.
15. Fleischer, T. and Grunwald, A., "Making nanotechnology developments sustainable. A role for technology assessment." *Journal of Cleaner Production*, 16 (2008), 889–898.
16. Williams, G.I. and Wool, R.P., "Composites from natural fibers and soy oil resins." *Applied Composite Materials*, 7, no. 5-6 (2000), 421–432.
17. Alamri, H. and Low, I.M., *Composites Part A*, 44 (2013), 23–31.
18. Assaedi, H., Shaikh, F.U.A., and Low, I.M., "Effect of nano-clay on mechanical and thermal properties of geopolymer." *Journal of Asian Ceramic Societies*, 4, no. 1 (2016), 19–28. doi:10.1016/j.jascer.2015.10.004, Journal of Asian Ceramic Societies
19. Huawen, Han, Rafiq, Muhammad Khalid, Zhou, Tuoyu et al. 2019. "A critical review of clay-based composites with enhanced adsorption performance for metal and organic pollutants." *Journal of Hazardous Materials*, 369, no. 5, (May 2019), 780–796. doi:10.1016/j.jhazmat.02.003
20. Ikarashi, Y., H. Mimura, and T. Nakai et al., "Selective cesium uptake behavior of insoluble ferrocyanide loaded zeolites and development of stable solidification method." *Journal of Ion Exchange*, 2014, no. 25 (2014), 212–219.
21. Iorio, M., Pan, B., Capasso, R., XingB. "Sorption of phenanthrene by dissolved organic matter and its complex with aluminum oxide nanoparticles." *Environmental Pollution*, 156 (2008), 1021–1029.
22. Ishizaki, M., S. Akiba, A. Ohtani et al., "Proton-exchange mechanism of specific Cs+ adsorption via lattice defect sites of Prussian blue filled with coordination and crystallization water molecules." *Dalton Transactions*, 2013, no. 42 (2013), 16049–16055.
23. Ismadji, S., F.E. Soetaredjo, A. Ayuchitra et al., *Clay Materials for Environmental Remediation*. Springer International Publishing: Berlin, Germany, 2015. ISBN978-3-319-16711-4.
24. Jiuhui, Q., "Research progress of novel adsorption processes in water purification: A review." *Journal of Environmental Sciences*, 20 (2008), 1–13.
25. Zinoviadou, K.G., Gougouli, M., and Biliaderis, C.G., "Innovative biobased materials for packaging sustainability." in: *Innovative Strategies in the Food Industy: Tools for Implementation*. Academic Press, 2016. doi:10.1016/B978-0-12-803751-5.00009-X
26. Lee, K.Y., Aitom€aki, Y., Berglund, L.A., Oksman, K., and Bismarck, A., "On the use of nanocellulose as reinforcement in polymer matrix composites." *Composites Science and Technology*, 105 (2014), 15e27. doi:10.1016/j.compscitech.2014.08.032.

27. Kuchibhaatla, S.V.N.T., Karakoti, A.S., Bera, D., and Seal, S. "One dimensional nanostructured material." *Progress in Materials Science*, 52 (2007), 699–913.
28. Kumar, Asheesh, Anshuman, S., "Preparation and mechanical properties of jute fiber reinforced Epoxy composites." *Industrial Engineering & Management*, 6, no. 234 (2017). doi:10.4172/2169-0316.1000234.
29. Laurent, S. , D. Forge, M. Port et al., "Magnetic iron oxide nanoparticles: Synthesis, stabilization, vectorization, physicochemical characterizations, and biological applications." *Chemical Reviews*, 2008, no. 108 (2008), 2064–2110.
30. Li, D. , Lyon, D.Y., Li, Q., and Alvarez, P.J.J. "Effect of soil sorption and aquatic natural organic matter fullerene water suspensions: Effect of nC60 concentration, CHAPTERS exposure conditions and shelflife." *Water Science and Technology*, 57, no. 10 (2008), 1533–1538.
31. Mohiuddin, M., Kumar, B., and Haque, S., "Biopolymer composites in photovoltaics and photodetectors." In *Biopolym*er, pp. 459–486. Elsevier, 2016. doi:10.1016/B978-0-12-809261-3.00017-6.
32. Zuber, M., Zia, K.M., and Barikani, M., "Chitin and chitosan based blends, composites and nanocomposites." 55e119, 2013. doi:10.1007/978-3-642-20940-6_3.
33. Reddy, M.M., Vivekanandhan, S., Misra, M., Bhatia, S.K., and Mohanty, A.K., "Biobased plastics and bio nanocomposites: Current status and future opportunities." *Progress in Polymer Science*, 38 (2013), 1653e1689. doi:10.1016/j.progpolymsci.2013.05.006
34. Ooshiro, Masaru, Kobayashi, Takaomi, Uchida, S.et al., "Fibrous zeolite-polymer composites for decontamination of radioactive waste water extracted from radio-Cs fly ash." *International Journal of Engineering and Technical Research*, 2017, no. 7 (2019), 6.
35. Morgada, M.E., Levy, I.K., Salomone, V., Farias, S.S., Lopez, G., Litter, M.I. "Arsenic (V) removal with nanoparticulate zerovalent iron: Effect of UV light and humic acids." *Catalysis Today*. https://doi.org/10.1016/j.cattod.2008.09.0382009.
36. Nakamoto, K., Ohshiro, Masaru, Kobayashi, Takaomi et.al. "Mordenite zeolite – Polyether sulfone composite fibers developed for decontamination of heavy metal ions." *Journal of Environment and Chemical Engineering*, 2017, no. 5 (2017), 513–525.
37. Osacky, M., M. Geramian, D.G. Ivey et al., "Influence of non swelling clay minerals (Illite, Kaolinite, and Chlorite) on nonaqueous solvent extraction of bitumen." *Energy Fuels*, 2015, no. 29 (2015), 4150–4159.
38. Visakh, P.M. andThomas, S., "Preparation of bionanomaterials and their polymer nano composites from waste and biomass." *Waste and Biomass Valorization*, 1, no. 1 (2010), 121–134.
39. Pan, B. and B.S. Xing, "Adsorption mechanisms of organic chemicals on carbon nanotubes." *Environmental Science and Technology*, 2008, no. 42 (2008), 9005–9013.
40. Salehi, R., Arami, M., Mahmoodi, N.M., Bahrami, H., and Khorramfar, S., "Novel biocompatible composite (chitos and zinc oxide nanoparticle): Preparation, characterization and dye adsorption properties." *Colloids Surf. B: Bio interfaces*, 80, no. 1 (2010), 86e93. doi:10.1016/j.colsurfb.2010.05.039.
41. Mishra, R.K., Sabu, A., and Tiwari, S.K., "Materials chemistry and the futurist ecofriendly applications of nanocellulose: Status and prospect." *Journal of Saudi Chemical Society* 2018 (2018). doi:10.1016/j.jscs.2018.02.005.
42. Mishra, Raghvendra Kumar, Ha, Sung Kyu, Verma, Kartikey, and Tiwari, Santosh K., "Heir engineering applications: An overview." *Journal of Science: Advanced Materials and Devices* (2018). doi:10.1016/j.jsamd.2018.05.003

43. Kalia, S., Kaith, B. S., Sharma, S., and Bhardwaj, B., "Mechanical properties of flax-gpoly(methyl acrylate) reinforced phenolic composites." *Fibers and Polymers*, 9, no. 4 (2000), 416–422.
44. Kalia, S., Vashistha, S., and Kaith, B. S., "Cellulose nanofibers reinforced bioplastics and their applications." in *Handbook of Bioplastics and Bio composites Engineering Applications*, Pilla, S., Ed., chapter 16, Wiley Scrivener Publishing, New York, NY, 2011.
45. Shabtai, I.A. and Y. G. Mishael, "Poly cyclodextrin–clay composites: Regenerable dual–site sorbents for bisphenol A removal from treated wastewater." *ACS Applied Material and Interfaces*, 2018, no. 10 (2018), 27088–27097.
46. Song, W., Li, G., Grassian, V.H., and Larsen, S.C. "Development of improved materials for environmental applications: Nanocrystalline NaY zeolites." *Environmental Science and Technology*, 39 (2005), 1214–1220.
47. Srinivasan, R., "Advances in application of natural clay and its composites in removal of biological, organic, and inorganic contaminants from drinking water." *Advances in Material Science and Engineering*, 2011, no. 2011 (2011), 1–17.
48. Stafiej, A. and Pyrzynska, K., "Adsorption of heavy metal ions with carbon nanotubes." *Separation and Purification Technology*, 58 (2007), 49–52.
49. Zimmermann, T. , Bordeanu, N., and Strub, E., "Properties of nano fibrillated cellulose from different raw materials and its reinforcement potential." *Carbohydrate Polymers*, 79, no. 4 (2010), 1086–1093.
50. Tambach, T.J. and E.J.M. Hensen, "Molecular Simulations of Swelling Clay Minerals." *Journal of Physical Chemistry B*, 2004, no. 108 (2004), 7586–7596.
51. Tan, L., Wang, S., and Du, W., "Effect of water chemistries on adsorption of Cs(I) onto graphene oxide investigated by batch and modeling techniques." *Chemical Engineering Journal*, 2016, no. 292 (2016), 92–97.
52. Uddin, M.K., "A review on the adsorption of heavy metals by clay minerals, with special focus on the past decade." *Chemical Engineering Journal* 308, (2017), 438–462.
53. Stelte, W. and Sanadi, A. R., "Preparation and characterization of cellulose nanofibers from two commercial hardwood and softwood pulps." *Industrial and Engineering Chemistry Research*, 48, no. 24 (2009), 11211–11219.
54. Tang, X.Z., Kumar, P., Alavi, S., and Sandeep, K.P., "Recent advances in biopolymers and biopolymer-based nanocomposites for food packaging materials." *Critical Review in Food Science and Nutrition*, 52, no. 2012 (2012) 426e442. doi:10.1080/10408398.2010.500508.
55. Xu, Z., Gan, L., Jia, Y., Hao, Z., Liu, M., and Chen, L. "Preparation and characterization of silica-titania aerogel-like balls by ambient pressure drying." *Journal of Sol-Gel Science and Technology*, 41, no. 3 (2007): 203–207.
56. Zhang, X., Du, A.J., Lee, P., Sun, D.D., and Leckie, J.O., "TiO_2 nanowire membrane for concurrent filtration and photocatalytic oxidation for humic acid in water." *Journal of Membrane Science*, 313 (2008), 44–51.
57. Zubair, M., Daud, M., McKay, G., Shehzad, F., and Al-Harthi, M.A., "Recent progress in layered double hydroxides (LDH)-containing hybrids as adsorbents for water remediation." *Applied Clay Science*, 143 (2017), 279–292.
58. Lagadic, I.L., Mitchell, M.K., & Payne, B.D. (2001). Highly Effective Adsorption of Heavy Metal Ions by a Thiol-Functionalized Magnesium Phyllosilicate Clay. *Environmental Science & Technology*, 35, no. 5, 984–990, https://doi.org/10.1021/es001526m.
59. Jaber, M., Miehé-Brendlé, J., Michelin, L., & Delmotte, L. (2005). Heavy Metal Retention by Organoclays: Synthesis, Applications, and Retention Mechanism. *Chemistry of Materials*, 17, no. 21, 5275–5281, https://doi.org/10.1021/cm050754i.

60. Lal, B., Badshah, A., Altaf, A.A., Khan, N., & Ullah, S. (2011). Miscellaneous applications of ferrocene-based peptides/amides. *Applied Organometallic Chemistry*, 25, 843–855, https://doi.org/10.1002/aoc.1843.
61. Zaghiyan, K.N., Mendelson, B.J., Eng, M.R., Ovsepyan, G., Mirocha, J.M., & Fleshner, P. (2019). Randomized Clinical Trial Comparing Laparoscopic Versus Ultrasound-Guided Transversus Abdominis Plane Block in Minimally Invasive Colorectal Surgery. *Diseases of the Colon & Rectum*, 62, no. 2, 203–210, doi: 10.1097/dcr.0000000000001292.

5 Impact of Nanoelectronics in the Semiconductor Field

Past, Present and Future

G. Boopathi Raja
Velalar College of Engineering and Technology,
Erode, Tamil Nadu, India

5.1 INTRODUCTION

Moore's Law, which predicts that computer processing performance will improve as miniaturization increases, has essentially determined advances in semiconductor technology over the last few decades, [1–3]. Electronics must continue to shrink to maximize processing speed and minimize per-bit manufacturing costs; however, as critical electronics proportions reach an atomic size, quantum tunneling and other effects become increasingly prohibitive.

As a result, scientists are seeking more extreme approaches to technological advancements, such as nanoscale physics. The world of nanoelectronics seems to be on the brink of revolutionizing electronics and emerging technologies.

The word "nanoelectronics" refers to a broad variety of technologies and materials that rely on atomic-scale interactions and quantum mechanical properties to work. At the nanoscale, various forces have a larger effect than at the macroscopic stage. For example, quantum tunneling and atomistic instability are critical issues for nanoelectronics researchers.

Nanoelectronics can boost electronics' capabilities while reducing their size, weight and power usage. Cutting power requirements, thus lowering display weight and density, will increase display screens. Researchers are also developing a nanoscale memory chip that can store one terabyte or more of data per square inch.

It corresponds to a large number of devices and resources. They both share one thing in common: they are so small that the interactions between atoms and quantum mechanical properties must be carefully studied. Other possibilities include a hybrid of electronic/semiconductor electronics, one-dimensional nanotubes/nanowires (e.g. silicon nanowires, carbon nanotubes [CNTs]) and sophisticated electronics. Nanoelectronic devices have critical dimensions ranging from 1 nm to 100 nm. The state now incorporates the latest silicon MOSFET technologies, such

DOI: 10.1201/9781003220350-5

as the following 22 nm nodes and the next 14 nm, 10 nm and 7 nm generation fin field-effect transistors (FinFETs).

Logic faults may occur in memories. These faults may be detected and corrected by specific digital circuits. Some of the techniques are Built-in Self Test (BIST), majority logic, etc. [4, 5].

5.2 TECHNOLOGY SCALING MOTIVATION

The market for portable battery-powered devices is growing every day, with a variety of applications, such as hearing aids, smartphones and laptops. Less space, lower power usage and less expensive construction are the "simple specifications" of such applications. As the battery capacity provided by portable devices is small, power distribution is important. Unfortunately, battery technology is unlikely to increase battery capacity by more than 30% every five years. It does not fulfill the requirements to meet the growing demand for power on mobile devices.

Gordon E. Moore estimated an important theory regarding the miniaturization of chips, called Moore's law, in 1965. It states that the number of transistors in the integrated circuit (IC) would double every two years. With the reduction of transistors, some circuits can be made of wafer silicon, which reduces circuit costs. The shorter channel length provides faster switching operation since it may take minimal time to pass from the drain to the source.

In another way, a smaller resistor results in lower capacitance. Transistor delay is reduced as a result of this. The power demand is reduced when dynamic power is proportional to the capacitance. "Scaling" is the term for reducing the size of transistors. When a transistor is scaled up, we refer to it as a new technology node. The node of the technology is the minimum channel length of the transistor. It may be 0.18 μm, 0.13 μm, 90 nm and so on. With each new generation of technology, the shrinking of transistor size improves cost, efficiency and energy efficiency.

5.3 ISSUES DUE TO SMALL DIMENSIONS

Electric field lines on long channel devices link directly with the channel surface everywhere. The power of the gate and the power behind the gate control these electric fields. On the other hand, short channel systems have source and drain structures that are nearer to the channel as the longitudinal electric field is considered. The electric field is regulated by the drain-source voltage, and it is perpendicular to the direction of current flow. The device is known as a short channel device since the dimensions of the channel are small when compared with the dimensions of the source and drain depletion regions.

The drawbacks of the high electric fields and 2D potential propagation in the short channel are discussed below:

5.3.1 Hot-Carrier Effect

The electric field increases above the drain for narrow geometric dimensions. Due to this, the charge carriers gain enormous energy, which is termed "hot carriers".

These carriers may accumulate with sufficient energy to induce ionization near the drain region. This in turn permits the formation of new electron-hole pairs. As a consequence, it leads to the flow of a drain-to-body current. A few hot electrons will tunnel through the oxide layer and accumulate at a gate region. Any hot carriers will weaken the structure by destroying the oxide.

5.3.2 Punch-through Effect

An extreme case of barrier lowering is punch-through. The depletion area around the drain would extend farther toward the source as the drain bias is increased, resulting in the merger of two depletion areas. This is referred to as a punch-through case.

In these cases, the drain current increases rapidly as the gate voltage loses controllability, and in turn, again the drain current exponentially increases. The punch-through effect gets stronger as the channel length gets shorter. Hence, it is difficult to switch off the system because of punch-through.

5.3.3 Carrier Velocity Saturation and Mobility Degradation

At the channel with the low electric field, the electron drift velocity is always proportional to the electric field. These velocities may saturate in a heavy electric field. It is known as velocity saturation. This leads normally to a rise in the longitudinal electric field of short-channel systems. Velocity saturation occurs in high electric fields, impacting MOSFET's I-V characteristics. The saturation mode of MOSFETs can now be achieved with lower drain-source voltages and lower saturation currents while maintaining the same gate voltage.

Due to an extremely intense electric field, the charge carriers in the channel may be distributed in the oxide interface. The drain current and carrier mobility all decrease as a result of this.

5.3.4 Drain-Induced Barrier Lowering (DIBL)

The drain-induced barrier lowering (DIBL) effect is another important short-channel phenomenon. It is nothing but an extremely high drain voltage; it leads to lower threshold voltage. If the applied voltage at the gate terminal is insufficient to provide inversion operation, then the charge carriers in the channel will encounter a potential barrier. This potential hurdle may be eliminated by increasing gate potential. On the other hand, the voltages Vgs and Vds monitor such a major constraint in short-channel devices.

The depletion layer of the drain body extends as the drain-to-source voltage (Vds) increases and grows steadily below the gate. As a result, even though Vgs is less than Vt, the channel's potential barrier lowers, allowing carriers (electrons) to flow between the source and drain. Drain dropping the channel barrier and raising the threshold voltage is referred to as DIBL. Vt roll-off refers to the decrease in threshold voltage as channel length increases. Subthreshold current is the current that flows under those conditions (off-state current). In saturation mode, DIBL allows the drain current to increase as drain bias rises.

5.4 METHODS FOR REDUCING SHORT-CHANNEL EFFECTS (SCE)

Short-channel effects (SCE) become unacceptable when channel duration is limited relative to depletion areas, as we saw in the previous segment. This reduces the amount of gate length reduction that can be achieved. To minimize these consequences, the width of the depletion area should be decreased along with the channel length. Increases in channel doping concentration, gate capacitance, or a mixture of the two may be used to do this. The gate control over the channel can be determined by the gate capacitance. Increasing gate oxide thickness improves gate capacitance. A system with a thinner gate oxide has been found to have a smaller depletion diameter and hence better SCE characteristics.

For the past 30 years, the oxide layer of Intel's process nodes can be approximately scaled down based on the channel dimension to reduce SCEs.

5.5 POST TRADITIONAL SCALING INNOVATIONS

5.5.1 Mobility Booster: Strained Silicon Technology

Mobility degradation is one of the major issues faced by nanoscale transistors. It is caused by a higher vertical electric field. Several methods are available to improve efficiency and mobility. The utilization of thin germanium film in the channel is one of the approaches so germanium has greater carrier mobility. The alternate way is the handling of stressed silicon in the channel by providing physical strain.

Strain silicon processing involves compressing the silicon crystal based on different techniques to maximize electron or hole mobility and the performance of the transistor. For example, whenever the channel is stretched compressively, the carrier mobility of PMOS (i.e. holes) may be enhanced.

To generate this pressure, during epitaxial growth, both the drain and source regions of the silicon channel must be filled with Si-Ge film. It is a silicon-germanium alloy with a typical composition of 80% silicon and 20% germanium respectively. The quantity of Silicon and Germanium atoms in the final product is the same as the number of Si atoms at the outset. The atoms in germanium are heavier than those in silicon. As a result, when pressure is applied, it allows the channel to move forward and increases the carrier mobility (i.e. holes). Increased semiconductor mobility increases drive current and transistor efficiency.

In 2003, Intel used stressed silicon techniques for MOS transistors for the first time in their 90 nm process technology. Hence, the Si-Ge source-drain structure was preferred as PMOS transistors lead to raising current by 25%. A high-stress Si3N4 capping film is used to apply NMOS strain to the transistor, which increases current by 12%.

5.5.2 Gate Leakage Reduction: High-K Dielectric

The SiO2 layer is considered as the oxide layer or insulator or dielectric medium. The thickness of this dielectric region should be equal to the tube volume.

An effective thickness of the oxide layer is approximately 2.3 nm is needed for the 65 nm node. Direct carrier tunnelling would become dominant if the thickness of the oxide region is minimized further. Gate leakage has achieved an unacceptably high degree as a result of all of this. As a result, the oxide thickness limit, which is calculated by gate-to-channel tunnelling leakage, is approximately 1.6 nm (also called quantum mechanical tunneling).

Using a dielectric material with a high dielectric constant is the only way to increase the oxide capacitance. The width of the dielectric layers can be increased, resulting in a higher gate-oxide capacitance. Because of the thicker layer, carrier tunnelling is minimized. The dielectric constant of the silicon dioxide (SiO2) layer is 3.9.

In 2007, Intel introduced the High-K dielectric material based on hafnium (HfO2) in its 45 nm high volume manufacturing phase, marking a watershed moment in the gate oxide industry. The dielectric constant of Hafnium is roughly 25, which is 6 times greater than SiO2.

5.6 JOURNEY OF MOS TRANSISTORS IN SEMICONDUCTOR FIELD

MOSFETs are also known as metal–oxide–silicon transistors. These are a kind of field-effect transistors made of semiconductors. These devices belong to the family of insulated-gate field-effect transistor (IG-FET). It is formed by oxidizing a semiconductor material, typically silicon or germanium, with utmost care. The electrical conductivity of a system is influenced by the voltage applied at the gate. This potential is used to adjust conductivity based on the applied voltage.

The concept of MOS transistors was initially proposed by Mohamed M. Atalla and Dawon Kahng of Bell Laboratory. It was first implemented in the year 1960. It was considered the popular device in history and the most basic component of modern electronics.

Transistors have the feature of both optical and analog ICs. It is just a tiny transistor that has been miniaturized and bulk manufactured for various applications. These devices act an important part in revolutionizing the electronics market and the global economy and playing a critical role in the new era, industrial age, and the contemporary planet.

Since the 1960s, MOSFET scaling and miniaturization have pushed electronic semiconductor manufacturing to exponential levels, allowing for more-density chips, such as microprocessors and storage devices. These devices are known as the workhorse of the semiconductor field. In comparison to bipolar junction transistors (BJTs), a MOSFET has the advantage of providing no input current to power the load current.

5.7 NECESSITY OF NANOELECTRONICS

The requirement of nanotechnology is considered as an essential one in most of the electronics field, and it could not be an unavoidable case in the following areas:

5.7.1 Spintronics

The research and application of electron spin, as well as its related magnetic moment and electric charge, is known as spintronics. Since spintronics is focused on the spin of individual electronics rather than the charges of several electrons, systems based on this area of research are supposed to have significantly higher computing efficiency and lower power consumption. Spintronics is also important in a variety of technologies that use quantum behavior for computation [3].

5.7.2 Optoelectronics

Electronic devices that emit, perceive and control light are known as optoelectronics. Nanoscale optoelectronics is gradually being used as a means to solve one of contemporary technology's most pressing issues: the utilization of energy. In optoelectronics, nanomaterials such as carbon nanofibers and CNTs are used, and atom-thick graphene has shown great potential for use in optoelectronic technology. One strategy for increasing the speed of data transmission among circuits is to use silicon nanophotonics elements in complementary metal-oxide-semiconductor (CMOS) circuits.

5.7.3 Display Technology

Nanowire-based electrodes could allow flat panel displays to be both versatile and thinner than currently available flat panel displays. CNTs could be used to send electrons to pixels, making for a millimeter-thick display that is light and small.

Quantum dots (QDs) are a kind of nanomaterial that could eventually replace fluorescent dots in display technology. QD displays are also expected to be cheaper to manufacture and use less fuel.

5.7.4 Wearable and Flexible Devices

As demonstrated by the exponentially expanding range of smartwatches and next-generation personal wellness products, the wearable electronics revolution has arrived. Many operating on these systems are looking for a blend of physical versatility, a quick manufacturing procedure, and low power consumption. Cadmium selenide nanocrystals were mounted on plastic sheets to build lightweight electronic circuits in one promising study.

Wearable nanoelectronics will be much more than optical watches and armbands in the future. Wearable, lightweight nanoelectronics, for example, may be incorporated in textiles, allowing for a wide range of smart clothes' forms, sizes and applications.

5.7.5 Energy Technology

Nanoelectronics is expected to play a significant role in energy technology, with solar cells and supercapacitors proving to be the most promising applications.

Alternative photovoltaic cells, such as thin-layer and polymer cells can benefit greatly from nanoelectronic technologies. Because of their low materials and fabrication costs, as well as their versatility, polymer solar cells are expected to have a lot of potentials, particularly in the field of portable electronic devices. Lithium-ion batteries are supposed to benefit from nanotechnologies in terms of power and protection.

5.8 ROLE OF NANOELECTRONICS—PAST, PRESENT AND FUTURE

Table 5.1 describes some important milestones of nanotechnology achieved in the field of electronics.

5.8.1 III–V CMOS FET

The InGaAs FET (indium gallium arsenide field-effect transistor) belong to III-V compound semiconductor-based FET. It was developed in the 1980s.

There are two main structures of InGaAs FETs. They are listed as follows:

- Insulated-Gate MOSFET and
- Schottky-gate high electron mobility transistors (HEMT) [7].

Figure 5.1 depicts the transconductance, gm, as a representative parameter for comparing and contrasting these two emerging device technologies. InGaAs correspond to any ternary composition ranging from pure GaAs to pure InAs. Following their first demonstration in 1980, HEMTs made steady progress. The characteristics of these devices have the highest transconductance. Also, the cut-off frequencies in the III–V FET band are being transformed into a highly competitive transistor technology.

In the 1980s and 1990s, on the other hand, InGaAs MOSFETs progressed steadily. This lasted until the mid-2000s, when the possibility of III–V CMOS began to pique researchers' attention around the world, resulting in a dramatic performance increase.

InGaAs MOSFET system advancement has been significantly improved credits to the III–V CMOS study. MOSFETs showed their promise in 2014, outperforming HEMTs in terms of transconductance. This analysis work resulted in the creation of a record InGaAs MOSFET computer, as well as a few others. Several III–V FET architectures, including HEMT and MOSFET, have been invented and studied since the mid-2000s. They were used to research process alignment and system physics in CMOS.

5.8.1.1 III–V HEMT as a CMOS Logic Device

The HEMT computer architecture is used in a variety of applications, including power amplifiers. These devices have also resulted in a remarkable shift in perception.

Figure 5.2 shows that HEMTs with the III–V channel can produce at least 2 times V_{inj} at lower drain voltages than strained silicon. V_{inj} enhancement is caused

TABLE 5.1
Milestone in the Development of Nanoelectronics Field

Year	Device	Feature
1989	1st FinFET Transistor	• It was known as a "Depleted Lean-channel Transistor," or "DELTA," and it was developed in Japan. • The semiconductor channel fin may lie on both sides or just on the sides. It can be protected and electrically contact by gate. It includes both double-gate transistor and tri-gate transistor. • Each side of a double-gate transistor may be wired to two separate terminals or contacts. This type of transistor is also known as a split transistor. This allows for more precise regulation of the transistor's output [6].
1998	17nm N-channel FinFET	
1998	The first demonstration of CNTFET was done	
1999	sub-50 nm P-channel FinFET	
2001	15 nm FinFET	
2002	10 nm FinFET	
	Omega FinFET	The shape of the gate is similar to the Greek letter "Omega" as it wraps around the TSMC-developed source/drain mechanism.
2004	High-κ/metal gate FinFET	
	Bulk FinFET Design	It was developed by Samsung. FinFET systems may be mass-produced in large quantities by analyzing DRAM using a 90-nm Bulk FinFET process.
2011	Tri-gate transistors	It was proposed by Intel. In contrast to planar transistors, the gate wraps around the channel on three sides, allowing for lower gate delay and hence improved performance.
2013	16 nm FinFET	Production started by TSMC and SK Hynix
	Devices at 10 nm	Production started by Samsung
2017	7 nm process	Fabrication of SRAM memory started by TSMC
2018	5 nm process	Initiated by Samsung
2021	The commercial production of a 3 nm GAAFET process	Samsung announced this plan in 2019

by increased InAs composition in InxGa1xAs. These findings affirm the material system's superior electron transport properties and demonstrate the benefit for possible greater performance and low-power CMOS applications.

The following configurations are not suitable for CMOS applications due to several issues. The first issue has to do with the gate leakage. The gate leakage in a

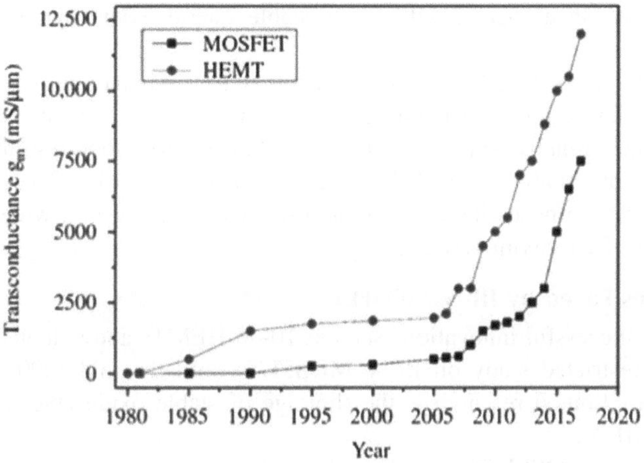

FIGURE 5.1 Transconductance—MOSFET versus InGaAs HEMT [8].

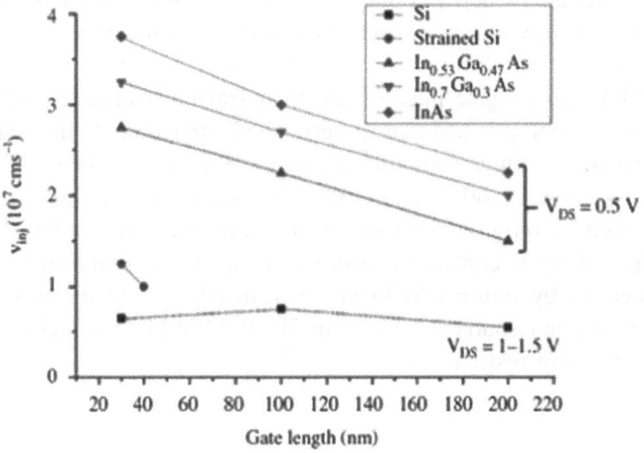

FIGURE 5.2 Source Injection Velocity versus Gate length—silicon, strained-silicon and InAs HEMTs FETs [3].

high-density circuit should be held below 1 A/cm^2. HEMT devices with the gate metal placed on top of a large bandgap semiconductor or barrier are used in "Schottky gate" technology (InAlAs). Carriers will quickly tunnel through the firewall if it is thin due to the narrowband offsets.

In this case, gate leakage in the ON-state will easily reach 100 A/cm^2. The gate leakage imposes a limit on the thickness of the barrier that can be scaled. As the device's lateral dimensions are reduced, this compromises SCE regulation. Secondly, "gate recess" refers to the elimination of the n-type semiconductor above the channel in the gate field. Selective wet etch is used to create a typical gate recess, resulting in a system with a noticeable lateral extent and a "underlap" architecture.

The recent CMOS "overlap" architecture has an n+ area marginally covered with the gate. The device footprint is limited by the underlap area, which also increases series resistance. Finally, since HEMTs do not have a large footprint, the gate and source/drain are usually fabricated using two different masks. The lithography alignment tolerance necessitates a long access area, resulting in a wide source-to-drain spacing of approximately 2 m.

5.8.1.2 Issues Faced by III–V MOSFET

Unlike more successful innovations such as III–V HEMTs and silicon MOSFETs, there was a restricted study on III–V MOSFETs until the mid-2000s. The main reason for this limited research is the shortage of stable oxide layer for the gate insulator on III–V.

The heart of a MOSFET is a high-quality gate stack. Recent advances in atomic layer deposition (ALD) and pulsed laser deposition (PLD) have provided the crucial enabling technologies required to solve this issue. Since the gate insulator is a deposited material, the oxide/III–V interface efficiency is a major concern in scaled transistors, as it directly affects current drive and SCE immunity.

High-density logic applications implement a pitch dimension criterion. Modern silicon MOSFETs have a gate pitch of less than 100 nm, with three active regions: one silicide contact, one gate and two spacers. These are similar to the active regions, gate, ohmic contacts—source and drain, as seen in Figure 5.3. The CMOS front-end method does not allow lift-off or Au for CMOS compatibility and manufacturability reasons. It is essential to improve nanoscale contact technologies for III–V transistors.

The features of III-V contact possess low contact resistivity and low film resistivity, as needed by future CMOS systems, in addition to the process specifications. The drain and source contacts for III–V MOSFETs remain a significant challenge due to these restrictions.

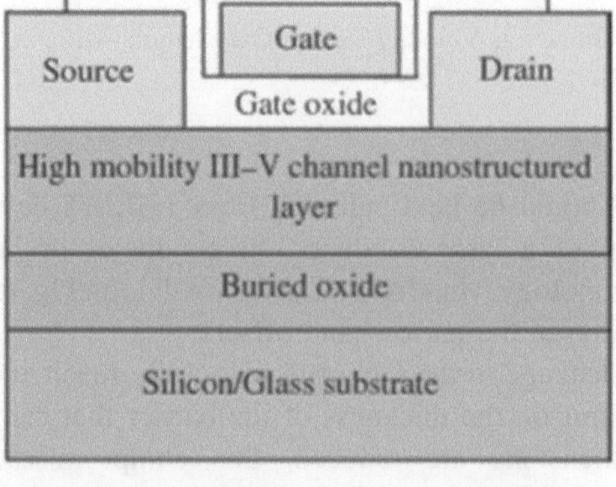

FIGURE 5.3 Planar Logic III–V MOSFET—cross-sectional View [7].

5.8.2 Quantum Dot

QDs are nanometer-sized semiconductor particles with optical and electronic properties that vary from those of larger particles due to quantum mechanics. It plays a vital role in nanotechnology. As UV light illuminates QDs, an electron in the QD may be excited to a higher energy state. This mechanism causes the transition of an electron from the valence band to the conductance band in a semiconducting ED. The excited electron will return to the valence band and release its energy by light emission. The differential energy between the valence band and the conduction band determines the color of the light.

These are called artificial atoms by highlighting their singularity, which includes bound, isolated electronic states similar to those found in natural atoms or molecules. The electronic wave functions of QDs were found to be similar to those of real atoms. An artificial molecule may be created by combining two or more of these QDs, displaying hybridization even at room temperature.

QDs are a hybrid of bulk semiconductors and isolated atoms or molecules in terms of properties. They have different optoelectronic properties depending on their size and form. Longer wavelengths, such as orange or red, are emitted by larger QDs with a diameter of 5–6 nm.

Single-photon sources, lasers, LEDs, Solar cells, single-electron transistors (SET), second-harmonic generation, medical Imaging, research in cell biology, microscopy and quantum computing are all possible uses for QDs.

5.8.3 FinFET

A fin field-effect transistor (FinFET) is a multi-gate MOSFET device, and it is a nonplanar, multi-gate MOS structure mounted on an SOI substrate. It is mounted on a substrate with the gate wrapped around the channel or on two, three or four sides of the channel, forming a multi-gate configuration. FinFET is a generic term used for these devices since the source/drain fields on the silicon surface are shaped like fins. The attractive features of FinFET are a maximum current density and marginally faster switching than planar CMOS [9].

FinFET belongs to the family of nonplanar transistors, and it is also known as a 3D transistor. It serves as the foundation for the fabrication of advanced nanoelectronic semiconductor chips. The FinFET devices were first commercialized during the year 2000, and they quickly established themselves as the most popular gate architecture at 22 nm, 14 nm, 10 nm and 7 nm process nodes. To maximize the drive strength and performance of a single FinFET, it is necessary to arrange more fins on all sides and all are protected by the same gate.

5.8.4 CNTFET

A carbon nanotube field-effect transistor (CNTFET) is another type of FET. In this device, a single or a group of CNTs is used as a channel instead of bulk silicon material [10]. Figure 5.4 shows the structure of a single-walled CNT.

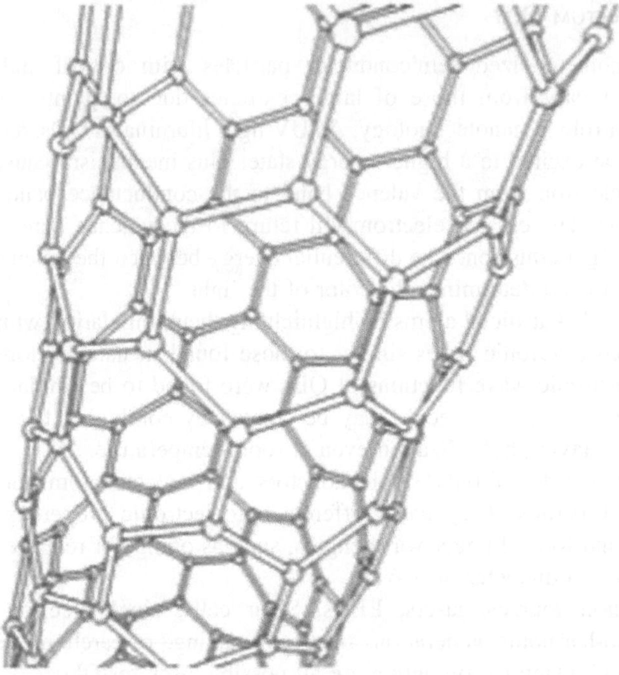

FIGURE 5.4 Structure of carbon nanotube.

The chiral angle and diameter of a CNT have a strong impact on its bandgap. CNTs may be a potential candidate for upcoming nanoscale transistor devices if those properties can be regulated. Furthermore, boundary scattering is absent in CNTs due to the absence of borders in their perfect and hollow cylinder structures. CNTs belong to quasi-1D materials. These devices permit forward and backscattering and elastic scattering to ensure that free paths in CNTs are long, usually on the order of micrometers.

CNTs are chemically stable and can hold huge quantities of electricity. CNTs are also almost as good at conducting heat as diamond or sapphire. The CNTFET can transform efficiently using much less power than a silicon-based system due to their miniaturized measurements [11].

The advantages of preferring CNTFET devices are high transconductance, improved electron mobility, greater current density, maximum linearity, good threshold voltage, better subthreshold slope and excellent controllability.

A temperature rise of several hundred kelvins will cause the current to drop and the CNT to burn. Due to various heat dissipation processes, the self-heating effect in a semiconducting CNTFET is much less extreme than in a metallic one. The channel dissipates a small portion of the heat produced in the CNTFET. The heat is spread unevenly, with the maximum values on the source and drain sides of the channel. As a result, the temperature near the source and drain regions drops dramatically. As compared to silicon, the temperature increase has a slight impact on the I-V properties of semiconducting CNT.

The disadvantages of using CNTFET devices are degradation in lifetime, reliability, challenges in mass production and cost.

5.8.5 Graphene Nanoribbon-Based Devices

A graphene nanoribbon is otherwise known as a GNR, nano-graphene ribbons, or nano-graphite ribbons. It is made up of graphene strips with a diameter of fewer than 100 nm. Mitsutaka Fujita et al. investigated the nanoscale size effect in graphene material by using GNRs as a theoretical model [12,13]. Figure 5.5 shows the 2D structure of the arrangement of a GNR.

Based on their 2D structure, reduced noise, and high thermal and electrical conductivity, GNRs are considered as the alternative material to copper for IC interconnects. At room temperature, GNRFETs with an Ion/Ioff ratio $>10^6$ have been produced. It might be a viable technological substitute for silicon semiconductors. It can keep microprocessor clock speeds in the 1 THz range with FETs that are less than 10 nm thick [14,15].

Graphene strips with a high length-to-width ratio are known as a GNR. GNRs have been the subject of extensive theoretical and laboratory studies. Because of their attractive properties, such as charge carrier mobility, band-gap or energy-level coordination, recent research, especially those focusing on the synthesis and characterization of new atomically specific GNR structures, is constantly being analyzed. The thickness of the GNR system, its edge orientation or termination, and the presence of heteroatoms all affect its properties. Furthermore, a few structural changes have a major effect on the nanoribbon's electrical properties.

FIGURE 5.5 Structure of graphene nanoribbon.

5.8.6 Nanowire

A nanowire is a 1 nm diameter nanoparticle (10^{-9} m). It is also regarded as having a greater-than-1000 length-to-width ratio. The dimensions of nanowires (i.e. the diameter or thickness) are about tens of nanometers or less [9].

Quantum mechanical effects are significant at these scales; hence, the word "quantum wire" was coined. Superconducting, metallic, semiconducting (e.g. InP, GaN, silicon nanowires) and insulating nanowires are only a few examples (e.g. TiO2, SiO2). Molecular nanowires are made up of organic or inorganic repeated molecular units.

Nanowires can be synthesized in any one of two ways, that is, either the bottom-up or top-down approach. A top-down technique uses lithography, milling or thermal oxidation to reduce a large piece of material into small parts. The nanowire is created by mixing constituent adatoms in a bottom-up approach. However, most of the synthesis techniques use bottom-up techniques due to self-limiting oxidation.

Vapor deposition, suspension, electrochemical deposition and VLSI growth are all typical laboratory techniques used to make nanowires. Ion track technology allows for the development of homogeneous and segmented nanowires with diameters as small as 8 nm. Thermal oxidation steps are often used to fine-tune the morphology of nanowires because their oxidation rate is regulated by their diameter.

MOSFETs can be made from nanowires. MOS transistors are widely used as fundamental building blocks of today's electrical circuits. Moore's law assumes that MOS transistors will continue to shrink until they reach the nanoscale. Maintaining strong channel gate access is one of the most challenging aspects of designing potential nanoscale MOS transistors.

5.9 PERFORMANCE COMPARISON OF VARIOUS NANOELECTRONIC DEVICES

For performance comparison, the standard 6T SRAM cell was considered an example. Based on the performance comparison of GNR-based designs to CMOS, FinFET and CNTFET-based designs, the behavior of GNR-based designs can be calculated [16–18]. The utilization of power, overall power dissipation, average delay and leakage current are some of the metrics used to compare efficiency [19,20]. The open-source model files are taken from Standford and nanohub [21].

Table 5.2 describes the performance comparison of 6T, 7T, 8T, 9T and 10T SRAM cells [22–24].

Table 5.3 compares the efficiency of GNRFET-based SRAM cells to CMOS, FinFET and CNTFET-based SRAM cells [25].

5.10 CONCLUSION

Although CMOS is irreplaceable, the power dissipated by GNRFET is somewhat less than that of CMOS. Beyond 32 nm, the power dissipated by CNTFET is also lower than that of CMOS. Beyond 32 nm, GNRFET-based circuits are more

TABLE 5.2
Performance Comparison of 6T, 7T, 8T, 9T and 10T SRAM Cell

SRAM Cell	6T SRAM	7T SRAM	8T SRAM	9T SRAM	10T SRAM
SNM	202	223	397	410	432
Write Margin	340	360	379	330	475
Dynamic Power	10.05	4.87	12.46	16.55	17.58
Delay	5.9	3.6	6.5	7.6	8.0

TABLE 5.3
Performance Comparison of Standard 6T-SRAM Cell at 32 nm technology node

Parameters Used	Average Power Consumption	Total Voltage Source Power dissipation	Average Delay
CMOS-Based design	16.61 nW	5.51 nW	0.29 us
FinFET-based design	10.23 nW	34.1 nW	4.65 ns
GNRFET-based design	8.23 nW	0.401 nW	3.75 ns
CNTFET-based design	6.21 nW	32.9 pW	2.50 ns

powerful than CMOS-based circuits. The potential of GNR in VLSI is exciting, paving the way for a breakthrough in MOSFET science. As a result, GNR can be used as an alternative to CMOS transistors beyond 32 nm.

AUTHOR BIOGRAPHY

Mr. G.Boopathi Raja received his B.E. (Electronics and Communication Engineering) from Anna University of Technology, Coimbatore, and his M.E. (Applied Electronics) from Anna University, Chennai. Currently, he is an Assistant Professor in the Velalar College of Engineering and Technology, Erode. He has published 10 papers in international journals and 4 book chapters. He is a lifetime member of IETE and ISSSE. His research interests include nanoelectronics, medical electronics and signal processing.

REFERENCES

1. *International Technology Roadmap for Semiconductors (ITRS)*. San Jose, CA: Semiconductor Industry Association, 2007.

2. Merritt, Rick. "Moore's law dead by 2022, Expert Says." *EE Times.* https://www.eetimes.com/moores-law-dead-by-2022-expert-says/.
3. Smith, Brett. "Definitions and applications of nanoelectronics." https://www.azom.com/article.aspx?ArticleID=18333.
4. Raja, G.B., and Madheswaran, M. "Logic fault detection and correction in SRAM based memory applications." *2013 International Conference on Communication and Signal Processing*, 2013, pp. 215–220, doi: 10.1109/iccsp.2013.6577046.
5. G. Boopathi Raja, and M. Madheswaran. "Design of improved majority logic fault dectector/corrector based on efficent LDPC codes." *International Journal of Advanced Research in Electrical Electronics and Instrumentation Engineering* 2 (2013), 3429–3437.
6. K Bindu Madhavi, Suman Lata Tripathi. "Strategic review on different materials for FinFET structure performance optimization." IOP Conference Series: Materials Science and Engineering, 2020.
7. John Chelliah, C., and Swaminathan, R. "Current trends in changing the channel in MOSFETs by III–V semiconducting nanostructures." *Nanotechnology Reviews*, 6, no. 6 (2017), 613–623. 10.1515/ntrev-2017-0155.
8. Del Alamo J.A. "Nanometer-scale InGaAs field-effect transistors for THz and CMOS technologies." 2013 Proc. ESSCIRC IEEE 2013, 16–21.
9. Monica Taba, and Gerhard Klimeck. "Investigation of the electrical characteristics of triple-gate FinFETs and silicon-nanowire FETs." (2006) https://nanohub.org/resources/1715.
10. Javey, A., J. Guo, D. Farmer, et al. "Carbon nanotube field-effect transistors with integrated ohmic contacts and high-κ gate dielectrics." *Nano Letters*, 4, no. 3 (2004), 447–450. 10.1021/nl035185x.
11. S. Chilstedt, C. Dong, and D. Chen. *Carbon Nanomaterials Transistors and Circuits, Transistors: Types, Materials and Applications*. Nova Science Publishers, 2010.
12. Natarajamoorthy, M., Subbiah, J.Alias, N., and TanM. "Stability improvement of an efficient graphene nanoribbon field-effect transistor-based SRAM design." *Journal of Nanotechnology*, 2020 (2020), 1–7. 10.1155/2020/7608279.
13. Chen, Y., Sangai, A., Rogachev, A., Gholipour, M., Iannaccone, G., Fiori, G., and Chen, D. "A SPICE-compatible model of MOS-type graphene nano-ribbon field-effect transistors enabling gate- and circuit-level delay and power analysis under process variation." *IEEE Transactions on Nanotechnology*, 14, no. 6 (2015), 1068–1082. 10.1109/tnano.2015.2469647.
14. W. A. de Heer, C. Berger, E. Conrad, P. First, R. Murali, and J. Meindl, "Pionics: The emerging science and technology of graphene-based nanoelectronics." *2007 IEEE International Electron Devices Meeting*, 2007, pp. 199–202, 10.1109/IEDM.2007.4418901.
15. Son, Y., M. Cohen, and S. Louie. "Energy gaps in graphene nanoribbons." *Physical Review Letters*, 97, no. 21 (2006). 10.1103/physrevlett.97.216803.
16. Deng, J., and Wong, H. "A compact SPICE model for carbon-nanotube field-effect transistors including nonidealities and its application—Part I: Model of the intrinsic channel region." *IEEE Transactions on Electron Devices*, 54, no. 12 (2007), 3186–3194. 10.1109/ted.2007.909030.
17. Deng, J., and Wong, H. "A compact SPICE model for carbon-nanotube field-effect transistors including nonidealities and its application—Part II: Full device model and circuit performance benchmarking." *IEEE Transactions on Electron Devices*, 54, no. 12 (2007), 3195–3205. 10.1109/ted.2007.909043.
18. J. Deng, and H.-s. Philip Wong. "A circuit-compatible SPICE model for enhancement mode carbon nanotube field effect transistors." 2006 International Conference on

Simulation of Semiconductor Processes and Devices, 2006, pp. 166–169, 10.1109/SISPAD.2006.282864.
19. "Predictive technology model for 32 nm CMOS technologies." http://www.eas.asu.edu/~ptm.
20. "Stanford University CNTFET Model Website." http://nano.stanford.edu/model.php?id=23.
21. nanoHUB. http://nanohub.org.
22. Joshi, S., and U. Alabawi. "Comparative analysis of 6T, 7T, 8T, 9T, and 10T realistic CNTFET based SRAM." *Journal of Nanotechnology*, 2017 (2017), 1–9. 10.1155/2017/4575013.
23. BoopathiRaja, G., and Madheswaran, M. "Design and performance comparison of 6-T SRAM Cell in 32nm CMOS, FinFET and CNTFET technologies." *International Journal of Computer Applications*, 70, no. 21 (2013), 1–6. 10.5120/12188-7751.
24. Raja, G., and Madheswaran, M. "Design and analysis of 5-T SRAM Cell in 32 nm CMOS and CNTFET technologies." *International Journal of Electronics and Electrical Engineering, Vol. 1, 4* (2013), 256–261. 10.12720/ijeee.1.4.256-261.
25. Boopathi Raja G, and Madheswaran, M. "Performance comparison of GNRFET based 6T SRAM cell with CMOS, FinFET and CNTFET technology." *International Journal of Innovative Research in Science and Engineering, Vol. 2, 5* (2016), 197–204.

6 Nanoelectronics

S. Dwivedi
S.S. Jain Subodh P.G. (Autonomous) College,
Jaipur, Rajasthan, India

6.1 INTRODUCTION

Nanotechnology is defined as the technology that deals with the tailoring of at least one of the dimensions of materials and devices below 100 nm [1–5]. The nanometer scale regime typically depicts sizes below one billionth the size of a human hair. Nanotechnology is a collective term that encompasses all associated domains of nanomaterials, nanocomposites, nanoelectronics, nano devices, nanomagnetism, nano-optics and even nano-biotechnology. Out of these different domains, nanoelectronics is the study of electronic properties of materials at the nanoscale. The electronic properties of nanomaterials are patterned in a particular manner by computational methods and experimental control and tailoring of properties at the orders of few nanometers such that highly specific novel physical phenomena guide the whole mechanism of interaction and operation. As a result, the largely used word of "microelectronics" is modified in the modern world to "nanoelectronics" because of the nanoscale regime. Earlier, computational devices used micron-scale technologies, and hence the word "microelectronics" came into use but has now become a misnomer because of the development of nanoscale technology. These microelectronic devices are an integral part of modern-day life, and it has become impossible to imagine life without these components. Some of the examples include (1) smart watches, (2) tablets, (3) smartphones consisting of small processors that act as mini-computers, (4) palmtops, (5) laptops that have become thinner and compact with touch-screen and voice-operability options possessing enhanced storage capacities in smaller volumes with integrated circuits (IC) along with much faster responding processors and clock speeds, (6) optical fiber-based circuitry for faster Internet speeds that enable use of Internet banking, (7) mobile app-based electronic commerce or e-commerce, (8) credit and debit card-based commerce, (9) modern toasters, (10) heat controllers, (11) air conditioners, (12) driver-less automobiles, (13) robot-enabled machineries and (14) the latest edition of artificial intelligence (AI)-based equipment, instruments, public user railway facilities and air ticketing systems [1–5].

The story of microelectronics started with the fabrication of bipolar junction transistor (BJT) by Bell Laboratories in 1948 [6,7]. This was followed by the discovery of insulated-gate field-effect transistor (IG-FET) in a famous research

DOI: 10.1201/9781003220350-6

work by M. M. John Atalla, along with Dawon Kahng in 1960 at the Bell Laboratories [8–10]. They removed the limitation experienced by Shockley, including other researchers by eliminating the "surface-states" that hinder electric field lines to pass through the semiconductor materials. It was discovered that these so-called "surface states" were greatly diminished at the surface-interface formed between semiconductor silicon (Si) and its thermally grown oxide, silicon-di-oxide (SiO_2) in a sandwich type of configuration consisting of metal (M at the gate), oxide (O – insulator) and silicon (S – semiconductor). This sandwich was given the acronym of MOSFET, standing for metal-oxide-semiconductor field-effect transistor that eventually came to be known as MOS. The semiconductor-based transistor was designed for telephonic systems of that time and was bit slower. Realizing its potential, Karl Zaininger and Charles Meuller at RCA were successful in fabricating a MOS transistor [11], whereas C. T. Sah at Fairchild fabricated a MOS-operated tetrode device in 1960 [12]. In 1962, an integrated circuitry of 16 MOS transistors was fabricated at RCA by Fred Heiman and Steven Hofstein [13]. A turning point in the semiconductor industry happened in 1964 when MOS-based transistor devices were available for commercial use. General Microelectronics and Fairchild fabricated commercially successful p-type doped channel devices for switching and logic applications, while RCA commercialized n-type doped channel devices for amplification applications. The MOSFET-based device technology has outphased BJT-based technology because of nanometric-sized architecture and lower power consumption, creating modern equipment and IT-based revolutions. In 1965, the founder of Intel, Gordon Moore, presented an analytical view that the microelectronics industry has been growing exponentially, and thus the typical size of a transistor becomes half of its then-size in every 18 months [14]. This typical analysis is known as Moore's Law and carries very high potential to shape economies. Reduction of size of the transistors leads to significantly smaller gates so that the switching action of the transistor becomes faster in comparison to the previous one which subsequently reduces the power consumption. Moreover, enhancement of number of transistors along with increased manufacturing yields cuts the cost of manufacturing of single unit of transistor, which leads to enhanced commercial gains by further scaling of transistor sizes to smaller dimensions.

6.2 INTRODUCTION TO FIELD EFFECT TRANSISTOR (FET)

FET can be defined as the type of transistor in which output can be tailored with the help of an applied electric field [15]. FETs consist of three terminals, source, drain and the gate. Generally, a transistor (transfer + resistor) is defined as a device with three operational terminals in which channel resistance between the two terminals of source and drain is precisely tailored by the top terminal of insulated gate electrode. The bottom or back side of the substrate serves as the fourth terminal for electrical contact to the substrate. A schematic of the FET with a capacitively coupled gate and metallized source, drain and gate electrodes is shown in Figure 6.1.

FETs can adopt two types of gate configurations, the back or bottom-gated and the top-gated device structure. Both types of gate configurations require a dielectric

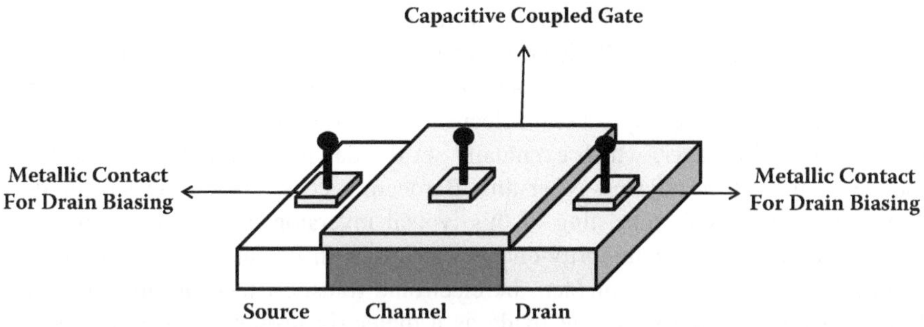

FIGURE 6.1 Schematic of the field effect transistor (FET) showing a capacitively coupled gate with metallized contacts on source, drain and gate electrodes.

metal oxide to be deposited followed by metallization, usually with noble metals to prevent oxidation of the surface layer. A voltage applied through the metallic contacts of the gate generates an electric field inside the channel area that controls the flow of charge carriers between the two electrodes by altering the conductivity. FETs are unipolar devices since they make use of one form of charge carriers only, either electron or hole, for their operation. A special property of FETs is that they produce impedances of higher degrees at low applied frequencies. MOSFET is the highest used configuration of the different types of FETs currently under application.

In 1925, Julius Edgar Lilienfeld, a physicist of Austro-Hungarian origin, was the first to patent the concept of FET, followed by Oskar Heil in 1934 [16]. However, both of them could not demonstrate a practical working semiconductor FET device based on the stated concept. In 1947, John Bardeen, along with Walter Houser Brattain, working at Bell Labs under the guidance of William Shockley, realized and explained the transistor action. Initial attempts were focused on modulation of conductivity of the semiconductor for obtaining the transistor effect. These attempts were ravaged by the incurring problems of dangling bonds, surface states and those associated with the materials of germanium (Ge) and copper (Cu). Further attempts led to the fabrication of a point-contact transistor by Bardeen and Brattain in 1947, followed by the development of BJT by Shockley in 1948 [17]. In 1945, Heinrich Walker patented the first junction-field effect transistor, or the JFET, which was the paramount FET device [18]. In 1950, engineers Jun-ichi-Nishizawa and Y. Watanbe, of Japanese origin, invented a variant of JFET consisting of a short channel, the static induction transistor (SIT) [19]. George F. Dacey, along with Ian M. Ross, constructed a practical JFET in 1953, which was based on the theoretical concepts of Shockley on JFETs [20]. However, JFETs were relatively heavy devices, difficult to produce at the mass-scale, limiting their practical applications. The IG-FET was presented as an option to the JFET. However, the major problem of penetration of externally applied electric field pervading the material persisted in this approach owed to the barriers presented by troublesome surface states. This resulted in a halt on the research on FET devices by the middle of the 1950s and researchers begin to focus their works on BJT technology. In 1946, Bardeen explained the failure of the FET device fabricated by Shockley in 1945 based on the existence of surface states. Bardeen presented the theoretical concept that the external electric fields of lines are blocked by the extra electrons lying at the surface states. As a result, those trapped extra electrons in the confined localized states form a typical inversion layer, which eventually set the dawn of the physics of surface phenomenon, surface physics. After this, Bardeen, successfully fabricated the device prototype IGFET, consisting of this typical inversion layer that precinct the transportation process of minority-charge carriers with enhanced modulation and conductivities. As a matter of fact, the electronic transport mechanism is heavily dependent on the quality of gate oxide as a dielectric insulator that is deposited over the inversion layer. The modern complementary metal-oxide semiconductor (CMOS) technology is heavily reliant on the conception of inversion layer and the patented technology of Bardeen. The goal of the scientists became to downsize the effects posed by surface states. It was finally concluded by the end of 1950 that there are two types of surface states, the fast and the slow surface states. The first one, i.e. the fast surface state, was discovered to be attached to the bulk along with the semiconductor-and-oxide surface-interface. The second one, the slow surface state, was numerous and was attached to the oxide layer due to the adsorption process of atoms, molecules as well ions from the surrounding environment or ambient along with possessing much longer relaxation times. Around this time, techniques to produce atomically clean semiconductor surfaces were introduced by many researchers.

In 1958, finally a breakthrough in the FET technology was obtained when Mohammed Atalla performed passivation of surface states by growing a thin SiO_2 layer over clean Si surfaces, called surface passivation. This technology became crucial to the semiconductor commercial technology, which eventually led to the mass-scale production of ICs on the semiconductor platform. In 1959, Mohamed Atalla, along with Dawon Kahng, developed the first-ever MOSFET that laid the foundation of modern CMOS technology [9]. MOSFET technology had superior characteristics of lower power consumption, higher density and high level of scalability that led to the fabrication of high-density ICs. In 1963, CMOS technology for fabrication of MOSFETs was produced by Chih-Tang Sah, along with Frank Wanlass, at Fairchid Semiconductor [21]. In 1967, Dawon Kahng, along with Simon Size, developed a floating-gate MOSFET. Toshihiro Sekigawa and Yutaka Hayashi presented the first double-gate MOSFET in 1984 [22]. In 1989, a 3D nonplanar multi-gated variant of MOSFET, known as fin field-effect transistor (FinFET), was developed by Digh Hisamoto at the Hitachi Central Research Laboratory [23].

MOSFET forms the fundamental building block of current electronics products and ranks top among all the manufactured devices in terms of number in human history of electronics manufacturing. According to a report, a total number of 13 × 1022, or 13 sextillion, MOSFETs were manufactured during the period 1960–2018 [24]. MOSFET devices are under intense scaling and miniaturization in the electronics manufacturing industry, driving the exponential growth through high-density IC to form microprocessors and memory chips for logic and memory operations. A MOSFET does not require an input current for controlling the load current.

6.2.1 Composition of FET

FET consists of the two electrodes of source and drain connected through an active channel in which active transport of charge carriers takes place. These charges carriers can be either electrons or holes. Thus, the three main components of a FET are:

i. Source (S): It is the electrode through which injection of charge carriers into the channel is performed.
ii. Drain (D): It is the output electrode at which number of charge carriers coming out of the channel are realized.
iii. Gate: The region enclosed between the source and drain electrodes is covered with an insulator oxide followed by metallization for biasing. This forms the gate region in the MOSFET and is the terminal through which channel conductivity can be modulated. Hence, the output or the drain current can be modulated by suitable applying bias at the gate terminal.

Thus, source, drain and gate are the three terminals of an FET that are comparable to emitter, collector and base of BJTs. Body or base or metallized bottom of the substrate of FETs form one more terminal for gate biasing the transistor in operational mode [25,26]. The horizontal distance confined between the two electrodes forms the gate length of the transistor that is channel. The length in a direction perpendicular to the horizontal gate length or cross-section is the width of the transistor. Width of the transistor is typically much larger in comparison to the gate length. Shorter gate lengths support higher frequencies of transistor operation. In a physical sense, the gate is equivalent to a switch that permits transportation of electrons or hinders their flow by formation or eradication of charge carriers between the two electrodes.

Out of the several versions of FETs, ***n***- and ***p***-channel devices are framed based on the type of carriers transported in the channel region. In the ***p***-type substrate, ***n***-channels are typically formed by electrons that become progressively conductive with the incrementing positive gate voltages. Oppositely, ***p***-channels are formed by holes in the ***n***-type body of substrate that become progressively conductive with enhancing negative gate voltages. Two different modes of operations of the transistor can be defined with reference to the application of zero or no gate biases. Thus, the FETs can be operated in depletion and enhancement modes [25,26]. In the enhancement mode of operation of the FET, or the normally-off state, the channel conductance is very low at zero applied gate biases so that the formation of a conductive channel requires application of a gate voltage. The oppositely configured mode is normally-on state, or depletion-mode, in which case the channel becomes conductive when no gate biases have been applied and a suitable gate voltage must be applied to bring the transistor in "off-state". All these different modes with their electrical characteristics have been shown schematically in Figures 6.2–6.5, given below [25]. As a customary, in case of a ***p***-channel device, holes are the charge carriers when oppositely oriented voltages are applied in the case of the ***n***-channel devices.

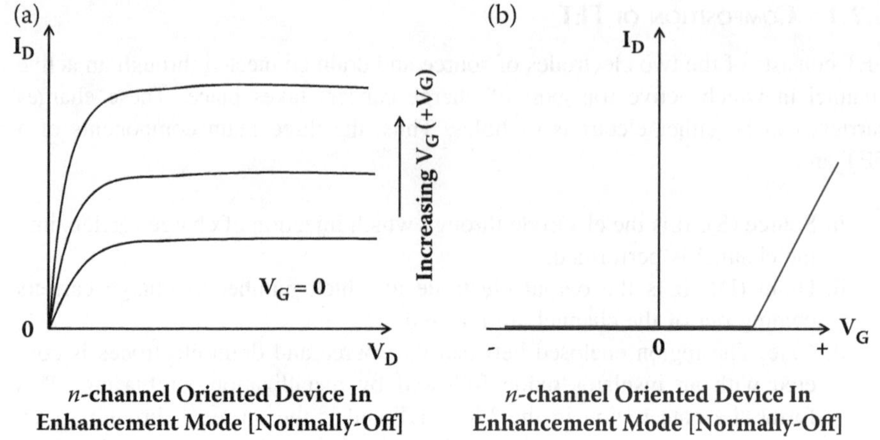

FIGURE 6.2 (a) *n*-channel enhancement mode of operation of the FET, and its (b) transfer characteristic, i.e. I_D*vs.* V_G electrical plot [25].

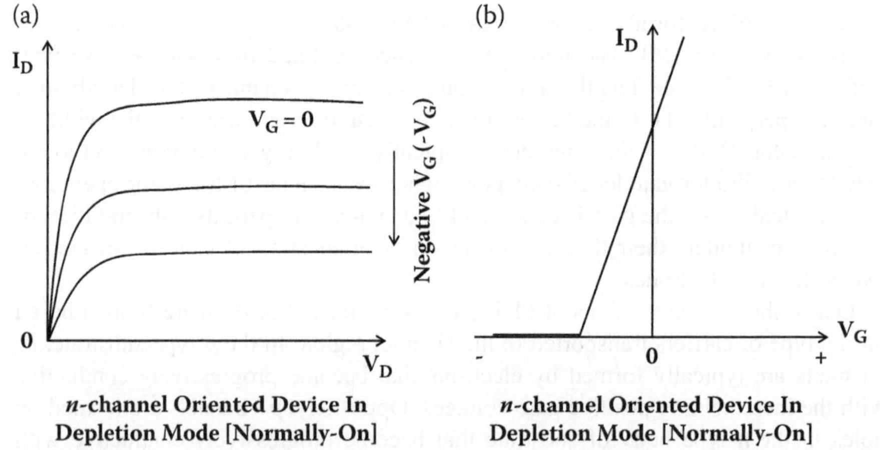

FIGURE 6.3 (a) *n*-channel depletion mode of operation of the FET, and its (b) transfer characteristic, i.e. I_D*vs.* V_G electrical plot [25].

6.2.2 Nature of Channels

Channel region can be defined as that formed by the surface inversion layer and bulk-buried layer that defines the nature of the channel. The first type of channel formed of surface inversion layer is a 2D sheet of charge carriers that is typically of the thickness of 5 nm order. In contrast, buried layered channel possesses a thickness that is much higher and is quite equivalent to the order of depletion width. It is owed to the fact that the channel region is completely encroached by the surface-based depletion layer in case when the transistor is in

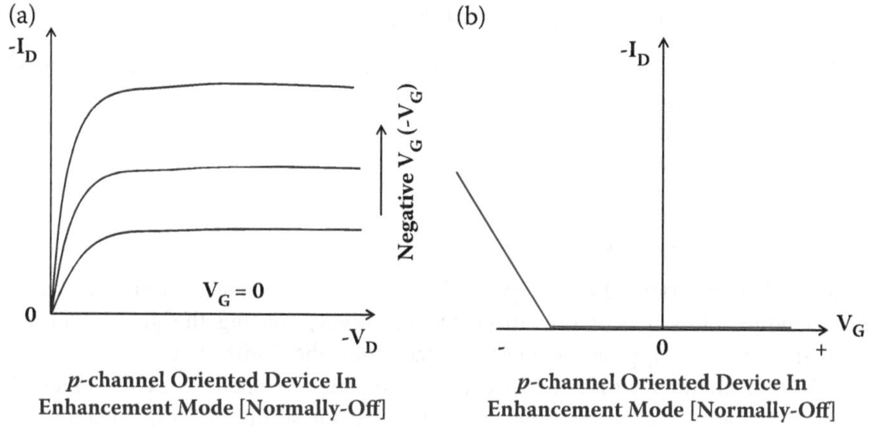

FIGURE 6.4 (a) *p*-channel enhancement mode of operation of the FET, and its (b) transfer characteristic, i.e. $I_D vs. V_G$ electrical plot [25].

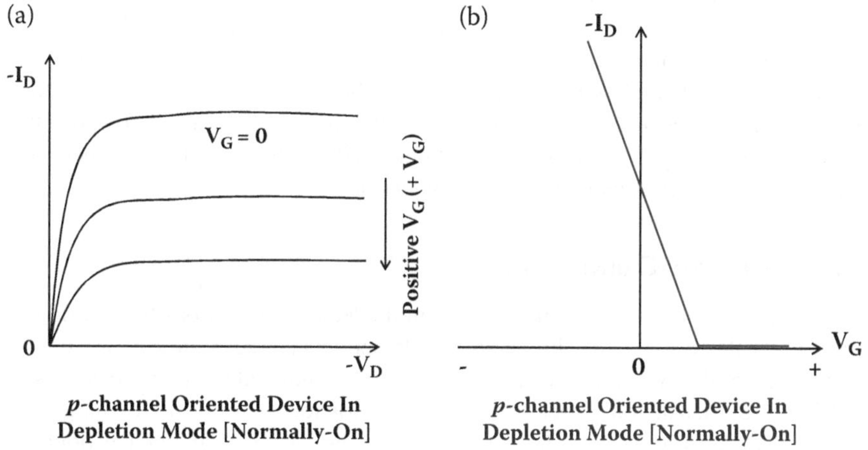

FIGURE 6.5 (a) *p*-channel depletion mode of operation of the FET, and its (b) transfer characteristic, i.e. $I_D vs. V_G$ electrical plot [25].

"off-state". Buried channel devices rely on bulk states for conduction and are usually devoid of surface defects and scattering fostering smoother carrier mobility. It is customary to apply buried channel configuration for depletion-mode devices. However, theoretically, the same functionality can be achieved by selecting an insulating gate material possessing appropriate work function tuning the threshold voltage to the desired operational levels.

Out of many FETs, MOSFET, MODFET and MISFET are mostly surface-channel device; however, MOSFETs and MISFETs can have the two kinds of channels parallely. JFETs and MESFETs are always buried-channel devices.

6.2.3 Functions of Gate of the MOSFET

The FET performs transistor operation or acts as a switch by controlling the flow of electrons in the channel region contained by the source and drain. For its operation, gate and source terminals are suitably biased to form a conductive channel of desired shape and size.

6.2.4 *n*-Channel-Oriented FET

Let an ***n***-channel oriented device in depletion mode be considered first for basic understanding of transistor operation. On negatively biasing the gate-to-source terminals, depletion region enhances in size along the width that narrows down the effective channel width. In case this active region encroaches the channel completely, there will be no flow of electrons from source to drain, leading to infinite resistance across the channel so that the FET becomes turned off, acting in the form of a switch. This phenomenon is called pinch-off, and the voltage that corresponds to this specific point is called "pinch-off voltage". In a reverse mode of operation of positive gate-to-source biasing of the FET, a conductive channel is formed by attraction of electrons by the positive voltage that counters the foreign ions doped into the body of the FET, forming a region devoid of mobile charge carriers, known as the depletion region. The corresponding voltage at which point this conductive channel starts acting is called the threshold voltage. On further increasing the gate-to-source voltage, more electrons are attracted towards the gate region, enabling formation of the channel with conductive features between source-to-drain electrodes that is called inversion.

6.2.5 *p*-Channel-Oriented FET

In the case of a ***p***-channel oriented device in depletion mode, a positive biasing is done from gate to body that broadens the depletion layered region. It forces the electrons towards the gate-to-insulator or gate-to-semiconductor interface. As a result, only carrier-devoid space region of acceptor ions with positive charge are exposed that are largely immobile. Similarly, in the case of a ***p***-channel oriented device operating in the enhancement mode, negative voltages are applied to create a conductive channel as generally the conductive channel does not exist.

6.2.6 *Drain-to-Source Biasing for Conductive Channel*

The FET can be activated in either enhancement or depletion modes depending on the specific requirements [25]. The FET can be gate-to-source biased for realization of the field effect or drain-to-source biased for output current. In the case of gate-to-source bias being significantly higher than the drain-to-source bias, channel conductivity or resistance can be tuned by alteration of the gate bias. In that case, the drain or output current is directly proportional to the drain voltage in reference to the source voltage. It indicates the linear or ohmic mode of operation of the FET and behaves similar to a variable resistor.

On enhancing the drain-to-source bias, potential takes the form of a gradient in a region from source to drain that develops an asymmetrical shape of the channel region. On the side of the drain electrode of the channel, the inversion region takes the shape similar to that of the pinched-off region [25, 26]. On increasing the drain-to-source bias further, the pinched-off region starts moving in a direction towards the source that is defined as the "saturation mode" or even "active mode". In the case when an amplified output signal is required, the region between ohmic and saturation modes is utilized. Such regions are sometimes regarded as a portion of the linear or ohmic region, even in the case when drain current does not carry an approximately linear relationship with the drain voltage.

A unique point is that the channel resistance becomes much higher in the saturation mode of operation of the FET; however, charge carriers are not hindered from transportation in the region. As an example, if an ***n***-channel oriented device operated in the enhancement mode is considered, the ***p***-type body or substrate consists of a depletion region that surrounds the conductive channel and source and drain electrodes. On application of the bias directed from the drain towards the source, the drain attracts the electrons out of the conductive channel via the space-charge confined depletion region. The main characteristics of the depletion region are that it is devoid of charge carriers along with a resistance comparable to that of silicon (***Si***). Enhancement of the drain voltage directed towards the source enhances the distance of the drain and the pinch-off point that leads to higher resistance of the depletion region that is directly proportional to the drain voltage directed towards the source. However, this proportional relationship does not have an impact much on the drain-to-source current that remains largely constant, reflecting an independent behavior out of the changes made in drain-to-source voltages. Hence, the FET works in the constant-current source configuration in the saturation mode that can be effectively employed as a device that amplifies the voltage. The gate-source bias is the parameter that ascertains the degree of constant current that transports through the channel.

The different types of FETs are MOSFETs (as described above), dual-gate MOSFET (DGMOSFET), consisting of double insulated gates, insulated-gate bipolar transistor (IGBT), metal-nitride-oxide-semiconductor transistor (MNOS), ion-sensitive field-effect transistor (ISFET), BioFET or biologically sensitive field-effect transistor, DNA-FET, junction field effect transistor (JFET), heterostructure insulated-gate field-effect transistor (HIGFET), modulation-doped field-effect transistor (MODFET), tunnel field-effect transistor (TFET), high electron mobility field-effect transistor (HEMT), carbon nanotube field-effect transistor (CNT-FET), organic field-effect transistor (OFET), quantum field-effect transistor (QFET), FinFETs and even spin field-effect transistor (*spin*-FETs).

6.2.7 Applications and Advantages of FETs

FETs are the potential candidates for analog switching applications, microwave amplifiers, high-input impedance amplifiers and the modern digital integrated circuitry for high-end smart-device applications. FETs are high-input impedance devices compared to BJTs that match easily with those of the standard microwave devices. FET is a highly stable device with a negative temperature coefficient at the higher degrees of currents, which means that current decreases as a function of temperature increment. This specific characteristic enforces a highly uniform distribution of temperature over the whole of the device area, preventing the device from thermally induced breakdown. This is highly specific to the FETs in the sense that the device does not suffer from thermal breakdown, even in the case of larger active area of the device or if several devices have been connected in parallel. BJT is highly prone to such problems, and breakdown happens easily. FETs do not support storage of minority charge carriers since there are no forward-biased ***p-n*** junctions, resulting in greater large-signal switching speeds.

6.3 *BASIC MOSFET CHARACTERISTICS*

The fundamental formation of a MOSFET is elucidated in Figure 6.6 that shows different parts involved in its operation. Throughout my discussion, I will focus on ***p***-type body of semiconductor substrate, consisting of electrons as the ***n***-channel carriers. MOSFET is a four-terminal or electrodes device in which two heavily doped ***n***-regions (written as n^+-region) have been formed by ion implantation technique that form the source and drain electrodes. The gate area of the channel length, enclosed between source and drain, is formed of insulated silicon-*di*-oxide (SiO_2) dielectric that is grown by thermal oxidation of silicon (***Si***) substrate or wafer to form a high-quality Si-SiO_2 interface. Above this insulating dielectric material, metallization is done to form an electrical biasing scheme that forms the gate structure. Thus, the gate consists of an insulating material over which metallization is done for electrical contacts. Gate electrode is usually formed of heavily doped polysilicon or silicide and polysilicon in conjunction with each other. Analytically, the basic parameters of the device are channel length (***L***), channel width (***M***), insulator thickness (***d***), substrate doping (N_A) and junction depth (t_j). The length of the part enclosed between the junctions formed by n^+ and ***p*** regions is called the channel length (***L***). In the discussion that follows, source electrode will be used as the voltage reference point. In the case of grounding or application of low voltage to the gate electrode, the channel is nonconductive. In this situation, source-to-drain junction

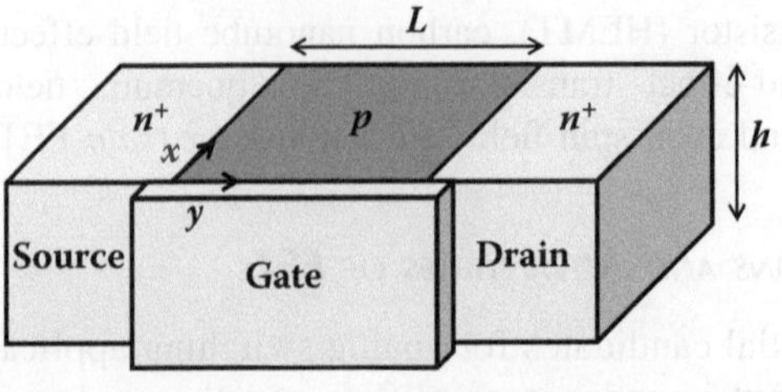

FIGURE 6.6 Schematic of the lateral view of the FET [25].

(*S-D*) electrodes are two different *p-n* junction entities that have a back-to-back connection. On biasing the gate to higher positive voltages, formation of a channel or surface inversion layer takes place between the two heavily doped n^+-regions. In this case, the channel becomes conductive, which facilitates large transport of negatively charged carriers. The bias applied through the gate voltage (V_g) generates an electric field that modulates the channel conductance by application on charge carriers. The back- or bottom-gated voltage or biased direct substrate electrical contact can be reverse-biased or may be taken as the reference voltage. This can have a bearing on the channel conductance as well; hence, this is named back- or bottom-gate electrode, constituting the "fourth terminal" of the MOSFET.

6.3.1 Inversion Charge in Channel

On biasing the source and drain contacts, height of minority carrier quasi-Fermi energy level (E_{Fn}) gets depressed in comparison to the equilibrium Fermi energy level (E_F). As a result, the metal-oxide-semiconductor (MOS) configurational structure acquires a non-equilibrium state that generates band-bending in the band structure [25,26]. Under equilibrium, the 2D flat-band type of band structure happens at $V_g = V_d = V_{bs} = 0$ [25,26]. Non-equilibrium condition happens under both types of gate and drain potentials, in which case formation of unique quasi-Fermi energy levels of electrons (E_{Fn}) and holes (E_{Fp}) takes place separately. Quasi-Fermi band energy level of holes or the E_{Fp} sustains itself at the bulk E_F level. On the other hand, quasi-Fermi band energy levels of electrons (E_{Fn}) get depressed to the side of the drain electrode. It can be inferred that the drain bias brings down the inversion layer charge at the drain electrode. Thus, the gate bias necessary for inversion layer formation at the drain electrode remains higher in comparison to the equilibrium case, whereby $\Psi_{s\ inversion} \approx 2\Psi_B$ signifies the case of weak inversion only. In the case of stronger inversion, $\Psi_{s\ inversion}$ is generally higher by a factor of a few thermal energy (kT). The process of lowering of space charge region of the inversion at the drain is guided by the lowering of E_{Fn} due to the applied drain bias. In this case, the surface potential follows the condition, $[E_{Fn} - E_{i(0)}] > q\Psi_B$, a necessary condition for the onset of the inversion layer. Here, $E_{i(0)}$ defines the intrinsic Fermi-band energy level (E_F) at $x = 0$.

Next, I focus on the equilibrium and non-equilibrium cases at the drain point for an inverted *p*-region. The discussion is framed around the variations in energy bands and charge distribution for the above case. The two situations for the inverted *p*-region can be outlined as follows,

i. In case of equilibrium condition, surface depletion region attains the maximum width (W_{DM}) at the point of inversion.
ii. In case of the non-equilibrium condition, the width of the surface depletion layer is deeper in comparison to W_{DM}, which changes with the applied drain bias (V_d). At the point of inception of strong inversion happening at the drain point, the surface potential $\Psi s\ (m)$ can be expressed by the following, nearly perfect approximation:

$$\Psi_{s\ \mathbf{inversion}} \approx V_d + 2\Psi_B$$

Under the non-equilibrium condition, characteristics of surface space charge have been derived based on the following two assumptions:

a. Since a p-region has been considered, majority charge carrier oriented quasi-Fermi band energy level E_{Fp} remains at the height similar to that of the wafer. This does not show any variation with respect to distance (x) from bulk in a direction towards the surface. This typical assumption ensures an almost error-free environment when the surface inversion happens. This is owed to the fact that only an insignificant part of the majority charge carriers make up the surface space charge.
b. The drain bias lowers the minority-carrier quasi-Fermi energy level (E_{Fn}) to the tune that depends on the y-position. This second assumption is true because minority charge carriers form a significant portion of the surface-situated space-charge region under surface inversion condition.

The 1D Poisson equation, based on the above assumptions, in the case of surface-oriented space-charge area at the drain electrode can be expressed as follows [25]:

$$\frac{d^2\Psi_p}{dx^2} = \frac{q}{\epsilon_s}(N_A - p + n)$$

$$N_A = p_{po} = \frac{n_i^2}{n_{po}}$$

$$p = N_A e^{-\beta\Psi_p}$$

$$n = n_{po}e^{(\beta\Psi_p - \beta V_d)}$$

where β is defined as, $\beta \equiv \frac{q}{kT}$

The charge in the inversion region constituted by the minority charge carriers is given by the following expression [25]:

$$|Q_n| \equiv q\int_0^{x_i} n(x)dx = q\int_{\Psi_s}^{\Psi_B} \frac{n(\Psi_p)d\Psi_p}{\frac{d\Psi_p}{dx}} = q\int_{\Psi_s}^{\Psi_B} \frac{n_{po}(e^{\beta\Psi_p - \beta V_d})}{(\sqrt{2kT/qL_D})F(\beta\Psi_p, V_d, n_{po}/p_{po})}$$

Here, x_i is the point at which $q * \Psi_p = E_{Fn} - E_i(x) = q * \Psi_B$ and function F is given as follows:

$$F\left(\beta\Psi_p, V_d, \frac{n_{po}}{p_{po}}\right) \equiv \sqrt{e^{-\beta\Psi_p} + \beta\Psi_p - 1 + \frac{n_{po}}{p_{po}}e^{-\beta V_d}\left[e^{\beta\Psi_p} - \beta\Psi_p e^{\beta V_d} - 1\right]}$$

The value of $\boldsymbol{x_i}$ is kept at smaller levels in the range of 3 nm to 30 nm to adhere to practical doping range profiles in silicon (Si) semiconductor [25,26].

After simplification with a range of semiconductor equations (outside the scope of current chapter), inversion charge $\boldsymbol{Q_n}$ collected at the drain electrode can be simplified as follows [25]:

$$\left|Q_n\right| \approx \sqrt{2}\,qN_A L_D\left[\sqrt{\beta\varphi_s + \left(\frac{n_{po}}{p_{po}}\right)e^{\left(\beta\varphi_s - \beta V_D\right)}} - \sqrt{\beta\varphi_s}\right]$$

Here, φ_s is surface potential in the above equation. A deficiency in the above solution is that it lacks a connection to the gate bias, $\boldsymbol{V_G}$, in addition to the difficulty in use since it becomes highly sensitive to surface potential φ_s under the condition of strong inversion.

6.3.2 Current–Voltage Characteristics

The basic output characteristics ($\boldsymbol{I_D vs.\ V_D}$ under different constant gate biases, $\boldsymbol{V_G}$) of a MOSFET-type transistor are shown in Figure 6.7.

At the low-valued sweep of $\boldsymbol{V_D}$, almost linear increment in $\boldsymbol{I_D}$ can be noticed as marked in the $\boldsymbol{I_D vs.\ V_D}$ curve with the help of a dashed line. Soon after this increment value of $\boldsymbol{V_D}$, the nonlinear region follows, after which locus of $\boldsymbol{I_D vs.\ V_D}$ can be seen after which $\boldsymbol{I_D}$ becomes saturated and does not show any vertical enhancement whatsoever at the gate voltage ($\boldsymbol{V_G}$). The nonlinear region is demarcated as the region contained between the two dashed lines.

Device operation can be understood in a better way with the help of a series of schematics given in Figure 6.8(a), (b) and (c) [25]. Suppose a big enough positive bias is applied at the gate that causes an inversion at the surface of the semiconductor. In the case of application of a small $\boldsymbol{V_D}$, the current starts flowing from the source to the drain electrode en route to the conducting region of the channel. On the lower side of $\boldsymbol{V_D}$ sweep, linear region happens so that $\boldsymbol{I_D}$ is directly proportional to $\boldsymbol{V_D}$. This situation is shown schematically in Figure 6.8(a). On increasing $\boldsymbol{V_D}$ to slightly large values, the $\boldsymbol{I_D}$ is diminished due to the effect of channel potential, and a deviation from linear relationship takes place such that the inversion charge at drain electrode eventually attains a point at which it becomes zero. This point at which inversion charge becomes zero is called the pinch-off point, shown in Figure 6.8(b). On increasing $\boldsymbol{V_D}$ beyond this point, since saturation current ($\boldsymbol{I_{D\ Saturation}}$) is already obtained, increment of $\boldsymbol{V_D}$ beyond this point drives the movement of pinch-off point in the direction of source electrode. However, at the pinch-off point, $\boldsymbol{V_D}$ remains the same and is $\boldsymbol{V_{D\ Saturated}}$. As a result, number of

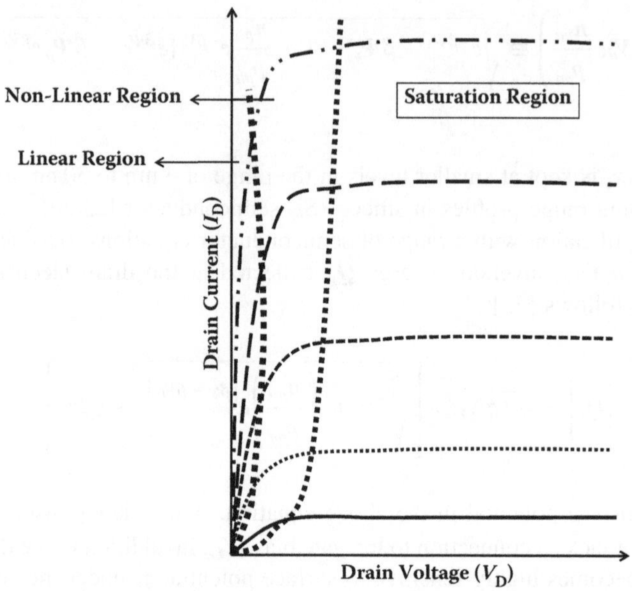

FIGURE 6.7 Output characteristics of a MOSFET showing I_D as a function of V_D sweep. Linear, nonlinear and saturated regions have been shown in the curve separated from each other by dashed lines.

charge carriers injected from the source at the pinch-off point is necessarily the same, which makes the flow of current in Figure 6.8(c). Additionally, a decrement in channel length happens at this point, giving rise to an effective channel length. In this situation, I_D will increase only when the cut-off part forms a considerable portion of the channel length, which is called short-channel effect.

In MOSFETs, the concept of ballistic transport is introduced in which charge carriers in the channel region do not undergo random scattering events in ultra-short channel lengths, which have the dimensions of the scale of or shorter in comparison to the mean free path [25,26]. The charge carriers do not lose energy by scattering with the lattice and keep on gaining energy from the electric field; thus, they acquire a velocity that is much higher in comparison to the saturation velocity. This mechanism is the ballistic transport that specifies that transconductance and current can be greater in comparison to the saturation velocity, which drives the device design with narrower channel lengths.

6.3.3 *MOSFET Scaling*

Since the inception of the CMOS-integrated circuit technology, device fabrication technology has reduced from 10 μm process technology node in 1971 produced by Intel Corporation and Radio Corporation of America (RCA) to as small as 7 nm process technology node chips in 2020 produced by Taiwan Semiconductor Manufacturing Corporation (TSMC). On downsizing the device dimensions to a

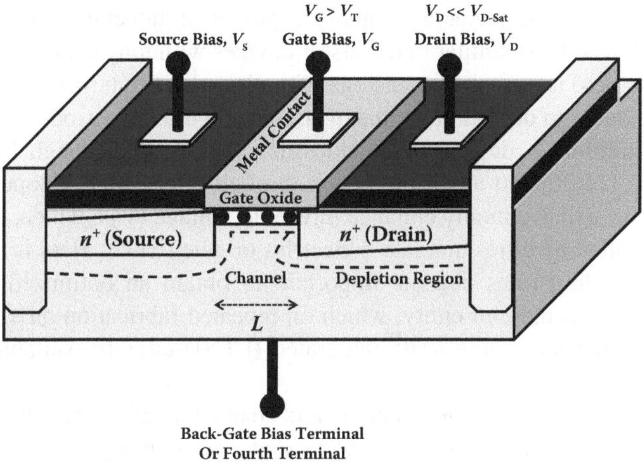
$V_G > V_T$
$V_D << V_{D\text{-Sat}}$
Source Bias, V_S
Gate Bias, V_G
Drain Bias, V_D
Metal Contact
Gate Oxide
n^+ (Source)
n^+ (Drain)
Channel
Depletion Region
L
Back-Gate Bias Terminal
Or Fourth Terminal

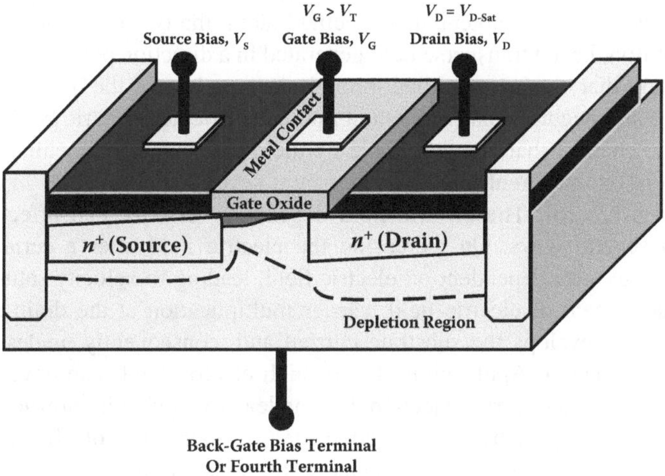
$V_G > V_T$
$V_D = V_{D\text{-Sat}}$
Source Bias, V_S
Gate Bias, V_G
Drain Bias, V_D
Metal Contact
Gate Oxide
n^+ (Source)
n^+ (Drain)
Depletion Region
Back-Gate Bias Terminal
Or Fourth Terminal

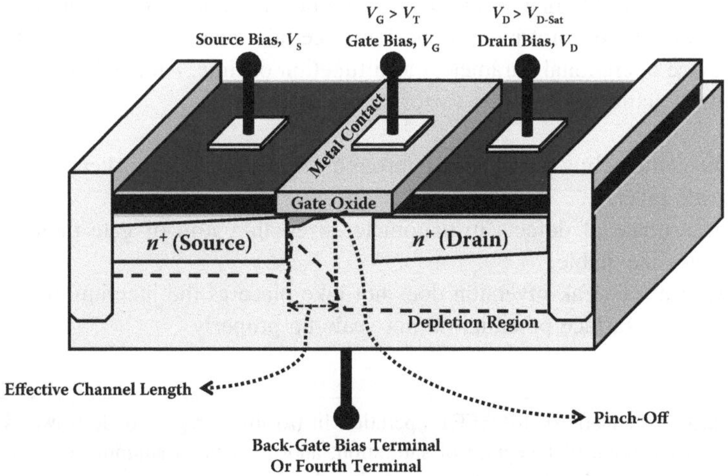
$V_G > V_T$
$V_D > V_{D\text{-Sat}}$
Source Bias, V_S
Gate Bias, V_G
Drain Bias, V_D
Metal Contact
Gate Oxide
n^+ (Source)
n^+ (Drain)
Depletion Region
Effective Channel Length
Pinch-Off
Back-Gate Bias Terminal
Or Fourth Terminal

few nanometers process technology node, a proper architecture needs to be designed to produce effects similar to transistor devices with long channels [25,26]. In the case of shrinked nanoscale dimensions with ultra-short channels, channel length is almost equivalent to depletion widths of source and drain electrodes. This type of parametric condition in device architecture develops punch-through between the two electrodes [25,26]. To avoid the punch-through effect, highly doped channels are required that will eventually enhance threshold voltage (V_T), which, in turn, can be better controlled by ultra-thin gate dielectrics or gate oxides. Here is the point at which device scaling rules become important to obtain an optimized device architecture that forms the unit entity, which on repeated fabrication on a large-sized ***Si*** wafer forms the ultra-large scale integrated (ULSI) circuitry suitable for commercial production.

On device scaling to only a few nanometers channel length, short-channel effects become inevitable, which are owed to the existence of high degree of electric fields inside the channel area and distribution of potentials in a **2D** geometry (along the two axis). Distribution of potential is specified along the two directions of electric field distribution, i.e. (i) transverse field generated in a direction perpendicular to the channel length that can be controlled through the gate bias or the back- or bottom-gated device architecture, and (ii) the longitudinally oriented electric field along the length of the channel that is controllable by the drain bias. This peculiar potential distribution in two different directions generates several abnormalities in electrical characteristics [25,26]. The electric field dependent behavior can affect channel mobility in several ways. On increasing the electric field after a certain value, channel mobility gets dependent on electric field, leading to velocity saturation. In further enhancement of electric field, carrier multiplication at the drain electrode happens, which develops the substrate current and, consequently, undesirable bipolar switching action. Application of very high electric fields can develop oxide charging effects by hot-carrier injection that can lead to a swing in threshold voltage and a subsequent degradation in transconductance. Such types of effects generate certain non-idealities, causing short-channel effects that are drain-current being not proportional to **1/*L***. Drain current does not acquire saturation with the drain voltage, threshold voltage does not remain constant a certain channel length, and degradation of device operational parameters as a function of time. An ideal device scaling is difficult to achieve due to the factors outlined below:

i. Direct tunneling of electrons through the ultra-thin gate dielectric or gate oxide layer.
ii. Occurrence of defects in nanometer-sized thin film of gate dielectric becomes inevitable.
iii. Arousal of weak-inversion does not take place as the junction built-in voltage and surface potential do not scale-up properly.

FIGURE 6.8 Schematic of MOSFET operation in (a) linear region under lower V_D bias sweep, (b) at the point of inception of saturation, and (c) beyond attainment of saturation level so that the effective channel length is reduced [25].

iv. Quantum-mechanical effect vitiates the total or gate capacitance because of the fact that charge carriers become finitely nanoconfined (≈ 1 nm) from the interface.
v. On decreasing $\boldsymbol{r_j}$, series resistance due to source and drain electrodes increases.
vi. It is not possible to increase the channel doping beyond a certain limit for control of punch-through because breakdown of ***p-n*** junction takes place.
vii. It is not possible to scale the threshold voltage because of off-current consideration.

The above limitations are responsible for showing up nonideal scaling factors, and, as a result, the electric field shows an increasing trend as a function of smaller gate lengths.

Channel length modulation is a scaling parameter in VLSI design technology that specifies change or decrement in channel length because of enhancement of drain-to-source bias in the saturation region [27]. As stated above in the previous section, pinch-off at the drain point happens in the channel in the confines of the saturation region. On the side of the saturation region, $\boldsymbol{I_D}$ becomes independent of $\boldsymbol{V_D}$ and increasing beyond does not create an effect on the shape of the channel ideally. However, in practice, on further increment in $\boldsymbol{V_D}$ in the saturation region, channel pinch-off point gets shifted towards the source point slightly. The pinch-off point gets deviated in a slight manner from the drain point because the drain electric field pushes it in the direction of the source. This leads to a decrement in effective channel length by $\boldsymbol{\Delta L}$ for a corresponding increment in $\boldsymbol{V_D}$. In this case, the channel does not touch the drain any longer, acquiring an asymmetrical shape that becomes thinner towards the drain point. It becomes important in shorter devices, and $\boldsymbol{I_D}$ increases on decreasing the channel length and increases as a function of $\boldsymbol{V_D}$ in the saturation region. The current-voltage curves are no longer flat for MOSFETs in the saturation region.

6.4 NANOSCALE DEVICES AND INTEGRATED CIRCUITS

With the advent of nanotechnology, MOSFETs based on nanoscale materials can be designed and scaled-up for commercial purposes. Semiconductor nanowires offer the integration capabilities as one of the components in MOSFETs. Semiconductor nanowires can be utilized as channel materials in MOSFETs that facilitate tremendous electrostatic gate control above the charge carriers in the channel for attainment of diminished short-channel effects. The size of MOSFETs has been reducing continuously to a few process technology nodes, down to 7 nm and even smaller than that. Moore gave the CMOS scaling laws that predicted the ever downsizing of transistors from micron-sized architecture to a few nanometer technology nodes. Similar scaling laws were put forth by Dennard et al. [28]. However, physics-related limitations and nanofabrication-related technological hurdles drove the deviation of these scaling laws that predicted ever-decreasing architectures for transistors. Down to scaling sizes that are closer to atomic dimensions, channel length modulation plays a major role in

MOSFET operation, along with heat dissipation and leakage current. As a result, novel innovative nanomaterials form the core for technological nanoelectronic advancements in device physics by presenting solutions to problems of high costs, reliability related issues, speed of device operation and power dissipation. Semiconductor nanowires offer large surface-to-volume ratio and comparatively much smaller channels that can form the gate-all-around (GAA) or gate-surrounding structure, along with efficient control over the nanometer-sized channel, showing quantum-confinement effects along with reduced dimension effects of the orders of atomic scale, leading to charge-carrier scattering. The GAA-structured nanowire-based FET design bears high resemblance to the FinFET technology in the currently famous 7 nm process technology node. Even GAA transistor technology provides a platform of extension of device scaling beyond the restrictions levied by FinFET technology in the sub-7 nm technological node. Such device technology offers higher on/off ratios, reduced leakage currents and smaller subthreshold slope.

Silicon (Si) nanowires belong to the class of semiconductor nanowires that can be grown along preferred crystal directions [29]. Smallest-sized Si nanowires synthesized by catalyst-assisted method grow in the direction of preferred <110> crystal orientation. On the other hand, Si nanowires with the largest diameter prefer to grow along the direction of <111> crystal orientation. It is pertinent here to shed light on modification of mechanical properties of Si crystalline material at the nanoscale in comparison to the bulk Si phase. Reportedly, <100> crystal oriented Si nanowires possess softened Young's modulus owed to the compressive surface stress that sharply declines on decreasing the diameter of the nanowire down to ~2.5 nm. In addition, mechanical property of Young's modulus of Si nanowires is highly directional in nature. As opposed to this trend, nanowires grown along the <110> crystal orientation possesses the highest mechanical property values. As regards the electronic properties, Si nanowires are direct band-gap material fulfilling the condition that the lowest dip of the conduction band and the extreme of the valence band lie at the similar point in *k*-space. As contrasted, bulk Si is an indirect band-gap material, with the maximum of the valence band lying at the Γ point, while the minimum of the conduction band lies at ~85% towards the X direction and the Γ point. Hence, direct band-gap materials are highly promising for photonics, photovoltaics and optoelectronics applications. Electronic properties of nanowires are a definite function of the preferred crystal orientation direction of nanowire growth, diameter and cross-section. On the other hand, band-gap of Si nanowire does not depend on shape of nanowire cross-section. An important parameter that can cause reduction of electrical conductance in nanowire transistors is the presence of surface roughness or surface defects generating enhanced electron scattering along the axis of the nanowire. Electronic transport along the preferred crystal orientation has a profound effect on conductance. For example, reportedly holes as charge carriers are transported smoothly along the <111> oriented nanowires, while electronic transport happens easily along <110> oriented nanowires bearing lessened effect of surface roughness in this particular direction.

6.4.1 Fabrication Technology of Si Nanowire-Based FET

Hao Zhu presented a self-alignment methodology of ***Si*** nanowire FET in his chapter on "Semiconductor Nanowire MOSFETs and Applications," which is given here in brief [29]. This fabrication technology consists of growth of Si nanowires on predefined locations with the help of Au catalysts. The perfect alignment of source and drain electrodes along with gate and gate metals is performed with photolithographic process. The first step of fabrication process involves the deposition of an ultra-thin layer (~1 nm) of Au catalyst over SiO_2/Si wafer, which is patterned and developed by photolithography and lift-off technological processes. Low-pressure chemical vapor deposition (LPCVD) process is used for the growth of Si nanowires from the Au catalyst that possess typical lengths of 20 μm and diameter of 20 nm. This is immediately followed by oxidation in a dry oxidation furnace to grow ~3 nm thick SiO_2. This helps in providing an efficient surface -interface between 1D-nanowire and the top-gated dielectric architecture. Source and drain contacts are then fabricated by photolithography and lift-off processes. This is followed by wet etching with 2% hydrofluoric (HF) acid for removing oxides lying over Si nanowires at the source-drain regions to expose contacts along with metal deposition. In the next step, gate-oxide is deposited by atomic layer deposition (ALD) technique, followed by subsequent deposition of 5 nm thin Al_2O_3 for improving interface with the aluminium (Al) top-gated structure. Thus, final fabricated devices were subjected to annealing in an environment of forming gas of composition 5% H_2 in N_2 at $\boldsymbol{T}$ = 325°C for reduction of interface trap density and improvement of contacts between Al metal and HfO_2 gate dielectric and Si nanowires. In these fabricated Si nanowire architectures, short-channel effects can be minimized with the minimum gate lengths for single-gated, double-gated and all-around gate architectures expressed by the following equations [29]:

$$\lambda_{Single-Gate} = \sqrt{\frac{\epsilon_{Si}\ t_{Oxide}\ t_{Si}}{\epsilon_{Oxide}}} \tag{6.1}$$

$$\lambda_{Double-Gate} = \sqrt{\frac{\epsilon_{Si}\ t_{Oxide}\ t_{Si}}{2\epsilon_{Oxide}}} \tag{6.2}$$

$$\lambda_{All\ Around-Gate} = \sqrt{\frac{2\ \epsilon_{Si}\ t_{Si}^2 \ln\left(1 + \frac{2t_{Oxide}}{t_{Si}}\right) + \epsilon_{Oxide} t_{Si}^2}{16\epsilon_{Oxide}}} \tag{6.3}$$

Here, $\boldsymbol{t_{Si}}$ and $\boldsymbol{\varepsilon_{Si}}$ are the thickness and permittivity, respectively, of ***Si*** wafer and $\boldsymbol{t_{Oxide}}$ and $\boldsymbol{\varepsilon_{Oxide}}$ are those of gate dielectric, respectively. The above-mentioned three equations specify that the all-around gate architecture offers the best control over short-channel effects.

6.4.1.1 *Carbon Nanotube-Based Field-Effect Transistors (CNT-FETs)*

The semiconductor fabrication technology has been following the trend, as predicted by Moore's law of miniaturization of semiconductor chips in a way such that the number of transistors forming the integrated circuitry gets doubled approximately in nearly two years' time. The rapid advancement in semiconductor fabrication technology has made possible to attain 7 nm fabrication process technological nodes. Short-channel effects, such as punch-through effect, direct tunneling of electrons from source to drain electrode undermining the electrostatic gate control of charge carriers, and enhancement in gate-leakage currents are detrimental to the transistor operation in the sub-10 nm nanoscale-regime. Semiconducting CNTs possess specific properties of large mean free path, better electrostatic control of charge carriers at the nanoscale due to their non-planar structure, excellent carrier mobility facilitating ballistic transport over hundreds of nanometers of length scales, and effective electrostatic gate-control owed to the absence of dangling bond-states over CNT surfaces including purely 1D electronic transport [30]. CNT-FETs possess the following characteristics [30]:

i. CNTs possess a 1D geometrical shape that significantly reduces the probability of electron scattering, thus facilitating ballistic transport.
ii. Passivation of the surface-interface formed between the CNT channel and gate-oxide or gate dielectric is not required since there are no dangling bonds forming interface states over the essentially CNT conducting surfaces, in which case all the chemical bonds are stable and saturated.
iii. Formation of nonlinear Schottky barrier takes place at the interface of CNT-metal contacts that act as an active switching element contained in inherent nanotube device architecture.

The different types of CNT-FET architectures are inclusive of bottom-gated CNT-FET, top-gated CNT-FET, wrap-around or all-around gated CNT-FET architecture and suspended CNT-FETs. CNT-FETs with schottky barrier (SB-CNT-FET) exhibit strong unipolar or ambipolar behaviour of current-voltage characteristics [30–32]. In the case of SB-CNT-FET, the probability of electron tunneling through the Schottky barrier at the source electrode in the conduction energy band is highly elevated for sufficiently high and positive gate biases. In the case of applied low negative biases, holes-based tunneling current increases through the Schottky barrier, created at the interface with drain in the valence band. As a matter of fact, in case of MOSFET-like CNT-FETs, only specifically sufficiently high positive gate biases can enhance the current due to lowering of barrier in the channel. This particular unipolar behavior of SB-CNT-FET restricts the application of this specific class of nano-transistors in traditional CMOS logic circuits. Another way of reducing the unipolar character is to fabricate the double-gated CMOS architectures while preserving the intrinsic behavior of CNTs.

6.4.1.2 *Gate Capacitance*

Gate capacitance in an important consideration in 1D geometry bearing SB-CNT-FETs that are alike MOSFETs. In such types of 1D geometrical architectures, gate capacitance is formed between the gate dielectric layer and the channel region, which is dependent on the density of states (DOS) and geometrical parameters. The electrostatic potential (V_S) applied on the channel has to be uniform so that potential drop above the gate dielectric is also homogeneous. In such a situation, the gate voltage is equal to the addition of potential drop at the gate dielectric and the channel potential. Thus, the total capacitance or gate capacitance can be thought of as a resultant of the series connection of two capacitances of the geometry-oriented electrostatic capacitance and the quantum capacitance or DOS capacitance. This has been modeled schematically, as shown in Figure 6.9.

Gate or total capacitance can be measured from the following mathematical expression [30]:

$$C_G = \frac{\delta Q}{\delta V_G} \tag{6.4}$$

Here, V_G is gate bias and Q denotes the total amount of charge accumulated on the gate that possess the magnitude similar to but opposite in sign of the total net charge spread over the CNTs. The oxide capacitance can be expressed by the following expression whereby gate dielectric or gate oxide is sandwiched to generate the electrostatic capacitance [30]:

$$C_{Oxide} = \frac{2\pi\epsilon L_G}{\ln\left\{\frac{t_{Oxide} + r_{CNT}}{r_{CNT}}\right\}} \tag{6.5}$$

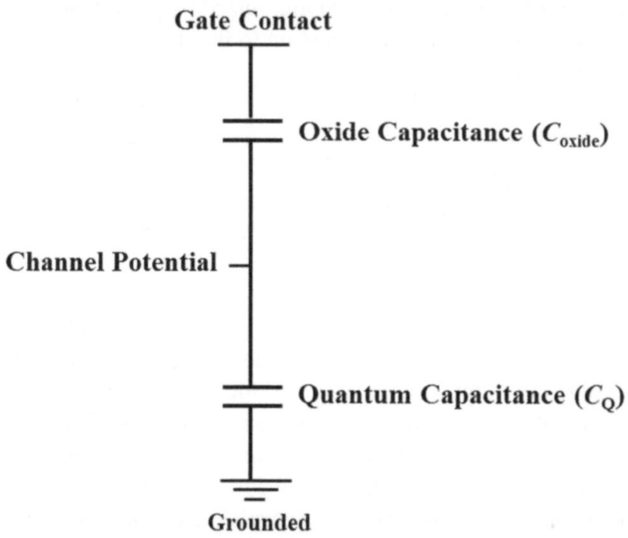

FIGURE 6.9 Schematic of oxide capacitance and quantum capacitance connected in series combination and the corresponding potential drop on the capacitances [30].

Here, ε denotes the dielectric constant belonging to the gate dielectric, t_{Oxide} denotes thickness of the gate oxide or gate dielectric, L_G denotes the length of the gate and r_{CNT} expresses the diameter of the CNT.

Now, if V_S is the channel potential, the quantum capacitance can be given as follows [30]:

$$C_Q = \frac{\delta Q}{\delta V_S} \tag{6.6}$$

In the above equation, charge (Q) on the channel can be determined from the Fermi level in the case of a given channel potential and DOS of the CNT. Here, it should be remembered that in the above case, the condition of V_S is not dependent on position, which holds strictly for the above relation to be true.

On downsizing the device dimensions down to 10 nm node, higher order of electrostatic coupling between source and drain poses severe challenges in terms of electrostatic control over charge carriers in the channel through the gate bias. An important parameter for efficient electrostatic gate control is to apply ultra-thin gate dielectrics. An ultra-thin layer of gate dielectric (~1 nm) suffers from excessive leakage currents through direct tunneling. A solution to the problem is presented by development of high-*k* gate dielectrics (usually hafnium-*di*-oxide or HfO_2), which can offer high total or gate capacitance, defying the thickness parameter.

Fringing capacitance is the parasitic capacitance that is generated as a result of fringing fields that exist between the gate metallic contact, and metallized source and drain electrodes. It becomes significant when the length of channel of a CNT-FET is condensed so that it is comparable to intrinsic device capacitances. Parasitic capacitances due to fringing fields must be minimized by proper choice of electrode geometries to make it negligible in comparison to the total or gate capacitance that modulates the conductance.

High-frequency operating efficiency of a transistor can be determined by the "unity current gain cut-off frequency". The "unity current gain cut-off frequency" is stated as that frequency at which point the current gain cascades to become unity. In another case of intrinsic CNT-FET, cut-off frequency can be stated as the intrinsic cut-off frequency, and when it is determined in presence of the parasitic capacitance as well, it is called the extrinsic cut-off frequency. If parasitic capacitance is small or negligible, extrinsic cut-off frequency approximates the intrinsic cut-off frequency. Quasi-static approximation based methodology can be exploited for the calculation of cut-off frequency. Quasi-static approximation produces the best result in the case of slow variation of the signal in comparison to the time constant calculated with the parameters of channel inductance and inherent gate capacitance. Based on the quasi-static approximation, a schematic in Figure 6.10 of the small-signal circuit model for CNT-FET omitting the equivalent inductive elements but inclusive of resistive elements and the equivalent capacitance is given as follows [30]:

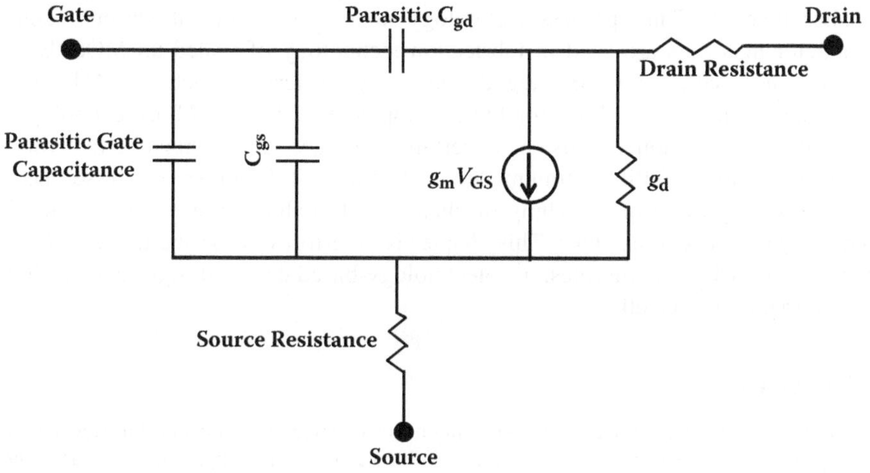

FIGURE 6.10 Schematic of small-signal circuit model for a carbon nanotube field-effect transistor [30].

Expression for cut-off frequency using the small-signal circuit model is denoted as follows [30]:

$$\frac{1}{2\pi f_T} = (R_S + R_D)C_{gd} + \frac{1}{g_m}(C_g + C_{gs} + C_{gd}) + \frac{g_d}{g_m}(R_S + R_D)(C_g + C_{gs} + C_{gd}) \tag{6.7}$$

Here, R_S and R_D denote the parasitic resistances arising from source and drain, respectively, g_m is transconductance, C_g is intrinsic gate or total capacitance, and C_{gs} and C_{gd} are parasitic capacitances belonging to source and drain, respectively. On omitting parasitic resistances, the above equation reduces to [30]:

$$f_T = \frac{1}{2\pi}\frac{g_m}{(C_g + C_{gs} + C_{gd})} \tag{6.8}$$

On omitting the parasitic capacitances, the intrinsic cut-off frequency is expressed as follows [30]:

$$f_T = \frac{1}{2\pi}\frac{g_m}{C_g} \tag{6.9}$$

6.5 CONCLUSION

Nanoelectronics and nanodevices are device architectures that are set to revolutionize electronic device manufacturing by reducing device dimensions even smaller than the 7 nm process technology node. Thin films of various semiconductor materials are used in fabrication technology of scaled-up MOSFETs. A profound challenge is to integrate nanoscale materials, such as CNTs and graphene in these MOSFETs, for the development of ULSI CMOS technology. Integration of nanomaterials can certainly reduce dimensions effectively. However, it must fulfill optimum device features of faster speed, lesser heat dissipation, high rate of switching mechanism, ultra-short in size, better control and other device scaling rules. This chapter is an effort to describe the MOSFET device along with scaling rules, nanotechnology-based device designs and scaling procedures in a nutshell.

REFERENCES

1. Shrivastava, S., and Dash, D. "Applying nanotechnology to human health: Revolution in biomedical sciences". *J Nanotech.* 184702 (2009), 1–14. http://repository.ias.ac.in/101266/1/184702.pdf.
2. Jeevanandam, J., Barhoum, A., Chan, S. Y., Dufresne, A. and Danquah, K. M. "Review on nanoparticles and nanostructured materials: History, sources, toxicity and regulations". *Beilstein J. Nanotechnol.* 9 (2018), 1050–1074. https://www.ncbi.nlm.nih.gov/pmc/articles/PMC5905289/pdf/Beilstein_J_Nanotechnol-09-1050.pdf.
3. Paul, D. J. "Nanoelectronics". In *Encyclopedia of Physical Science and Technology* (Third Edition), ed.R. A. Meyers, 285–301. Academic Press: Elsevier Science Ltd, 2004.
4. Tan, G. S. and Jalil, B. A. Mansoor. *Introduction to the Physics of Nanoelectronics.* Woodhead Publishing: Woodhead Publishing Series in Electronic and Optical Materials, 2012.
5. Ojha, A. K. 2004. "Nano-electronics and nano-computing: status, prospects, and challenges". Region 5 Conference: Annual Technical and Leadership Workshop. Norman, OK, USA, 85–91. doi: 10.1109/REG5.2004.1300168.
6. Brinkman, F. W. "A history of the invention of the transistor and where it will lead us". *IEEE J. Solid-State Circuits* 32. no. 12 (1997): 1858–1865. DOI: 10.1109/4.643644.
7. Huff, R. H. "John Bardeen and transistor physics". *AIP Conference Proceedings* 550. no. 3 (2001): 1–28. doi: 10.1063/1.1354371.
8. Kahng, D. 1960. "Electric field controlled semiconductor device". U. S. Patent No. 3,102,230 (Filed 31 May 31, 1960, issued August 27, 1963).
9. Atalla, M. M. "Stabilization of silicon surfaces by thermally grown oxides." *Bell System Technical Journal* 38 (1959). 749–783: 10.1002/j.1538-7305.1959.tb03907.x.
10. Bassett, K. R. *To the Digital Age.* Johns Hopkins University Press: Baltimore, Maryland, 2007.
11. "1960: Metal oxide semiconductor (MOS) transistor demonstrated." Computer History Museum. https://www.computerhistory.org/siliconengine/metal-oxide-semiconductor-mos-transistor-demonstrated/
12. Sah, C. T. "Evolution of the MOS Transistor." *Proceedings of the IEEE* 76. no. 10 (1988): 1293–1298.
13. Hofstein, S. R. and Heiman, F. P. "The Silicon insulated gate field effect transistor." *Proceedings of the IEEE* 51 (1963). 1190–1202: DOI: 10.1109/PROC.1963.2488.
14. Moore, G. E. "Cramming more components onto integrated circuits. Intel.com." *Electronics Magazine*, 1965. https://newsroom.intel.com/wp-content/uploads/sites/11/2018/05/moores-law-electronics.pdf.

15. Shockley, W. "A unipolar 'Field-Effect' transistor." *Proceedings of the IRE* 40. no. 11 (1952). 1365–1376: doi: 10.1109/JRPROC.1952.273964.
17. Bhattacharya, P., Fornari, R. and Kamimura, H. *Comprehensive Semiconductor Science and Technology*. Elsevier Science Ltd: Amsterdam, 2011.
16. "What is a FET: Field effect transistor." Electronics Notes. www.electronics-notes.com.
18. US 2673948 H. F. Mataré/H. Welker/Westinghouse: "Crystal device for controlling electric currents by means of a solid semiconductor." FR priority 13.08.1948.
19. McCluskey, P. F., Podlesak, T. and Grzybowski, R. *High Temperature Electronics*. CRC Press: ISBN 0-8493-9623-9, 1996.
20. Jun-Ichi N. "Junction field-effect devices." *Semiconductor Devices for Power Conditioning*. Springer. 241–272, 1982: doi:10.1007/978-1-4684-7263-9_11. ISBN 978-1-4684-7265-3.
21. "1963: Complementary MOS circuit configuration is invented." Computer History Museum. https://www.computerhistory.org/siliconengine/complementary-mos-circuit-configuration-is-invented
22. US Patent 3102230A, Dawon, K. 1961. "Electric field controlled semiconductor device." https://patents.google.com/patent/US3102230.
23. "The breakthrough advantage for FPGAs with tri-gate technology." Intel. 2014.
24. "13 sextillion & counting: The long & winding road to the most frequently manufactured human artifact in history". Computer History Museum. April 2, 2018.
25. Sze, M. S. and Ng, K. K. *Physics of Semiconductor Devices* (Third Edition). Wiley Interscience: Hoboken, NJ, 2007.
26. Neamen, A. D. *Semiconductor Physics and Devices: Basic Principles* (Third Edition). McGraw-Hill Higher-Education: New Delhi, 2003.
27. Chauhan, Y. S., Lu, D. D., Vanugopalan, S., Khandelwal, S., Duarte, P. J., Paydavosi, N., Niknejad, A. and Hu, C. *FinFET Modeling for IC Simulation and Design*. Academic Press: Elsevier Inc, 2015.
28. Dennard, R. H., Gaensslen, F. H., Yu, H., Rideout, V. L., Bassons, E. and LeBlanc, A. R. "Design of ion-implanted MOSFET's with very small physical dimensions". *IEEE Journal of Solid State Circuits*, SC-9 (1974), 256.
29. Zhu, H. "Semiconductor Nanowire MOSFETs and Applications". In *Nanowires – New Insights*, ed. Maaz, Khan, 101–131. IntechOpen, 2017, DOI: 10.5772/67446. Available from: https://www.intechopen.com/books/nanowires-new-insights/semiconductor-nanowiremosfets-and-applications.
30. Kordrostami, Z. and Sheikhi, H. M. "Fundamental physical aspects of carbon nanotube transistors". In *Carbon Nanotubes*, ed. Jose Mauricio Marulanda. IntechOpen, 2010, DOI: Available from: http://www.intechopen.com/books/carbon-nanotubes/fundamental-physical-aspects-of-carbonnanotube-transistors.
31. Abe, M., Ohno, Y. and Matsumoto, K. "Schottky barrier control gate-type carbon nanotube field-effect transistor biosensors". *Journal of Applied Physics*, 111. no. 034506 (2012), 1–7: doi 10.1063/1.3681902.
32. Svensson, J. and Campbell, E. B. E. "Schottky barriers in carbon nanotube-metal contacts." *Journal of Applied Physics,* 110. no. 111101 (2011), 1–17: doi 10.1063/1.3664139.

7 Evolution of Nanoscale Transistors

From Planner MOSFET to 2D-Material-Based Field-Effect Transistors

Dr. Kunal Sinha

7.1 INTRODUCTION

Metal-oxide-semiconductor field-effect transistor (MOSFET) is the fundamental component in modern complementary metal-oxide semiconductor (CMOS) technology and is widely used in many applications, across several fields. The concept of the transistor was first proposed by Lilienfield [1] in 1925, and later, in 1947, John Bardeen, Walter Brattain and William Shockley invented the first working transistors at Bell Labs, according to Gorton [2]. In the beginning of the 1960s, researchers started exploring how transistors worked by fabricating them in different laboratories [2–5]. The innovation of the metal-oxide-semiconductor (MOS) transistor overcame the challenges of bipolar junction transistors (BJTs) in terms of ease of bulk production and associated costs. It also helped the growth of the semiconductor industry and commercial electronics technology from the late 1960s onward [4]. In 1965, Gordon Moore predicted the growth of MOSFET fabrication in modern semiconductor technology: the number of transistors in any given area will double every 18 months [6]. This is known as Moore's law, which is an industry axiom that is still followed by designers.

One of the initial MOSFET devices was designed by Dennard et al., with dimensions of the order of 1 μ [7], and performance of the device has been studied thoroughly for various doping profiles, threshold voltage variations, current variations, etc. The purpose of the study was to predict the transistor performance for highly miniaturized integrated circuits (ICs). With the advancement of technology, the dimensions of transistors have reduced significantly. From micron technology, transistor dimensions have moved into the nanotechnology domain, and currently researchers and the industry are working with transistors that have dimensions in the deep nanoscale regime.

DOI: 10.1201/9781003220350-7

From the initial days of MOSFET devices, the dimensions of transistors have reduced significantly to accommodate more and more components in an IC and improve performance. However, with reduced dimensions, various challenges, in terms of both fabrication as well as device operation, came to the surface and hindered the growth of technology. However, designers worked hard and are still working tirelessly to overcome these difficulties. In this chapter, the gradual growth of the transistor is discussed. Different difficulties and their solutions, as reported by various scientists, are analyzed in brief to understand the growth of modern transistor technology from the initial days. In the beginning of the chapter, the conventional MOSFET device structure is discussed, followed by the development of fin-shaped field-effect transistor (FinFET) and tunnel field-effect transistor (TFET) devices. After that, how modern-day two-dimensional (2D) materials like graphene, MoS_2, etc. are used in FET devices is included. In all these sections, the gradual development of the devices, along with some of the latest research and various advantages and disadvantages of these devices, are explored. Finally, a summary of the chapter is presented, and some possible future work in this field of research is shared for interested readers.

7.2 METAL-OXIDE-SEMICONDUCTOR FIELD-EFFECT TRANSISTOR (MOSFET)

Although the idea of the FET was first introduced in 1925 [1], researchers found it difficult to implement the concept of controlling current with the help of electric field. The initial laboratory studies were performed on a thin film semiconductor in Bell Laboratory and reported in Shockley and Pearson [8]. The first fabricated MOSFET device structure was invented and patented by Mohamed M. Atalla and Dawon Kahng [9,10], and later the physics of MOSFET operation were reported in detail by several other researchers [11–13]. The schematic of the device from the patent [10] is shown below (Figure 7.1).

The initial challenge of fabricating a MOSFET device was to introduce the external electric field into the semiconductor region, and Atalla et al. showed in their work that by growing a high-quality Si-SiO_2 stack, this problem can be solved [14]. Later in the same year, Atalla proposed the idea of using MOSFET to fabricate the MOS-IC due to the ease of its fabrication [15]. However, during the initial period of MOSFET fabrication, it was relatively slower and less reliable when compared to the existing BJT devices [16], and various researchers from different leading laboratories worked together to overcome these challenges. Finally, in 1963 Steve R. Hofstein and Fred P. Heiman built the MOS insulated chip [12], and in 1968 Albert Medwin invented and patented the first CMOS-IC [17].

The initial application of the MOSFET device was done by using the p-MOSFET device for switching and logic operation, and it was performed by General Microelectronics and Fairchild in 1964. Later, RCA introduced the n-MOSFET device to amplify signals [18]. It was then observed that the n-channel device is faster than its p-channel counterpart; however, these devices are more difficult to fabricate. The tireless work of researchers in leading laboratories helped to overcome the initial challenges of MOSFET fabrications, and at the end of the 1960s, it became the

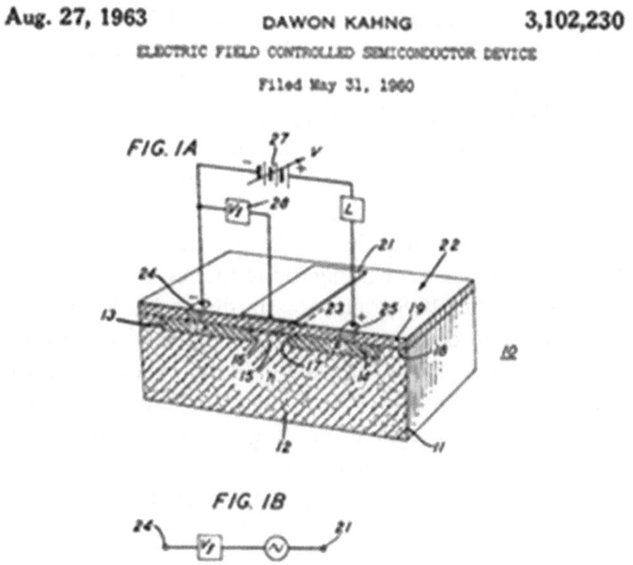

FIGURE 7.1 MOS device structure from Dawon Kahng Patent [10].

leading device in the semiconductor industry and outperformed the existing bulky BJT devices.

From the beginning of the 1970s, researchers studied various aspects of the MOS transistor and reported their findings in leading journals and conferences. The massive work on this device to perform more complicated tasks using the new MOSFET structure improved performance and lead to constant miniaturization of device dimensions. As a consequence, transistor density in an IC increased. Furthermore, researchers studied and found that the power consumption by this transistor was significantly lower than the bulky BJT device, and it further helped to sustain the growth of MOS devices in the semiconductor industry. The scaling of transistor dimensions was predicted by Moore [6]; however, in 1968, Robert Dennard invented the one transistor memory cell [19] and further established the scaling rule for supply voltage without compromising the current voltage relationship [20,21]. Dennard established that supply voltage scaling should be done less rapidly than dimension scaling so that the electric field within the device remains high [21]. On the other hand, in 1980, Brew et al. established the empirical relation between MOSFET gate oxide thickness, gate length and junction depth parameter [22] to retain the transistor behavior even in reduced dimensions. In this work, it was shown that all MOSFET device dimensions do not need to be reduced by the same scale factor. Following the scaling guidelines, when the 4K-DRAM cell of the 1970s is compared with the 90 nm technology transistor of 2003, it can be observed that gate oxide thickness has reduced from 50–100 nm to 1.2 nm effective oxide thickness (EOT), the gate length has reduced from 7.5 μm to 45 nm and junction depth has decreased from several microns to ~20 nm [23].

On the other hand, another critical parameter of MOSFET performance evaluation is the variation of CV - I curve, which indicates the speed of the transistor. As per the ITRS guidelines in 2001 [23], for N-MOSFET, the speed has become less than 1 p-sec, and for P-MOSFET it is approaching 1 p-sec. At the moment, with the advancement of technology, the speed is less than 1 p-sec for both N-MOS and P-MOS transistors. It indicates that with time, not only have the dimensions of transistors reduced but the performance has also improved significantly.

Scaling of transistor dimensions led to the rise of several challenges, called short channel effects. One such challenge is reduction of gate oxide thickness. During the initial period of MOS transistors in the early 1970s, silicon dioxide material was used as gate oxide with thickness 50–100 nm. With the gradual reduction of transistor dimensions, the gate oxide thickness has also reduced in proportionate ratio, and in the late 1990s, it became ~2 nm. Various researchers studied the impact of such a thin layer of gate oxide material on transistor performance [24, 25] and found that when the gate oxide (silicon dioxide) thickness becomes 2 nm or less, direct tunneling occurs between the gate and channel region, which results in a high amount of unwanted leakage current flow through the device. To overcome this challenge of leakage current, several scientists have studied possible alternatives. Finally, they found that high dielectric constant insulating materials can be used as an alternative to silicon dioxide [26–28]. In this regard, hafnium dioxide (HfO_2) material has been found to be the best possible alternative, with the available fabrication process facilities. Due to the high dielectric constant value of hafnium (almost 4 times more than silicon), the oxide thickness for HfO_2 is larger than that of SiO_2 for the same EOT, resulting in a significantly lower value of leakage current between the gate and channel region.

Another observation of researchers was that with the reduction of transistor dimensions, carriers from heavily doped source/drain regions enter into the channel with their saturation velocity and move toward the gate dielectric. This is known as hot-carrier effect, which deteriorates the transistor threshold voltage and increases the drain leakage current. To overcome this effect, various scientists proposed a lightly doped region between the channel and source/drain region, which is called lightly doped drain (LDD) region. Due to the presence of this LDD region, the electric field at the channel-source/drain interface region reduces. As a result, the generation of hot-carrier effects is small [29, 30]. The advantage of reduced hot-carrier effects comes at the cost of increased drain resistance, and as a result, reduced drain current. To overcome the reduction of drain current, researchers have explored possible alternatives to increase the drive current. In this regard, introduction of controlled strain in the channel region of a MOS transistor has been observed to have significant impact. The strain technology in the MOS transistor was first introduced by Manasevit et al. and People et al. [31, 32] in the early 1980s. These works showed that Si-SiGe hetero-structure introduced stain within the device that increased the mobility of carriers in the channel, and as a result, the performance of the transistor improved. Several research works have been published on strain engineering in the MOS transistor; some of these are summarized in Sinha [33].

With constant miniaturization of MOSFET dimensions, several other short channel effects have been observed when the dimensions reach the nanoscale domain. To overcome these effects, some researchers have studied other materials to find the alternative of popular silicon (Si) as a fabricating material so that these short channel effects can be avoided. In this regard, various III-V materials like GaAs, InGaAs, InP, etc. have been studied extensively, and the research work is still going on. The advantage of these materials is the high carrier mobility, low power requirement, etc., such that these materials can be an excellent alternative for Si, especially in the transistor channel region, with high-speed operation. However, the fundamental challenge with these materials is process compatibility with the current established fabrication set-up.

Furthermore, the gradual miniaturization of conventional planner MOSFET led to the restriction of device operation by the fundamental device physics. In the early 2000s, when the gate length of the transistor became 45 nm in 90 nm technology node design, the gate controllability over the carrier transport through the channel started to become very weak, and as a result, a high amount of leakage current started to flow through the device in OFF condition. To keep the leakage current in control, researchers started to modify the conventional MOSFET device structure, keeping the basic principle of operation constant. In 1967, H. R. Farrah and R. F. Steinberg proposed the concept of double gate thin film transistor [34], and in 1984 Sekigawa and Hayashi demonstrated the silicon-over-insulator device between two connected gate transistors [35]. From these two inventions, researchers later developed double-gate FET (DG-FET) and SOI-FET architecture to improve gate controllability. Furthermore, triple gate FET (tri-gate FET) is a 3D transistor introduced by the scientists in 1989 [36] that enhances gate control more than the DG-FET device structure. The purpose of such devices is to increase the number of gate terminals around the channel to increase the control of carrier transport through it. In this way, the growth of the semiconductor industry can keep going even in the era of nanoscale dimensional transistors.

In the next section, the growth of a very popular FET device structure, called FinFET, is discussed in brief. Although the device structure was not introduced long ago, within a very short time, it has become very popular and widely used by the industry in many applications. Currently, FinFET devices are used in the latest computer processors with excellent performance.

7.3 FIN CHANNEL FIELD-EFFECT TRANSISTOR (FINFET)

Although the idea of the first 3D MOSFET device with three gates was first introduced in 1989 [36], it was in 2000 when the term "FinFET" was first introduced by Hisamoto et al. [37] in their fabricated p-channel device. In this architecture, the conventional MOSFET structure is modified by introducing a 3D thin fin-shaped channel between the source and drain and on top of the substrate region. The schematic diagram of the device as reported in Hisamoto [37] is shown below (Figure 7.2).

In the above work, the Si Fin channel was 10 nm wide, with a poly-silicon gate wrapped from three sides around the channel. Since the channel is surrounded by

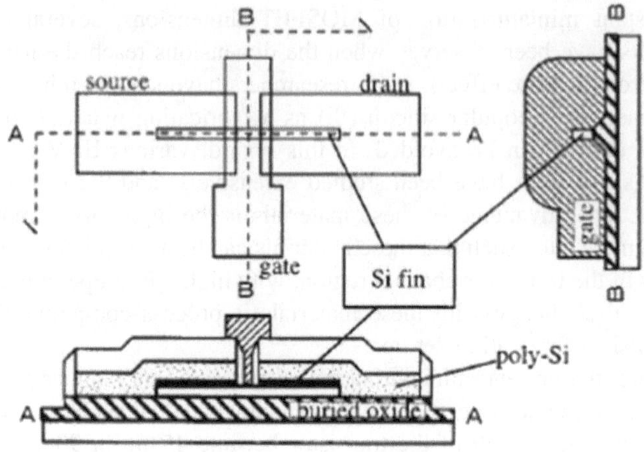

FIGURE 7.2 Schematic layout of the proposed FinFET device in Hisamoto [37].

the gate from top, back and front sides, the gate control over the carrier transport through it improves significantly and leakage current flow reduces. Also, in this work the authors have shown the steps to fabricate the device by using the established fabrication set-up. Thus, the proposed 3D FinFET device structure was found to be compatible with the conventional CMOS process flow.

The FinFET device structure got immense popularity among researchers and the industry. In 2001, Choi et al. showed from the transfer and output characteristics of a sub-20 nm FinFET device that it offers high drive current, keeping the leakage current significantly low [38]. The said characteristics as reported in Choi et al. [38] are shown below (Figure 7.3).

In 2002, TSMC (Taiwan Semiconductor Manufacturing Company) showed that a 25 nm transistor can operate by using 0.7 V supply, along with only 0.39 picosecond gate delay for N-FET and 0.88 picosecond gate delay for a P-FET device [39]. The data indicate that FinFET has the potential to work in low-power switching applications. In 2004, Samsung electronics demonstrated a DRAM cell by using a 90 nm FinFET device [39]. In 2011, researchers showed that FinFETs can have electrically independent gates, which gave the circuit designers flexibility to design a more efficient and low-power gate [40]. Later on, in 2013 the TSMC, and in 2014 the Global Foundries, announced their respective version of FinFET process technologies for 13 nm and 14 nm technology nodes design, respectively. Currently, the FinFET is used by most of the leading companies for their cutting-edge processors due to its advantages like superior performance, even in 10/7 nm technology nodes, compatibility with established process flow, etc.

The reason for the FinFET devices gaining such popularity is not only the fabrication process compatibility, but also excellent drain-induced barrier lowering (DIBL) and subthreshold slope values, along with high I_{on}/I_{off} ratio and low threshold voltage (V_{Th}). When the planner MOSFET devices are scaled down to 90 nm technology node and beyond, it has been observed that the subthreshold slope DIBL values are becoming very difficult to control. The ability to overcome

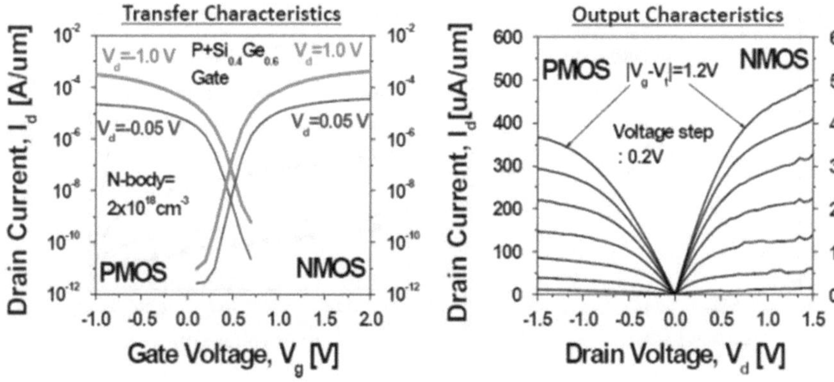

FIGURE 7.3 Transfer and output characteristics of a sub-20 nm FinFET, as published in Choi et al. [38].

such short channel effects made the FinFET architecture well accepted by all in 22 nm, 14 nm and even 7 nm technology node designs. During the last few years, researchers have explored various techniques to improve FinFET device performance further. In this regard, strain engineering is a popular technique that is well explored by several researchers in conventional planner MOSFET device structure. In Gupta et al. [41], it has been shown that strain technology is also effective for the 7 nm node FinFET structures and helps to improve the transistor performance in terms of drive current, I_{on}/I_{off} ratio, transconductance, speed, etc. when compared with an unstrained similar specifications device. In Sinha et al. [42], the authors introduced uniaxial compressive strain in the channel of a FinFET by using a $Si_{(1-x)}Ge_{(x)}$ embedded source/drain region, and they varied the channel strain by varying the mole fraction (x) of the source/drain composition. The studies clearly indicate that with the increase of channel strain, channel conductivity increases and drive current improves significantly. The variations of these parameters, as reported in Sinha et al. [42], are shown below (Figure 7.4).

Apart from strain engineering in a FinFET device, a few researchers have studied the impact of using other semiconductor materials like InGaAs and compared it with a conventional Si device with similar specifications [43, 44]. The studies show that in sub-20 nm technology node devices, where downscaling of device dimensions are posing a great challenge to keep the short channel effects within tolerable limits, these compound materials give an alternative for Si to sustain the growth of transistors with improved performance. In Park et al. [43], the authors compared the performance of InGaAs and strained-Si channel transistor devices and reported the comparison data in tabular form. The data clearly indicated that strained-Si channel devices give superior performance compared to the III-V semiconductor devices. On the other hand, in Vardi et al. [44], the authors fabricated a InGaAs FinFET device with sub-10 nm fin width. It has been found that in such devices, the threshold voltage has strong dependence on the fin width due to quantum confinement effect. This dependence was not found in Si devices, as the effective mass of electron in InGaAs is very low.

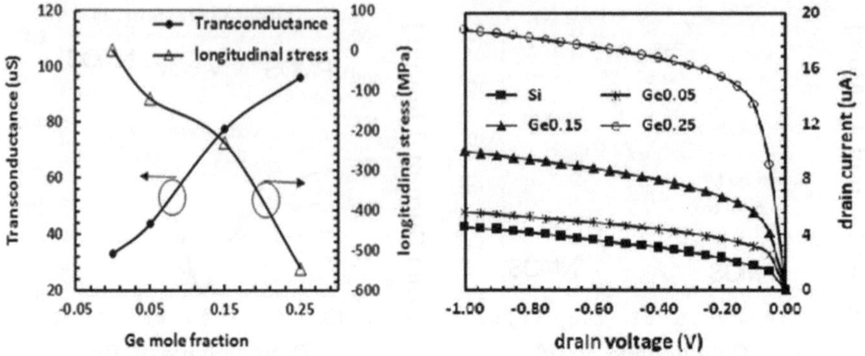

FIGURE 7.4 Impact of strain on channel transconductance and output characteristics of a sub-20 nm FinFET, as reported in Sinha et al. [42].

In this way, the FinFET device has found its place in the present sub-20 nm technology node domain. Both scientists and the industry are still working together to overcome the challenges of such low-dimensional devices. Strain engineering and material engineering are some of the alternatives that are being explored and were discussed in this section. In the next section, a new device is explored where scientists use a different physics of current flow in the transistor to sustain the growth of the semiconductor industry.

7.4 TUNNEL FIELD-EFFECT TRANSISTOR (TUNNEL FET OR TFET)

One of the major drawbacks of MOSFET devices is that their basic physics of operation does not allow them to work as excellent switching devices. For a good switching device, a transistor should be able to increase the current flow quickly below the threshold voltage of the device, called subthreshold swing. For any MOSFET device, researchers have proved that the subthreshold swing can never go below 60 mV/decade, i.e. beneath the threshold voltage of any MOS transistor, for 1 decade rise of drain current at room temperature, one needs to apply minimum 60 mV supply voltage. Due to this drawback as a switching transistor, researchers are exploring alternatives, since the state-of-the-art MOS transistors (DG-FET, Tri Gate FET, FinFET, etc.) are all having subthreshold swing more than 60 mV/decade. The tunnel field-effect transistor (TFET) has the ability to overcome this barrier, and researchers have shown that a TFET can have subthreshold swing value well below the 60 mV/decade. The transfer characteristics of a MOSFET, TFET and ideal switch are shown below (Figure 7.5).

In 1965, J. Appenzeller and his colleagues in IBM research lab proposed a new concept of current flow through a tunnel junction triode where current swing below 60 mV/decade is possible [45]. In this work, the authors have shown that instead of p^+ - i – p^+ (or n^+ - i – n^+) source – channel – drain MOS transistor, if the regions are designed with p^+ - i – n^- (or n^+ - i – p^-) type, then reverse-biased low-doped type region will act as a collector and the carriers will flow from source region, through the barrier by tunneling mechanism, to the collector side.

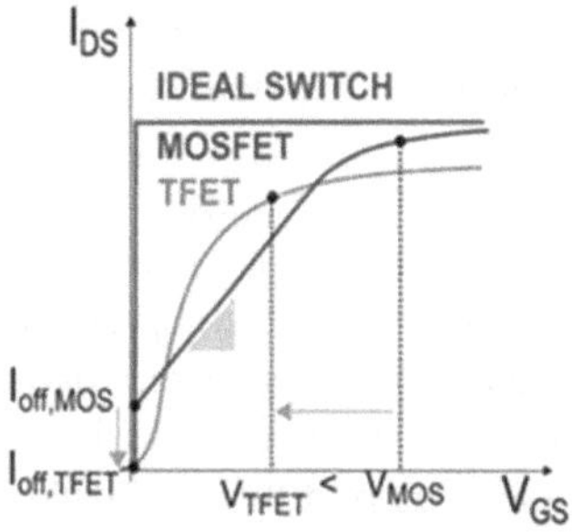

FIGURE 7.5 The transfer characteristics of a MOSFET, TFET and ideal switch.

Thus, unlike the thermionic carrier flow over the barrier in MOS transistors, the authors proposed a new quantum tunneling mechanism of carrier flow through the barrier. Due to this new concept of carrier transport, the Maxwell–Boltzmann statistics on carrier transport were no longer valid in this device, and the limitation of subthreshold swing to 60 mV/decade at room temperature was eliminated.

By using this concept, the authors of Appenzeller et al. [46] in 2004 created a transistor by using a carbon nanotube channel and reported the subthreshold swing as 40 mV/decade. In this work, a detailed study on band-to-band tunneling was presented, and how the current flow is controlled by the position of conduction and valence bands are elaborated.

The basic device structure of TFET is similar to that of MOSFET. The only difference is that the drain region is doped with the opposite type of source doping. The channel region is intrinsic in nature, and electrostatic potential of the channel is controlled by the gate voltage. The schematic of the device and corresponding band structure is shown below (Figure 7.6).

There are many reasons why researchers and the industry became highly interested in this new transistor. In addition to breaking the 60 mV/decade barrier of subthreshold swing, the TFET device showed better immunity to various short channel effects, high speed, lower threshold voltage and off-state leakage current compared to the conventional planner and 3D MOS family of transistors, which drew the attention of many people. Also, due to the extremely low level of off-state leakage current and low operating voltage to reach saturation current level, TFET

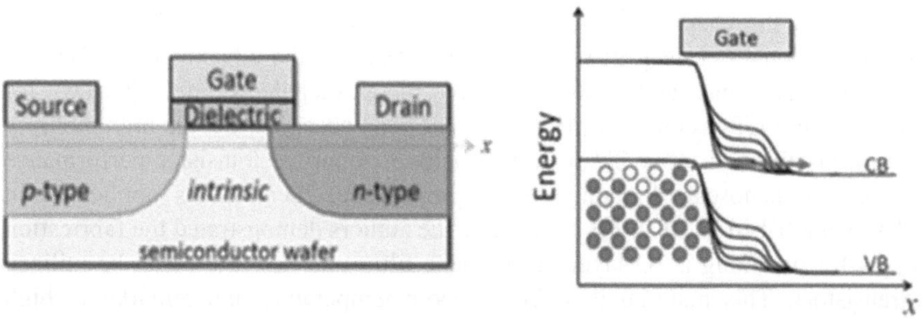

FIGURE 7.6 The tunnel field-effect transistor (TFET) device structure and the band diagram of the device along the channel (shown as dotted line in the device structure).

devices were widely used in low-power applications and low-power SRAM ICs. Due to these advantages, TFETs are currently considered to be a potential alternative for conventional MOS transistors [47, 48].

The primary challenge with TFET devices is the low level of drive current when compared with the conventional MOS transistors. Various researchers are currently working on different techniques to improve the ON-state current. R. Narang et al. studied the influence of structural modifications on TFET behavior and reported their findings [49]. Many researchers evaluated the TFET performance for various materials like Ge, SiGe and other III-V materials, and they reported their valuable observations in reputed journals [50–52]. The strain engineering discussed in the previous section for FinFET devices has also been explored in TFET devices by researchers, and large improvement in terms of drive current, turn ON voltage, etc. has been observed and reported [53, 54]. In a recent publication, researchers presented a scaled III-V hybrid TFET-MOSFET on silicon substrate [55]. In this work, both the advantages of TFET and MOSFET devices are utilized, and the architecture is designed in such a way that both devices can be optimized independently.

TFETs are currently explored by researchers for bio-sensing applications also. TFETs have high sensitivity when the dielectric constant in any of their regions changes. These changes are quantified by using the changes in the current level. This property is utilized in various TFET-based bio-sensing devices by various researchers, and their findings are published in reputed journals [56–58]. Since the TFET device is comparatively new, it is still being studied before being utilized for commercial purposes. Various people all over the world are currently exploring features of this device and how it can be utilized for different applications.

7.5 2D MATERIALS IN FIELD-EFFECT TRANSISTORS

The gradual development of the transistor, from planner MOSFET devices to the current industry standard 3D FinFET structure, has seen the reduction of its geometric dimensions from micron-size devices to the current few nanometer-size transistor devices. This reduction of transistor size gave rise to severe quantization effects in transistor performance. Research by several people indicates that by reducing the scattering of the transistor, the carrier flow through the channel can be streamlined from source to drain and then, not only will the speed of the device improve, but the channel conductivity and transistor efficiency will also be enhanced. In Houssa et al. [59] several authors reported the advantages of 2D materials like graphene and other transition metal dichalcogenides (TMDs) for application in FETs as channel material. It has been observed that the high mobility of electrons through these 2D materials helps to improve transistor performance in the current nanoscale domain. The first proposal for 2D materials' application in FET was published in 2004 [60, 61], where the authors demonstrated the fabrication process for obtaining a 2D layer of graphene films and how these can be utilized in transistors. This material is stable at room temperature and remarkably high quality with very small overlap between conduction band and valence band. Schwierz et al. [62] have shown the schematic of 2D material channel transistor and band diagrams of different 2D materials in their article. These were extremely

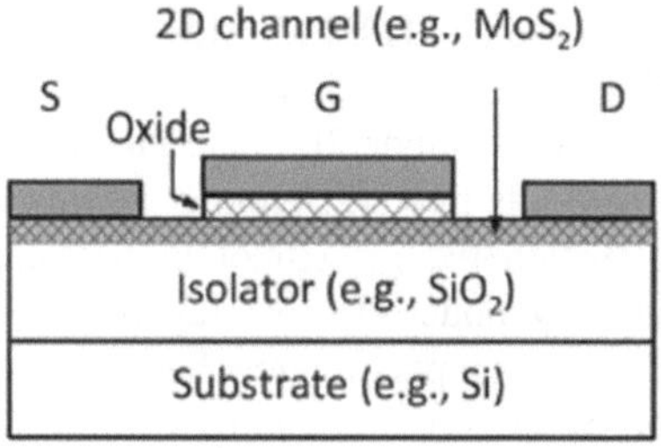

FIGURE 7.7 Generic structure of a FET with 2D channel material, as reported in Schwierz et al. [62].

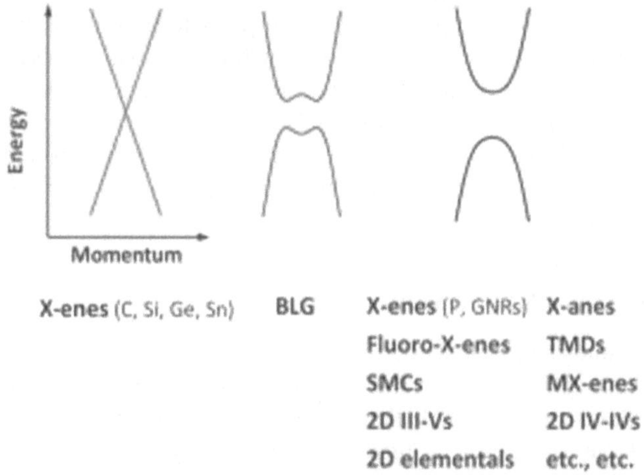

FIGURE 7.8 Schematic band structure of 2D materials relevant for transistors: BLG: bilayer graphene; GNR: graphene nanoribbon; TMD: transition metal dichalcogenide; SMC: semimetal chalcogenide, as reported in Schwierz et al. [62].

helpful to visualize the new materials and how they are placed in conventional transistor structure. The schematics of this device and corresponding material band structure, as reported in Schwierz et al. [62], are shown below (Figures 7.7 and 7.8).

The journey of 2D materials as a potential alternative for silicon in transistors started in 2004 in the form of graphene. The material received immense popularity from researchers and the industry due to its high mobility and strength. In 2007, the first graphene MOSFET was demonstrated by Lemme et al. [63]. Later in 2011, researchers published an integrated graphene circuit fabrication technique in Han et al. [64]. However, the advantages of high carrier mobility and conductivity of graphene did not last long since the band gap of pristine graphene, which is highly essential for any FET operation, is zero. Due to the absence of proper band gap, the OFF state current of the graphene channel FETs becomes very high and therefore could not be used for low-power applications. To overcome the challenge of band gap in graphene, various researchers worked hard, and in 2011, an alternative to graphene was found in MoS_2 material [65], which is 2D in nature with high carrier conductivity. It also has a high intrinsic band gap of 1.8 eV. The authors of Radisavijevic et al. [65] reported I_{on}/I_{off} ratio of 1×10^8. Furthermore, since the

MoS_2 material is direct band gap material, it has been used in TFETs for low-power applications [66, 67], optoelectronic devices [68], etc.

Apart from MoS_2, researchers have developed several other 2D materials and explored them for FET devices [69–71]. These materials have different band gap and are applied in various device structures to sustain the growth of transistor performance in the current state-of-the-art technology nodes. In this regard, researchers are working all over the world, and different compact models have been developed for 2D materials [72]. The fabrication process and challenges are reported [73, 74], and how these 2D materials can be used for different types of applications are also being explored [75].

Although the 2D materials possess high prospects to become an alternative to conventional 3D materials and sustain the growth of transistors, there are several gray areas that need to be explored in detail. Researchers are working all over the world to investigate more hidden features of these materials and how these can be used in a more fruitful way. There are many challenges already reported in reputed journals and conferences; however, how these shortcomings can be solved is currently under investigation.

7.6 SUMMARY AND FUTURE WORKS

Transistor technology has developed leaps and bounds since its inception in the early 1960s. During the early days, the dimensions of FETs were a few hundred microns, and there were many challenges to fabricate the device on a large scale with the available fabrication technology. That is why, although the theoretical concept of the FET device was patented in the late 1920s, it was first fabricated in 1960, and the device gained popularity from the late 1960s onward. Once the fabrication process of the MOSFET device was demonstrated, both academia and the industry welcomed it due to its easy fabrication and ability to fabricate on a large scale. Soon it replaced the existing bulky bipolar transistor, and the growth is still ongoing.

In the 1970s and 1980s, researchers tried to miniaturize transistor size so that more and more components could be included within a given IC chip. At the same time, improving transistor performance in the reduced dimension was also a focus of researchers. During this period, to improve transistor performance, strain engineering, material engineering, structural modifications, etc. were explored extensively by several researchers, and these are discussed in brief in the previous sections. Some of the relevant publications are cited during the discussion to create more interest among readers. The gradual reduction of transistor dimensions led to entry in the nanoscale domain from micron-level dimensions, and due to the miniaturization of transistors, several short channel effects started to creep in. To overcome the short channel effects, new device structures were explored by several researchers, and finally the FinFET device came into the picture. The operation of the FinFET is available in the cited literatures. Interested readers can review those references for a better understanding. In this article, the advantages and applications of this new device are discussed in brief, along with some relevant citations. Also, the impact of strain engineering on

FinFET structure are discussed from published literature. It has been demonstrated by the researchers that strain engineering helps to improve FinFET device performance, and this device is extensively used by the industry in the current 7 nm technology node architecture.

In the next section, another new device structure was discussed, TFET. The current transport in this device follows a different physics than what is followed in MOSFET devices. The advantages and published reports are discussed in brief, along with several other potential applications that researchers are currently exploring by using this TFET device structure. There are a few challenges researchers are facing while exploring TFET devices, and these are also mentioned during the discussion. After that, the latest research topics are analyzed; researchers are exploring various 2D materials for better performance in FET devices. Since 2004, various research papers have been published on this topic. Fabrication processes, various challenges and numerous potentials of these materials have been discussed. Some of the cited publications have been published recently in leading journals, which indicates the potential of research in this field.

Due to the limitation of pages, the topics are discussed in brief in this chapter. However, for any interested reader, there is ample opportunity to work in this field. Presently many researchers are trying to introduce strain technology in 2D materials in a precise, controlled manner; however, there are many unexplored areas that need to be studied before being implemented in fabrication. On the other hand, some of the 2D materials are being used in TFET devices for better performance. There are many challenges that have been observed by researchers, but there is enormous opportunity to work to overcome these challenges, along with working with some other new 2D materials that are yet to be explored for FET operation and that may have potential for future devices.

The semiconductor device is an ever-expanding field of research, where researchers all over the world are working on different topics. New material engineering, device structure modification, new physics of operation, etc. are currently being explored to sustain the growth. These new ideas are being explored by academic and industry researchers together as none of them can work independently without the help of each other. The present world is enjoying their joint work, which they did in the past, and is eagerly waiting for some more fascinating findings in the coming days.

REFERENCES

1. Julius Edgar Lilienfeld. "Method and apparatus for controlling electric current". US 1745175 filed in Canada, 1925.
2. Gorton, W.S.. Genesis of the transistor, Vol. 3, A History of Engineering and Science in the Bell System, 1949.
3. Bassett, Ross Knox. *To the Digital Age: Research Labs, Start-up Companies, and the Rise of MOS Technology*. Johns Hopkins University Press. p. 22, 2007. ISBN 978-0801886393.
4. Arns, R.G. "The other transistor: early history of the metal–oxide–semiconductor field-effect transistor". *Engineering Science and Education Journal*, 7, no. 5 (1998), 233–240.

5. Butrica, Andrew J. "NASA's role in the manufacture of integrated circuits". In *Historical Studies in the Societal Impact of Spaceflight*. Ed: Dick, Steven J. NASA. pp. 149–250 (239–42), 2015.
6. Moore, G.E. "Cramming more components onto integrated circuits". *Electronics*, 38 (1965), 114–117.
7. Dennard, Robert H., Fritz H. Gaensslen, Hwa-Nien Yu, et al. "Design of ion-implanted MOSFET'S with very small physical dimensions". *IEEE Journal of Solid-State Circuits,* Sc-9, no. 5 (1974), 256–268.
8. Shockley, W., and G.L. Pearson. "Modulation of conductance of thin films of semi-conductors by surface charges". *Physical Review*, 74 (1948), 232.
9. Kahng, D., and M.M. Atalla. "Silicon-silicon dioxide field induced surface devices". *IRE-AIEE Solid-state Device Research Conference* (Carnegie Institute of Technology, Pittsburgh, PA), 1960.
10. Kahng, Dawon. 1963. "Electric field controlled semiconductor device". U. S. Patent No. 3,102,230. Issued August 27 1963.
11. Ihantola, H.K.J., and J.L. Moll. "Design theory of a surface field-effect transistor". *Solid-State Electronics*, 7, no. 6 (1964), 423–430.
12. Hofstein, S.R., and F.P. Heiman. "The silicon insulated-gate field-effect transistor". *Proceedings of the IEEE*, 51, no. 9 (1963), 1190–1202.
13. Sah, C.T. "Characteristics of the metal-oxide-semiconductor transistors". *IEEE Transactions on Electron Devices*, 11, no. 7 (1964), 324–345.
14. Deal, Bruce E. "Highlights of silicon thermal oxidation technology". *Silicon Materials Science and Technology*. The Electrochemical Society. p. 183, 1998.
15. Moskowitz, Sanford L. *Advanced Materials Innovation: Managing Global Technology in the 21st century*. John Wiley & Sons. pp. 165–167, 2016.
16. Bassett, Ross Knox. *To the Digital Age: Research Labs, Start-up Companies, and the Rise of MOS Technology*. Johns Hopkins University Press. pp. 53–54, 2002.
17. Medwin, Albert H. "Semiconductor translating circuit". US3390314A. Filed by RCA Corp on 30.10.1964.
18. "1960: Metal oxide semiconductor (MOS) transistor demonstrated". Computer History Museum. https://www.computerhistory.org/siliconengine/metal-oxide-semiconductor-mos-transistor-demonstrated/ (accessed March 28, 2021).
19. Dennard, R.H. 1968. "Field-effect transistor memory". U.S. Patent No. 3,387,286, Issued 1968.
20. Dennard, R.H. "Evolution of the MOSFET dynamic RAM – a personal view". *IEEE Transactions on Electron Devices*, ED-31 (1984), 1549–1555.
21. Dennard, R.H. "Scaling challenges for dram and microprocessors in the 21st century". *Electrochemical Society Proceedings* 97–3 (1997), 519–532.
22. Brews J.R., W. Fichtner, E.H. Nicollian, et al. "Generalized guide for MOSFET miniaturization", *IEEE Electron Device Letters*, EDL-1 (1980), 2–4.
23. "International technology roadmap for semiconductors (ITRS)". http://www.itrs2.net/.
24. Lo, S.-H., D.A. Buchanan, and Y. Taur. "Modeling and characterization of quantization, poly silicon depletion and direct tunneling F3191 effects in MOSFETs with ultrathin oxides"., 43 (1999), 327–337.
25. Frank, D.J., R.H. Dennard, E. Nowak, et al. "Device scaling limits of Si MOSFETs and their application dependencies". *Proceedings of IEEE*, 89 (2001), 259–288.
26. Kingon, A.I., J.-P. Maria, and S.K. Streiffer. "Alternative dielectrics to silicon dioxide for memory and logic devices". *Nature*, 406 (2000), 1032–1038.
27. Wilk, G.D., R.M. Wallace, and J.M. Anthony. "High-k gate dielectrics: Current status and materials properties considerations". *Journal of Applied Physics*, 89 (2001), 5243–5275.

28. Huff, H.R., A. Agarwal, Y. Kim, et al. "Integration of High- K gate stack systems into planar CMOS process flows". *International Workshop on Gate Insulator (IWGI2001)*, 2–11, 2001.
29. Takeda, E., H. Kume, T. Toyabe, et al. "Submicron MOSFET structure for minimizing hot-carrier generation". *IEEE Transactions on Electron Devices*, 29 (1982), 611–618.
30. Liou, J.J. 1994. *Advanced Semiconductor Device Physics and Modeling*. Artech House, Inc.
31. Manasevit, H.M., I.S. Gergis, and A.B. Jones. "Electron mobility enhancement in epitaxial multilayer Si-Si1−xGex alloy films on (100) Si". *Applied Physics Letters*, 41 (1982): 464–466.
32. People, R., J.C. Bean, and D.V. Lang, et al. "Modulation doping in GexSi1-x/Si strained layer heterostructures". *Applied Physics Letters*, 45 (1984), 1231.
33. Sinha, Kunal. "Strain engineering in modern field effect transistors". In *Electrical and Electronic Devices, Circuits and Materials*. Ed. Suman Lata Tripathi, Pervej Ahmad Alvi, and Umashankar Subramaniam. pp. 1–18. Scrivener Publishing, Wiley, 2021.
34. Farrah, H.R., and Steinberg, R.F. 1967. "Analysis of double-gate thin-film transistor". *IEEE Transactions on Electron Devices*, 14 (2): 69–74.
35. Sekigawa, Toshihiro, and Y. Hayashi. "Calculated threshold-voltage characteristics of an XMOS transistor having an additional bottom gate". *Solid-State Electronics* 27, no. 8 (1984): 827–828.
36. Hisamoto, D., T. Kaga, Y. Kawamoto, and E. Takeda. "A fully depleted lean-channel transistor (DELTA) – A novel vertical ultra thin SOI MOSFET". *International Technical Digest on Electron Devices Meeting* (1989), 833–836.
37. Hisamoto, Digh, Chenning Hu, J. Bokor, et al. "FinFET – A self-aligned double-gate MOSFET scalable to 20 nm". *IEEE Transactions on Electron Devices*, 47, no. 12 (2000), 2320–2325.
38. Choi, Y., N. Lindert, P. Xuan, et al. "Sub-20 nm CMOS FinFET technologies." *International Electron Devices Meeting. Technical Digest* (Cat. No.01CH37224), 19.1.1–19.1.4, 2001.
39. Tsu-Jae King, Liu. "FinFET: History, fundamentals and future". University of California, Berkeley. Symposium on VLSI Technology Short Course, June 11, 2012.
40. Rostami, M., and Mohanram, K. "Dual independent-gate FinFETs for low power logic circuits". *IEEE Transactions on Computer-Aided Design of Integrated Circuits and Systems*, 30, no. 3 (2011), 337–349.
41. Gupta, Suyog, Victor Moroz, Lee Smith, et al. "7-nm FinFET CMOS design enabled by stress engineering using Si, Ge, and Sn". *IEEE Transaction on Electron Devices*, 61, no. 5 (2014), 1222–1230.
42. Sinha, Kunal, Hafizur Rahaman, and Sanatan Chattopadhyay. "Investigation of the impact of embedded SiGe source/drain induced uniaxial stress on the performance of Si p-channel 3D FinFETs". 6th International Conference on Computers and Devices for Communication (CODEC-2015), Kolkata, India, 2015.
43. Park, S.H., Yang Liu, Neerav Kharche, et al. "Performance comparisons of III–V and strained-Si in planar FETs and nonplanar FinFETs at ultrashort gate length (12 nm)". *IEEE Transactions on Electron Devices*, 59, no. 8 (2012), 2107–2114.
44. Vardi, A., X. Zhao, and J.A. del Alamo. "Quantum-size effects in sub 10-nm fin width InGaAs FinFETs". *2015 IEEE International Electron Devices Meeting (IEDM)*, (2015), pp. 31.3.1–31.3.4.
45. Hofstein, S.R., and Warfield, G.. "The insulated gate tunnel junction triode. 12 (2)". *IEEE Transactions on Electron Devices*, (1965), 66–76.
46. Appenzeller J., Y.M. Lin, J. Knoch, et al. "Band-to-band tunneling in carbon nanotube field-effect transistors". *Physical Review Letters*, 93, no. 19 (2004), 196805.

47. Khatami, Y., and K. Banerjee. "Steep subthreshold slope n- and p-type tunnel-FET devices for low-power and energy-efficient digital circuits". *IEEE Trans. Electron Devices*, 56, no. 11 (2009), 2752–2761.
48. Asra, R., M. Shrivastava, K.V.R.M. Murali, et al. "A tunnel FET for VDD scaling below 0.6 V with a CMOS-comparable performance". *IEEE Transactions on Electron Devices*, 58, no. 7 (2011), 1855–1863.
49. Narang, R., M. Saxena, R.S. Gupta, et al. "Assessment of ambipolar behavior of a tunnel FET and influence of structural modifications". *J. Semiconductor Technology and Science*, 12, no. 4 (2012), 482–491.
50. Kao, K.-H., A.S. Verhulst, W.G. Vandenberghe, et al. "Direct and indirect band-to-band tunneling in germanium-based TFETs". *IEEE Transactions on Electron Devices*, 59, no. 2 (2012), 291–301.
51. Singh, G., S.I. Amin, S. Anand, et al. "Design of Si0.5Ge0.5 based tunnel field effect transistor and its performance evaluation". *Superlattices and Microstructures*, 92 (2016), 143–156.
52. Convertino, C., C.B. Zota, H. Schmid, A.M. Ionescu, and K.E. Moselund. "III–V heterostructure tunnel field-effect transistor". *Journal of Physics: Condensed Matter*, 30, Number 26 (2018).
53. Nayfeh, O.M., J.L. Hoyt, and D.A. Antoniadis. "Strained-Si{1 - x}Ge{x}/Si band-to-band tunneling transistors: Impact of tunnel-junction germanium composition and doping concentration on switching behavior". *IEEE Transactions on Electron Devices*, 56, no. 10 (2009), 2264–2269.
54. Zhao, Q., Simon Richter, Christian Schulte-Braucks, et al., "Strained Si and SiGe nanowire tunnel FETs for logic and analog applications". *IEEE Journal of the Electron Devices Society*, 3, no. 3 (2015), 103–114.
55. Convertino, C., C.B. Zota, H. Schmid, et al. "A hybrid III–V tunnel FET and MOSFET technology platform integrated on silicon". *Nature Electronics*, 4 (2021), 162–170.
56. Barbaro, M., A. Bonfiglio, and L. Raffo. "A charge-modulated FET for detection of biomolecular processes: Conception, modeling, and simulation". *IEEE Transactions on Electron Devices*, 53, no. 1 (2006), 158–166.
57. Gao, X.P.A., G. Zheng, and C.M. Lieber. "Subthreshold regime has the optimal sensitivity for nanowire FET biosensors". *Nano Letters*, 10, no. 2 (2010), 547–552.
58. Kanungo, Sayan, Sanatan Chattopadhyay, Partha Sarathi Gupta, et al. "Study and analysis of the effects of SiGe source and pocket-doped channel on sensing performance of dielectrically modulated tunnel FET-based biosensors". *IEEE Transactions on Electron Devices*, 63, no. 6 (2016), 2589–2596.
59. Houssa, Michel, Athanasios Dimoulas, and Alessandro Molle. *2D Materials for Nanoelectronics*. CRC Press, ISBN 9781498704175, 2016.
60. Novoselov, K.S., A.K. Geim, S.V. Morozov, et al. "Electric field effect in atomically thin carbon films". *Science*, 306 (2004), 666–669.
61. Berger, C., Z. Song, T. Li, et al. "Ultrathin epitaxial graphite: 2D electron gas properties and a route toward graphene-based nanoelectronics". *Journal of Physical Chemistry B*, 108 (2004), 19912–19916.
62. Schwierz, Frank, Jörg Pezoldt, and Ralf Granzner. "Two-dimensional materials and their prospects in transistor electronics". *Nanoscale*, 7, no. 18 (2015), 8261–8283.
63. Lemme, M.C., T.J. Echtermeyer, M. Baus, et al. "A graphene field-effect device". *IEEE Electron Device Letters*, 28 (2007), 282–284.
64. Han, S.-J., A. Valdes-Garcia, A.A. Bol, et al. "Graphene technology with inverted-T gate and RF passives on 200 mm platform". *International Electron Devices Meeting*, 2011 (2011), 19–22.

65. Radisavljevic, B., A. Radenovic, J. Brivio, et al. "Single-layer MoS2 transistors". *Nature Nanotech* 6 (2011), 147–150.
66. Szabo, A., S.J. Koester, and M. Luisier. "Metal-dichalcogenide hetero-TFETs: Are they a viable option for low power electronics? English", *Device Research Conference,* Conference Digest (2014), 19–20.
67. Gong, C., H. Zhang, W. Wang, et al. "Band alignment of two-dimensional transition metal dichalcogenides: Application in tunnel field effect transistors". *Applied Physics Letters* 103 (2013), 53513.
68. Singh, Eric, Pragya Singh, Ki Seok Kim, et al. "Flexible molybdenum disulfide (MoS2) Atomic layers for wearable electronics and optoelectronics". *ACS Applied Materials and Interfaces*, 11, no. 12 (2019), 11061–11105.
69. Wang, Q.H., K. Kalantar-Zadeh, A. Kis, et al. "Electronics and optoelectronics of two-dimensional transition metal dichalcogenides". *Nature Nanotechnology*, 7 (2012), 699–712.
70. Butler, S.Z., S.M. Hollen, L. Cao, et al. "Progress, challenges, and opportunities in two-dimensional materials beyond graphene". *ACS Nano*, 7 (2013), 2898–2926.
71. Schwierz, F., Pezoldt, J., and Granzner, R. "Two-dimensional materials and their prospects in transistor electronics". *Nanoscale*, 7 (2015), 8261–8283.
72. Das, Biswapriyo, and Santanu Mahapatra. "A predictive model for high-frequency operation of two-dimensional transistors from first-principles". *Journal of Applied Physics* 128 (2020), 234502.
73. Liu, Y., S. Zhang, J. He, et al."Recent progress in the fabrication, properties, and devices of heterostructures based on 2D Materials". *Nano-Micro Letters*, 11 (2019), 13.
74. Chun-Li, Lo, Benjamin A. Helfrecht, Y. He, et al."Opportunities and challenges of 2D materials in back-end-of-line interconnect scaling". *Journal of Applied Physics*, 128 (2020), 080903.
75. Lemme, Max, L. Li, L., Tomás Palacios, et al. "Two-dimensional materials for electronic applications". *MRS Bulletin*, 39 (2014). 711–718.

8 Memory Design Using Nano Devices

Deepika Sharma, Shilpi Birla, and Neha Mathur
Manipal University Jaipur, Jaipur, India

8.1 INTRODUCTION

A study conducted at nanoscale from 1 nm to 100 nm in the field of science, engineering and technology is defined as nanotechnology. The applications of nanotechnology increased after the 2000s in the field of commercial products, though most of these are limited to the bulk use of passive nanomaterials.

Nanoelectronics is the branch of nanotechnology in the field of electronics. It includes all devices and materials with common characteristics that have small interatomic interactions and quantum mechanical properties. The size of nanoelectronic devices varies between 1 nm and 100 nm. Recently, the generation of silicon metal-oxide-semiconductor field-effect transistor (MOSFET) technology lies in this regime, including complementary metal-oxide-semiconductor (CMOS) for 22 nm, 14 nm and 10 nm nodes and FinFET for 7 nm nodes. Gordon Moore in 1965 observed that with the rapid development of technology, the size of silicon transistors was continuously scaling, and later his observations were summarized into a law called Moore's law. Moore's observations reduce the size of transistors to the nanometer range as of 2019.

8.1.1 Nanotechnology Approaches for Nanoelectronics

Two types of methods are defined for manufacturing nanomaterials in nanoelectronics:

- Top-down approach
- Bottom-up approach

In the top-down process, nano-dimensional particles are obtained by sequential cutting of bulk materials, whereas bottom-up approaches include storage of materials from the bottom, such as atom by atom, molecule by molecule or cluster by cluster [1]. Both the approaches differ in the process of obtaining nanoparticles, with the top-down process starting from the top and the bottom-up process starting from the bottom. Top-down approaches are faced with the problem of defectiveness of surface structure and significant crystallographic impairment of processed patterns [1]. This causes an extra problem in design and construction of the device.

DOI: 10.1201/9781003220350-8

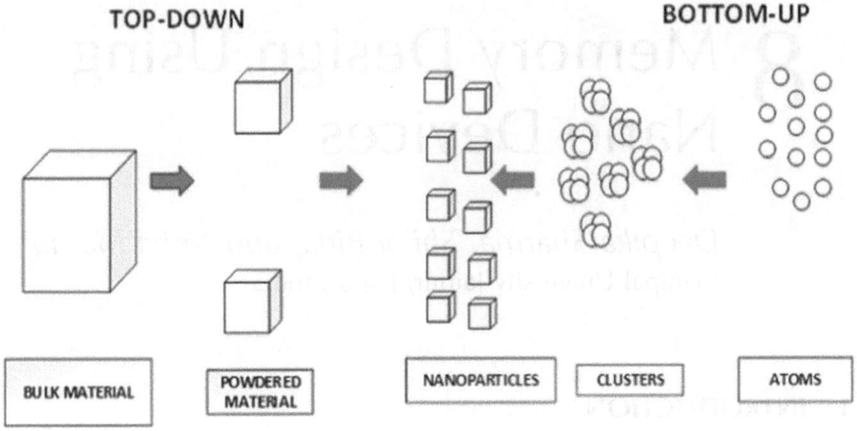

FIGURE 8.1 Top-down and bottom-up approaches [1].

Top-down processes include bulk fabrications of nanomaterial, whereas in bottom-up processes flaws are produced. The bottom-up approach is the effective technique for nanostructures with fewer defects when compared to the top-down approach, which includes stress, surface defects and contaminations.

For the design of integrated circuit (IC) technology, the top-down process is used. Various procedures are used (etching, oxidation, film deposition) to design the interconnects and devices. In nanotechnology, bottom-up approaches have several advantages, such as low power dissipation, high package density and high functionality, which are in high demand for portable devices and are unattainable with top-down approaches. Figure 8.1 shows both types of approaches.

Table 8.1 shows the basic differences between these two technologies.

By considering the advantages and disadvantages of these two approaches, one can conclude that bottom-up technology is more effective than top-down technology for the fabrication of different materials in nanoelectronics.

8.1.2 Materials Used in Nanoelectronics

Materials used in nanoelectronics are of two types:

- Inorganic nanocrystals
- Organic molecular components

Nanotubes and nanowires are named after the shape obtained from the process of growth kinetics. All nanotubes and nanowires come under the category of inorganic nanocrystals, whereas all molecular wires, monolayers and supramolecules come in organic molecule components. An organic component includes molecular wires, single molecules, molecular monolayers and supramolecules with different schemes. Figure 8.2 shows steps to generate nanochips.

TABLE 8.1
Comparison between two approaches in nanoelectronics

Top-Down Technology	Bottom-Up Nanotechnology
It goes from larger to smaller	It goes from smaller to larger
Method uses pattern and etch	Method uses synthesis and self-assembly
High production cost	Low production cost
Low accessibility	High accessibility
Design tools are developed	Design tools are open
Material selection is obstinate	Material selection is flexible

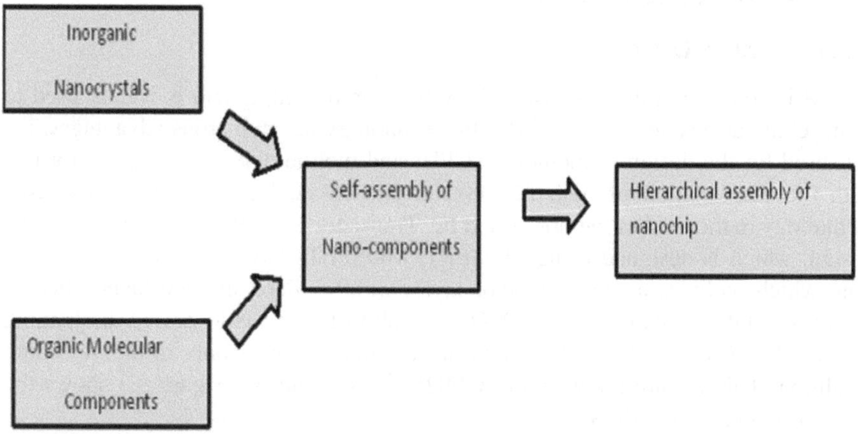

FIGURE 8.2 Nanoscale components for next-generation nanochips [1].

8.2 HISTORY OF NANOSCALE DEVICES

At Texas Instruments in 1958, by using only two transistors, the IC flip-flop was developed. Nowadays, there are billions of transistors integrated into chips [2]. Today's mobile devices, which are used by half the population of the world and for which memory is an important part, are supported by the whole company's system. This rapid development in technology occurred because of transistor scaling and advancement in the manufacturing process of silicon.

The electronics industry started with the invention of vacuum tubes. In these devices, electron flows were controlled in a vacuum. After some time, these devices were not working effectively because of increasing complexity, power dissipation and the number of components [2]. This resulted in decrement in the performance of the device, i.e. it was no longer successful.

After the failure of vacuum tubes, the bipolar junction transistor (BJT) was developed in 1950 by Shockley. BJTs are more efficient when compared to vacuum tubes with the features of having less area, low power consumption and greater speed. Transistors behave as a switch with three terminals, in which one terminal is a control terminal. When current flows from the control terminal, the transistor acts as a closed switch; otherwise, it will act as an open switch. The BJT faces the problem of having large standby power dissipation, meaning circuit dissipated power switches to the OFF state. This basically limits a greater number of transistors from integrating into circuits.

To eliminate the problems of static power in BJTs, the first logic gate was developed in 1963 in which both n-channel and p-channel transistors were used [2]. These transistors were connected in pull-up and pull-down configuration to form CMOS devices. They draw negligible static power dissipation. The excellent features of CMOS technology replace all the technology for digital applications.

8.3 NANOSCALE DEVICES

8.3.1 CMOS Device

CMOS is considered the best technology for chip designing and is widely used to form IC in various applications [3]. This technology has numerous advantages, i.e. it is used for the design of memories, CPUs and mobile devices. CMOS is formed with the combination of both PMOS and NMOS devices. It is the most used technology in the field of memories and IC. Transistor is the basic component for IC design, which is designed using MOSFET. MOSFETs have sandwich-like formation, which includes a semiconductor layer, wafer, silicon dioxide layer, slice of silicon crystal and metal layer. CMOS has almost zero static power dissipation, unlike NMOS and BIPOLAR. CMOS has dissipated power only in the switching condition. This feature allows more CMOS gates on an IC. Figure 8.3 shows the basic structure of a CMOS device.

8.3.1.1 NMOS

NMOS is a three-terminal device. This substrate is formed by p-type, and the source and drain is formed by n-type [3]. Electrons are the majority charge carriers. NMOS is conducted when voltage is applied on it and remains OFF when there is no voltage. NMOS is faster than PMOS because NMOS has electrons as the majority

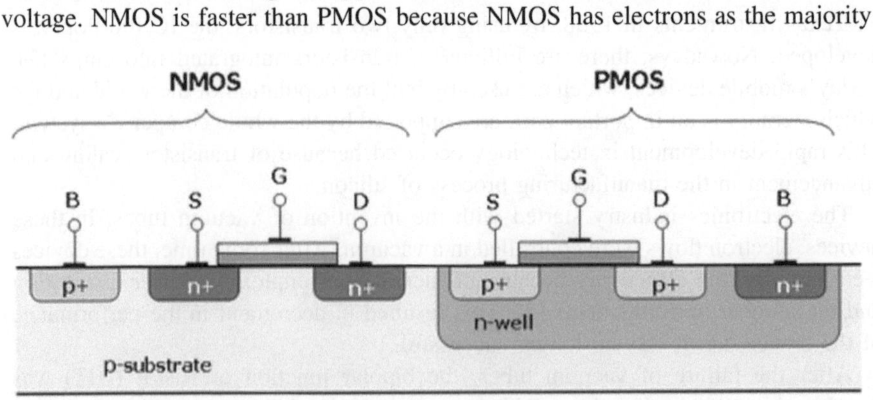

FIGURE 8.3 CMOS device structure [3].

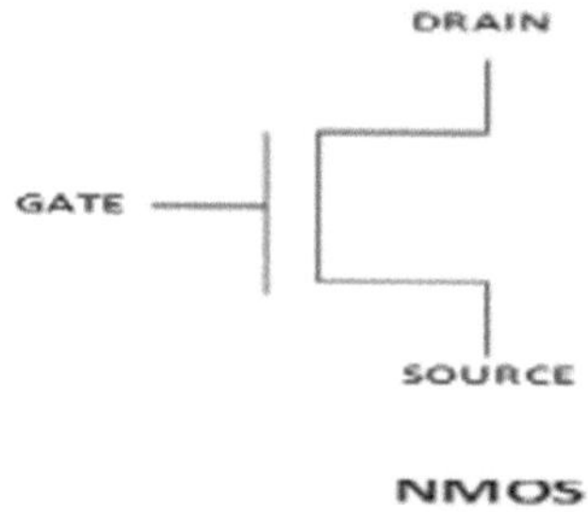

FIGURE 8.4 NMOS [3].

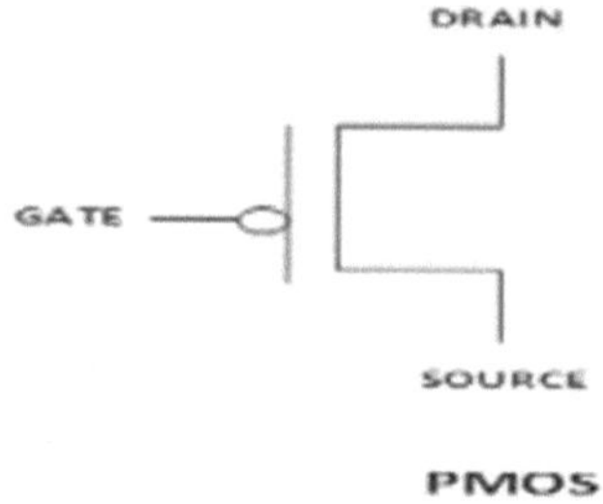

FIGURE 8.5 PMOS [3].

charge carriers, which travel twice as fast than PMOS. Figures 8.4 and 8.5 show the symbol of NMOS and PMOS switches.

8.3.1.2 PMOS

The formation of PMOS is almost the same as NMOS except source and drain are of p-type formed on n-type substrates. PMOS has holes as a majority charge carrier. It will remain ON when a negative supply is given to the gate terminal and will remain OFF with the positive supply. The noise immunity power of PMOS is more than NMOS.

8.3.1.3 Working

Logic functions are designed with the help of a CMOS device by applying input to the pull-up and pull-down network. The input signal is applied on the gate of NMOS and PMOS in such a way that it will turn ON one network and turn OFF another network. In this device, n-type MOSFETs are connected in a pull-down network between the output node and ground, whereas PMOS is connected in a pull-up network between supply and output [3]. The connections of a network are arranged in such a fashion so that ON of one transistor will OFF the other for any type of input pattern. In both the states, CMOS retains its properties and operates for various combinations of input patterns [3]. CMOS batteries have a life span of 10 years, but they can be changed according to their uses and surroundings.

8.3.1.4 Advantages

- CMOS has good noise immunity and less power dissipation over TTL.
- It has high input impedance.

- It has has simple gates.
- It uses a single power supply.
- It has zero static power dissipation.
- It has high fan out.
- It has temperature stability.
- It has large logic swing.
- It is more compact.

8.3.1.5 Disadvantages

- As the processing steps increase, the cost also increases.
- CMOS has low package density.
- It covers more area than NMOS.
- It has increased short channel effects (SCEs).

8.3.1.6 Applications

- CMOS is used in computer memories and CPUs.
- It is used for designing microprocessors.
- It is used for designing flash memory chips.
- It is used for designing application-specific integrated circuits (ASICs).

8.3.2 FinFET Device

FinFET is categorized as a multi-gate MOSFET. Chenming Hu and his teams at the University of Berkley, California, designed the first FinFET [4]. FinFET is a multigate transistor with more than one gate in a single device. In this device, a thin film of silicon is enfolded around the channel. The structure of FinFET looks like a set of fins, i.e. its name is derived from "fin". The channel length of the device is determined by its thickness. FinFET is a nonplanar device that is built either on silicon-on-Insulator or on silicon wafers. The features for wrapping the gate around the conducting channel help in reducing leakage currents. Figures 8.6 and 8.7 show the double gate and basic structure of FinFET.

8.3.2.1 Necessity of FinFET

As the technology node is scaled in CMOS devices, there are various types of unwanted side effects, also called short channel effects. These effects reduce the performance, modeling and reliability of the device [4]. Planar MOSFETs suffer from various problems below 22 nm. FinFET solves the issues of SCE better than MOSFET, hence making transistor scaling more effective. These effects are categorized as follows:

8.3.2.1.1 DIBL (Drain-Induced Barrier Lowering)

This effect occurs when the depletion region of the drain extends to the source and the two depletion layers merge. This causes punch through [4]. This can be minimized by thin oxides, long channels, shallower junctions and high doping on substrates.

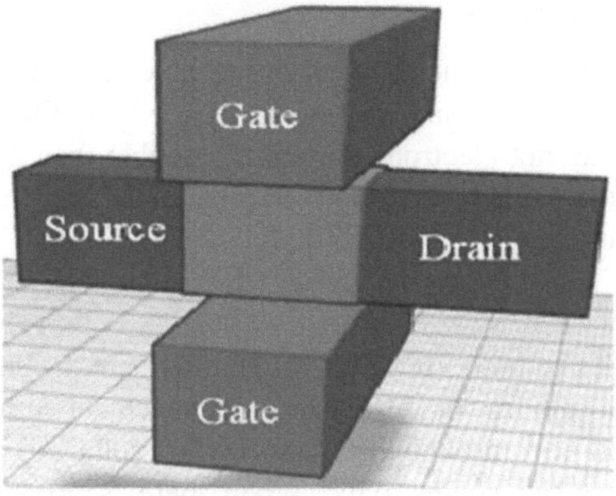

FIGURE 8.6 Double gate structure [2].

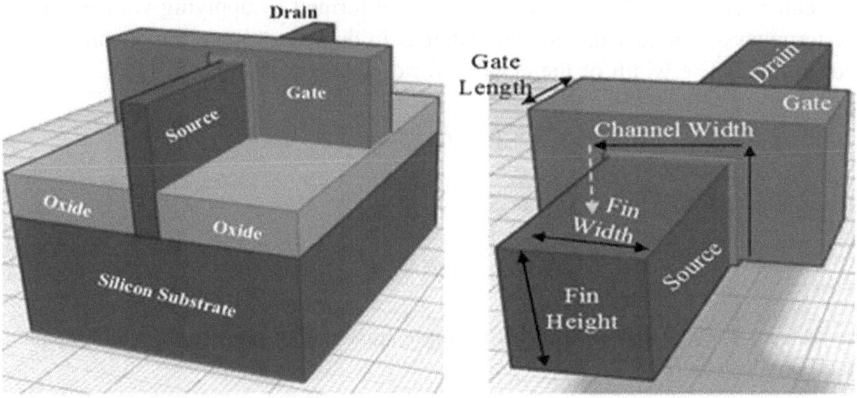

FIGURE 8.7 FinFET structure [2].

8.3.2.1.2 Surface Scattering

It is nothing but the collisions experienced by the electrons, which travel toward the interface. Surface scattering reduces the mobility of charge carriers, and electrons face difficulty moving parallel to the interface.

8.3.2.1.3 Velocity Saturation

The velocity of charge carriers is directly proportional to the electric field that drives them. When the value of the electric field is above the critical field, then carrier velocity reaches its maximum value and is saturated at this velocity. When this happens, the transistor enters the state of velocity saturation [4]. This causes an increase in transit time of carriers through the channel.

8.3.2.1.4 Impact Ionization

This effect generally occurs due to generation of electron-hole pairs by impact ionization caused by high velocity electrons in the presence of a high electric field. These effects cause degradation in the performance of devices.

8.3.2.1.5 Hot Electron Effect

In the presence of high electric fields, electrons or holes gain high kinetic energy in the semiconductor device, causing a hot electron effect. Electrons have higher mobility than holes, i.e. hot electrons are more probable than hot holes. Hot electrons travel through the oxide layer and get trapped in the gate, causing undesirable behavior of the device.

8.3.2.2 Working

The working principles of both MOSFET and FinFET are the same. There are two types of conduction modes in MOSFET: enhancement mode and depletion mode [4]. Current will flow through the channel when voltage is applied on the gate terminal. This reduces the conductivity. When there is no flow of current, the channel shows the maximum value of conductance. Figure 8.8 explains the basic symbol of n- and p-channel FinFET.

In enhancement mode, the channel has to be formed by applying voltage on the gate terminal to conduct the device, whereas in depletion mode the channel is already formed. The width of the channel for double gate FinFET is [4]:

$$Width = 2 * Fin\ \ Height + Fin\ \ Width$$

For triple gate FinFET it is [5]:

$$W = 2 * Hfin + T$$

The transistor width increases with multiple fins. If the number of fins is n, then effective width is [5]:

$$Weff = n * W$$

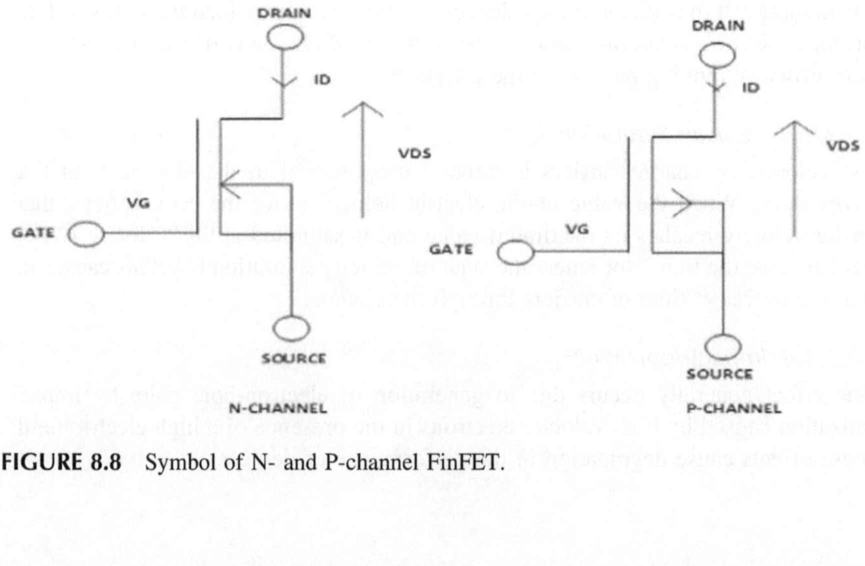

FIGURE 8.8 Symbol of N- and P-channel FinFET.

By increasing the Fin height, the channel width also increases, which in turn increases the current flowing through the channel. The second way to increase the current is to increase the number of parallel fins.

8.3.2.3 Classification of FinFET

FinFET is categorized in two ways:

- Shorted gate (SG)
- Independent gate (IG).

SG is three-terminal, while IG is four-terminal [6]. SG FinFET has both the front and back gates physically short, while IG FinFET has both gates independent of each other. SG FinFET offers higher values of ON and OFF currents than IG FinFET because both gates combine control of the channel, while in IG two different signals or voltages are applied to two gates to provide flexibility. IG FinFET faces the problem of area overhead. Figure 8.9 indicates the structure of SG and IG FinFET.

8.3.2.4 Advantages

- FinFET has good controlling power over the channel.
- It has reduced SCE.
- It has lower static leakage current and power consumption.
- It has higher switching speed.
- It has increased drain current.

8.3.2.5 Disadvantages

- FinFET's fabrication cost is high.
- It has high value of capacitance.

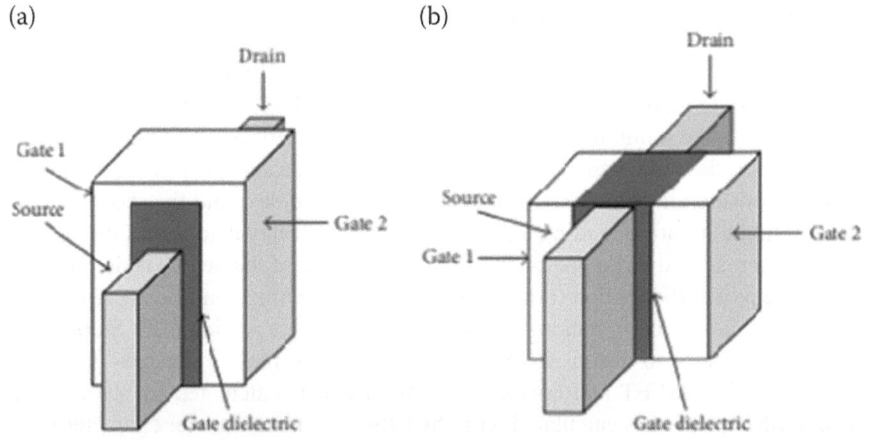

FIGURE 8.9 (a) SG (b) IG FinFET.

- It is highly parasitic.
- Users must manage dynamic Vth.
- It has corner effect.
- Designers face difficulty making fractions of fins.

8.3.2.6 Applications

- FinFET 14 nm technology is used in Samsung smartphone.
- Apple, Intel and TSMC also use the FinFET chip.

8.3.3 CNTFET

In 1998, the first carbon nanotube field-effect transistors (CNTFETs) were introduced. They are designed by depositing single-wall CNTs on oxidized Si wafers. These wafers are prepatterned with gold or platinum electrodes. These electrodes form the source and drain contact connected by the nanotube channel. The gate terminal is formed by doped Si substrate [7]. The CNTFETs channel is formed by semiconducting carbon nanotubes. The electrical properties basically depend on chirality or the direction of distortion. Depending on chirality, CNTs can be metallic or semiconducting. The structure of CNTFETs is almost similar to MOSFETs, i.e. CNTFETs are the best replacement of MOSFETs. Both devices have the same I-V and transfer characteristics. However, CNTFETs have some different properties from MOSFETs, such as being one-dimensional, having an absence of dangling bonds, and the Schottky barrier behaves as an active switching element. CNTFETs have two types: single-walled carbon nanotube (SWCNT) and multiple walled carbon nanotube. A SWCNT is formed by rolling the single sheet of graphene [8]. It can be made either metallic or semiconducting by changing the chirality vector. The diameter of CNTFETs decides the threshold voltage of transistors. Both diameter and chirality vectors are related to each other. Figure 8.10 shows the planar and coaxial structure of CNTFETs.

8.4 DESIGN OF SRAM USING CMOS

SRAM occupies a large area of a chip because a number of transistors are required to design a single memory cell. Therefore, for high package density the size of the transistor should be minimum in order to design the cell [9]. As the device is scaled from past decades, the size of SRAM cells are also scaled [10]. There are two important considerations in SRAM: one is propagation delay, and the second is the power dissipation during reading and writing the data. Delay determines the speeds of the cell in read and write operations [11]. Battery life of the device is determined by dynamic power dissipation (power dissipated in switching conditions). There are lots of challenges faced in nanometer design due to device scaling [12]. Stability is the important factor to design the SRAM cell. Stability mainly depends upon the aspect ratio of MOSFET and operating conditions. It is calculated by static noise margins (SNM). SNM is calculated with the help of voltage transfer characteristics of the cell [13]. SNM is the amount of noise that can be tolerated by the cell without

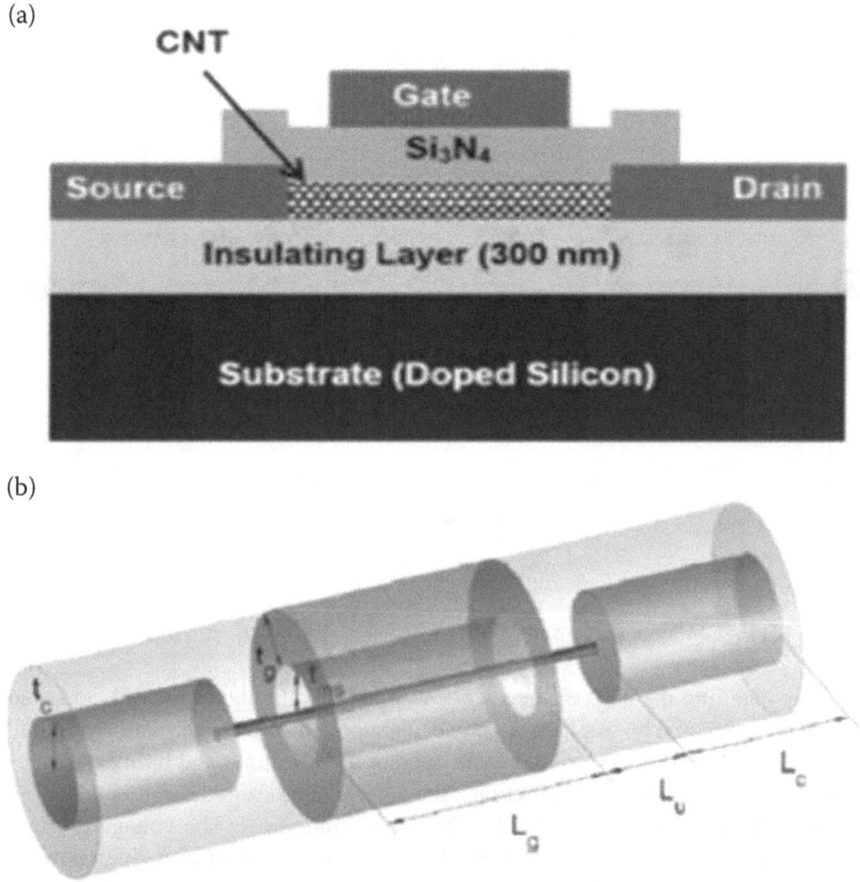

FIGURE 8.10 The CNTFET structures: (a) planar, (b) coaxial [8].

changing its state [14]. During standby mode, the data of the cell should not change or new data can't be written in the cell [15]. The overall power dissipation in SRAM cells is dynamic and static dissipation. Dynamic occurs during active mode of operation, while static occurs in OFF mode [16].

Here, the applications of nanoscale devices in memories are studied. The design of conventional 6T SRAM cells using CMOS, FinFET and CNTFET devices are discussed. The design of the 6T SRAM cell using CMOS is shown in Figure 8.11. The operation of SRAM cells for all three nanoscale devices is similar expect for minor differences in device orientation. One of the major differences between CNTFETs and MOSFET is that the source and drain terminal of CNTFETs are not interchangeable with each other, as is the case for MOSFET, i.e. care must be taken when designing the cell. The cell consists of two inverters cross-coupled to each other (P1-N1 is first inverter, and P2-N2 is second inverter) to store the bit, and two access transistors N3-N4 are used to read and write data from the cell. Access transistors are ON and OFF with the help of wordline. There are two bitlines to read

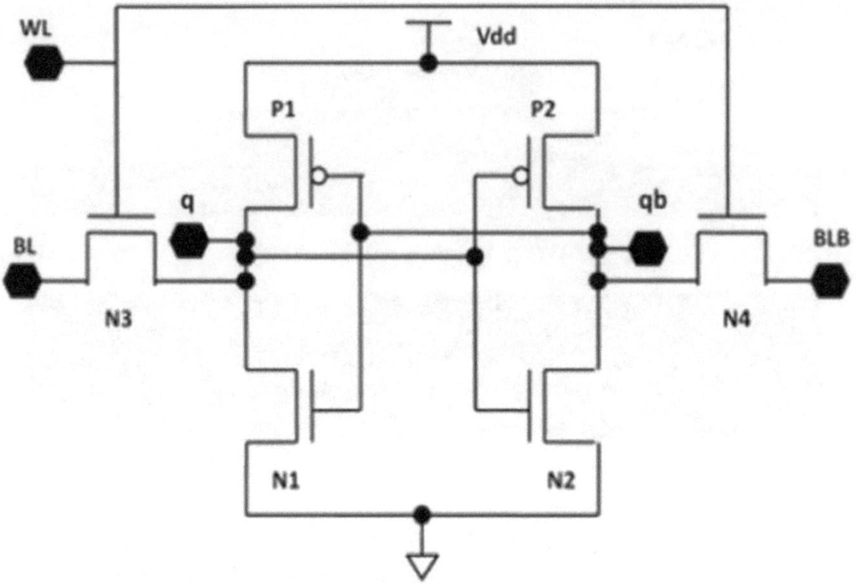

FIGURE 8.11 CMOS-based cell [19].

and write data into the cell. Bitlines work as an input and output in write and read operations, respectively. As long as power is supplied to the cell, data are stored into the cell [16]. To start read and write operations, wordlines and bitlines are pre-charged to Vdd [17]. In read operations, the stored value from the cell is taken out by activating the wordlines to ON for the access transistors. The voltage drop on bitlines is sensed by the sense amplifiers. In write mode, the data are written into the cell. In hold mode, the wordlines are deactivated, which disconnects the access transistors from the cell and data remain stored into the cell [18]. Figures 8.12 and 8.13 show the design of conventional 6T SRAM using FinFET and CNTFET devices.

8.5 OPERATION OF 6T SRAM CELL

8.5.1 Standby Mode

In this mode, wordline is deactivated, which turns OFF the access transistors. The cell cannot be accessed at this time because there are no connections between cell and access transistors. The two cross-coupled inverters will continue feedback to each other as long as power is given to them, and data will remain stored into the cell [21, 22].

8.5.2 Read Mode

During read operations, the wordline is activated, which turns ON the access transistors and the cell connected to it. The value stored in nodes q and qb is

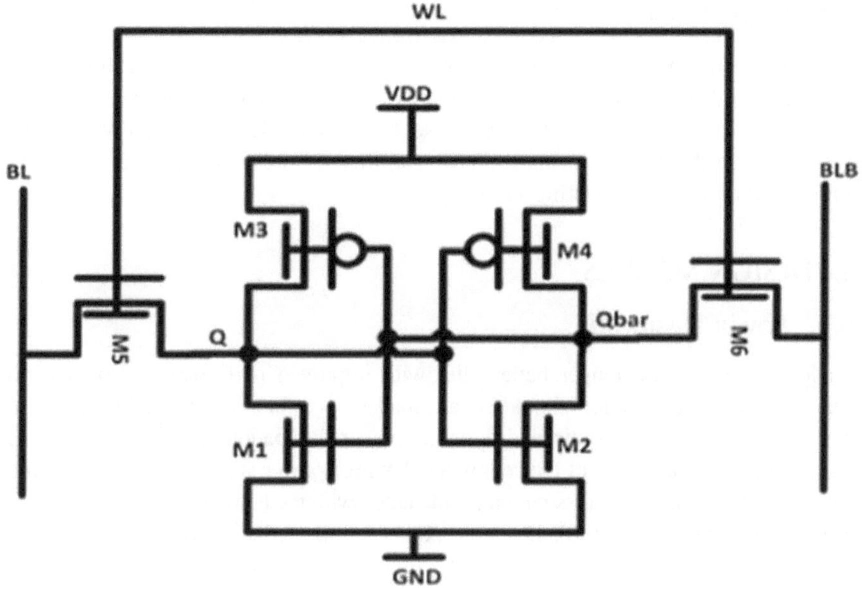

FIGURE 8.12 FinFET-based cell [20].

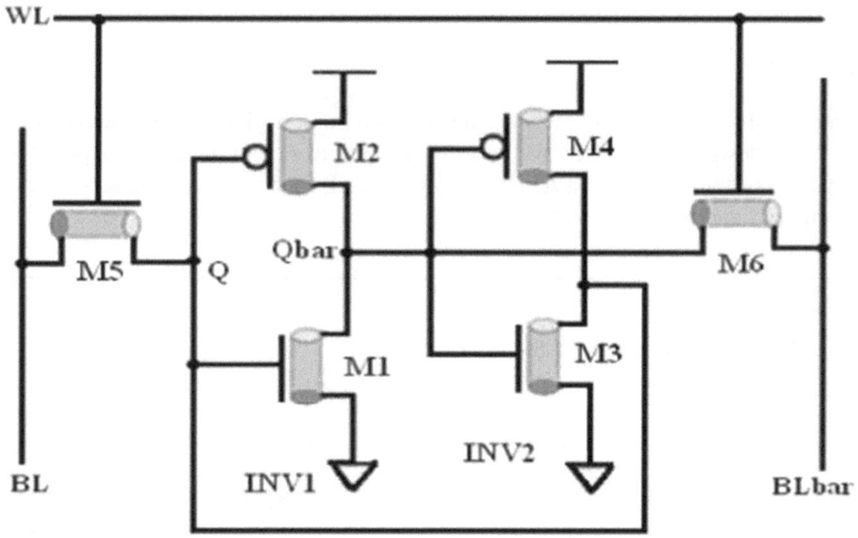

FIGURE 8.13 CNTFET-based cell [8].

transferred to the bitlines. If the node q stores 1, then the BLB will discharge through driver transistors N2 and load transistor P1 pull up to Vdd. The read stability of the cell is an important parameter. It means the data cannot be changed while reading the data [23, 24].

8.5.3 Write Mode

In this mode, both bitlines are charged or discharged according to the value. Write into the cell and the wordline always remains ON. If the cell already stored 1 and we want to write 0 into the cell, then BL is set to 0 V and BLB to 1 V and vice-versa to write 1 when the cell is storing 0. Write stability should be high so that the data cannot be changed while writing [25–28].

8.6 DESIGN METRICS

8.6.1 Power Dissipation

Portable devices require longer battery life with improved performance. For a good battery life, power dissipation should be as small as possible. There are two types of power dissipation in cells: static and dynamic power dissipation [29]. Static power is consumed in OFF conditions of the cell, while dynamic power is consumed in switching conditions of the cell. Power dissipation is calculated with the help of current and voltage consumed from the source [30]. Dynamic power can be reduced with the technology node, while static power is increased in the subthreshold region. In this chapter static power is calculated for all devices by using 32 nm PTM model files. All cells are simulated by using HSPICE tools. Table 8.2 shows the values of static power for all cells at 32 nm technology nodes. From the results, it can be revealed that FinFET-based cells have less static power compared to MOSFET-based and CNTFET-based cells.

8.6.2 Delay

Delay is defined as the difference in time at which input is given and output is taken out. While designing a system, the delay of the system should be as low as possible so that the system becomes faster. In the SRAM cell, speeds are calculated with the help of read and write access time.

TABLE 8.2
Static Power Analysis for 6T SRAM Cell

Parameter	SRAM Cells		
Supply Voltage	CMOS	FinFET	CNTFET
900 mV	32.54 μW	14.79 nW	0.90 μW
800 mV	20.41 μW	11.31 nW	0.85 μW
700 mV	10.69 μW	7.52 nW	0.72 μW
600 mV	4.92 μW	5.08 nW	0.61 μW
500 mV	1.73 μW	2.95 nW	0.59 μW
400 mV	245.54 nW	2.23 nW	0.34 μW
300 mV	21.06 nW	1.05 nW	0.15 μW
200 mV	1.26 nW	526.45 pW	0.07 μW

8.6.3 Power Delay Product (PDP)

PDP is calculated by multiplying the delay and average power consumption. It also calculates the energy consumed in the charging and discharging process of the cell. Systems with a low value for PDP are considered to be energy efficient.

8.6.4 Static Noise Margin

Stability of the SRAM cell is defined by the parameter SNM [21, 22]. SNM is the minimum amount of noise that can be tolerated by the cell without changing its state [23]. It is calculated by two methods. One is the butterfly curve method, and the second is the N-curve method [21]. The butterfly curve method is considered to be the most used method. The voltage transfer characteristics (VTCs) of one inverter are superimposed on inverted VTCs of a second inverter to form a lobe and then find the maximum possible square between them [21, 22, 24].

8.6.4.1 Hold Margin

Hold stability is defined as the ability of a cell to store data while wordline voltage is OFF. Figure 8.14 shows the butterfly curve in hold mode of operation [10]. The dashed line in the curve indicates that there is no connection between cell and access transistors. Table 8.3 shows the values of hold SNM for all cells. From the

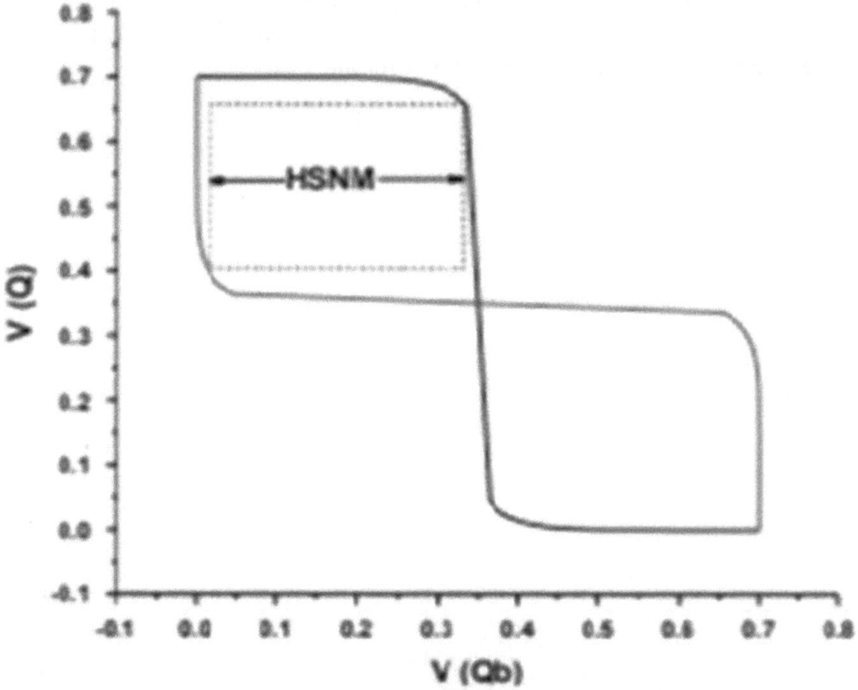

FIGURE 8.14 Hold SNM [29].

TABLE 8.3
Hold SNM Analysis for 6T SRAM Cell

Parameter	SRAM Cells		
Supply Voltage	CMOS	FinFET	CNTFET
900 mV	195 mV	240 mV	260 mV
800 mV	186 mV	200 mV	220 mV
700 mV	174 mV	190 mV	200 mV
600 mV	166 mV	170 mV	190 mV
500 mV	154 mV	150 mV	170 mV
400 mV	123 mV	130 mV	150 mV
300 mV	105 mV	70 mV	90 mV
200 mV	55 mV	60 mV	70 mV

results, it can be concluded that CNTFET-based cells have large values of SNM compared to the other two devices.

8.6.4.2 Write Margin

Write margin is defined as the minimum amount of voltage required to write new values into the cell [22]. It is the capability of the cell to replace new values in place of previously stored values. Figure 8.15 shows the butterfly curve for write mode. Table 8.4 indicates the values of write SNM for all cells.

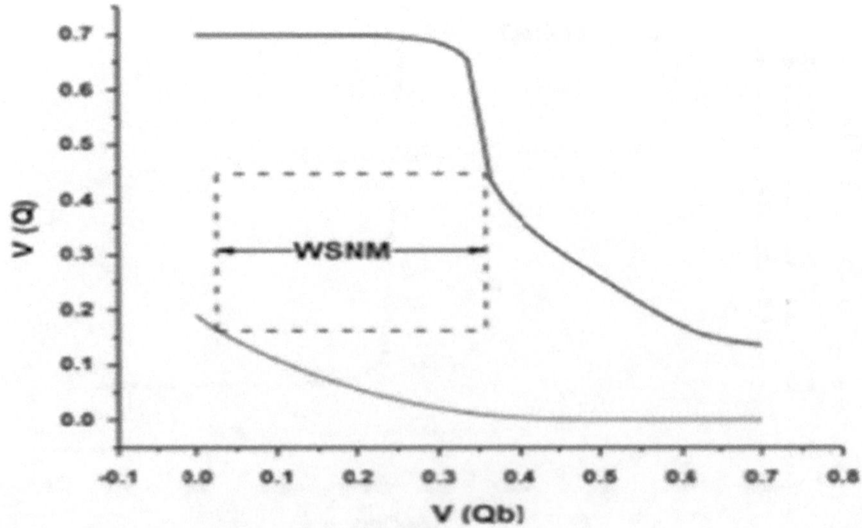

FIGURE 8.15 Write SNM [29].

TABLE 8.4
Write SNM Analysis for 6T SRAM Cell

Parameter	SRAM Cells		
Supply Voltage	CMOS	FinFET	CNTFET
900 mV	105 mV	155 mV	170 mV
800 mV	90 mV	135 mV	140 mV
700 mV	70 mV	128 mV	135 mV
600 mV	55 mV	120 mV	130 mV
500 mV	45 mV	95 mV	100 mV
400 mV	35 mV	65 mV	80 mV
300 mV	25 mV	45 mV	50 mV
200 mV	15 mV	25 mV	30 mV

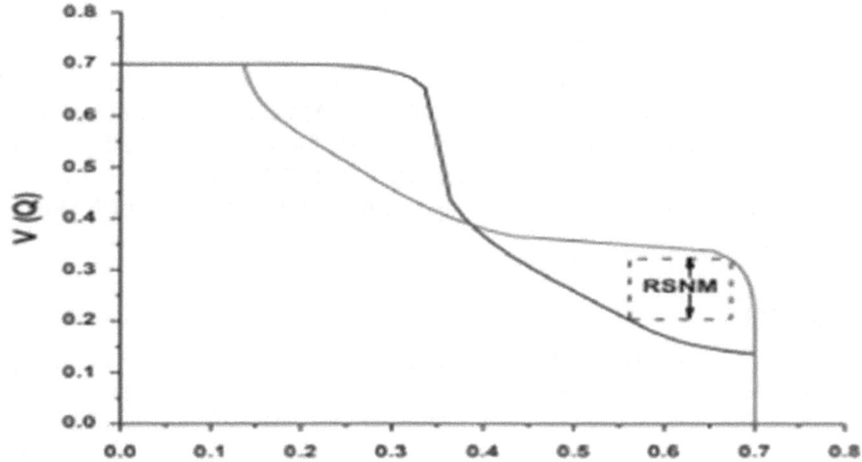

FIGURE 8.16 Read SNM.

8.6.4.3 Read Margin

Read stability of the cell is defined by read margin. Read margin is the ability of the cell to prevent data while the data are read [22]. Figure 8.16 shows the butterfly curve for write mode. The cell is more prone to noise in read operation. Table 8.5 indicates the values of read SNM for all cells.

8.7 CONCLUSION

Nanotechnology includes technology that covers the whole process of chip designing. Nanoelectronics is the branch of nanotechnology that refers to the use of

TABLE 8.5
Read SNM Analysis for 6T SRAM Cell

Parameter	SRAM Cells		
Supply Voltage	CMOS	FinFET	CNTFET
900 mV	155 mV	170 mV	200 mV
800 mV	140 mV	160 mV	170 mV
700 mV	130 mV	140 mV	150 mV
600 mV	115 mV	135 mV	140 mV
500 mV	100 mV	120 mV	130 mV
400 mV	80 mV	95 mV	110 mV
300 mV	50 mV	60 mV	85 mV
200 mV	30 mV	40 mV	60 mV

electronic components at nanoscale. Nanoelectronics include various types of devices at nanosclae, such as CMOS, FinFET and CNTFET. The main focus of this chapter is on nanoelectronic devices, including nanoelectronics and application of these devices in memories. The conventional 6T SRAM cell is designed at 32 nm by using CMOS, FinFET and CNTFET devices. Performance of the cell is taken by various parameters, such static power, HSNM, RSNM and WSNM. Results concluded that FinFET-based cells are superior to MOSFET-based and CNTFET-based cells.

REFERENCES

1. Pandey, Gaurav, Deepak Rawtani and Yadvendra Kumar Agrawal, "Aspects of nanoelectronics in materials development," *Intech Open Chapter*, July 2016.
2. Vora, Pavan H, Ronak Lad, "A review paper on CMOS, SOI and FinFET technology", *Einfochips Pvt. Ltd.* https://www.design-reuse.com/articles/41330/cmos-soi-finfet-technology-review-paper.html
3. "What is a CMOS: Working principle & its applications", https://www.elprocus.com/cmos-working-principle-and-applications/
4. Katakkar, T., All about FINFET", https://www.engineersgarage.com/article_page/all-about-finfet/
5. Mari, L., "What is a FINFET", https://eepower.com/technical-articles/what-is-a-finfet/#
6. Bhattacharya, Debajit and Niraj K. Jha, "FinFETs: From devices to architectures", *Advances in Electronics*, 2014, no. 365689, (2014), 21.
7. Busi, R., Swapna, P.,Babu, K., & Srinivasa, R. (March 2010). Carbon Nanotubes Field Effect Transistors: A Review. *International Journal of Electronics & Communication Technology*, 2, 204–208.
8. Rajendra Prasad, S., Prof. B.K. Madhavi and Prof. K Lal Kishore, "Design of low write-power consumption SRAM cell based on CNTFET at 32nm technology", *International Journal of VLSI design & Communication Systems (VLSICS)* 2, no. 4 (December 2011), 167–177.

9. Sharma, V., Francky Catthoor, andWim Dehaene (Eds.), "SRAM bit cell optimization", in *SRAM Design for Wireless Sensor Networks* (New York: Springer Science) pp 9–30, 2013.
10. Lim, W., H.C. Chin, L.S. Cheng, and M.L.P. Tan, "Performance evaluation of 14nm FinFET-based 6T SRAM cell functionality for DC and transient circuit analysis", *Journal of Nanomaterials*, 2014. doi:10.1155/2014/820763
11. Azizi-Mazreah, A., M.T.M. Shalmani H. Barati, and A. Barati, "Delay and energy consumption analysis of conventional SRAM", *International Journal of Electrical and Computer Engineering* 2 (2008), 74–78.
12. Kumar, Y. and S.K. Kingra, "Stability analysis of 6T SRAM cell at 90 nm technology", *International Journal of Computer Applications* 1 (2016), 32–36.
13. Singh, S.K., S.V. Singh, B.K. Kausik, C. Chauhan and T. Tripathi, "Characteriztion & improvement of SNM in deep submicron SRAM design", *International Conference on Signal Processing and Integrated Networks,* pp. 538–542, 2014.
14. Arandilla, Christiensen D.C., Anastacia B. Alvarez and Christian Raymund K. Roque, "Static noise margin of 6T SRAM cell in 90-nm CMOS", *International Conference on Modelling and Simulation*, pp 534–539, 2011.
15. Sung Mo, Kang, and Leblebici Yusuf, *CMOS Digital Integrated Circuits* (New York: McGraw-Hill), 1996.
16. Kundan, Vanama, Gunnuthula Rithwik and Prasad Govind, "Design of low power stable SRAM cell", *International Conference on Circuit, Power and Computing Technologies*, pp 1263–1267, 2014.
17. Roy, Kaushik and Sarat C. Prasad, *Low-Power CMOS VLSI Circuit Design* (New York: Wiley & Sons), 2000.
18. Rohit and G. Saini, "A stable and power efficient SRAM cell", *IEEE International Conference on Computer, Communication and Control*, 2015.
19. Saun, Shikha and Hemant Kumar, "Design and performance analysis of 6T SRAM cell on different CMOS technologies with stability characterization", *IOP Conference Series: Materials Science and Engineering*, 2019.
20. Chodankar, Prathamesh and Ajit Gangad, "Low power SRAM design using independent gate FinFET at 30nm technology", *First International Conference on Computational Systems and Communications (ICCSC)*, 17–18 December 2014.
21. Seevinck, F. J. List and J. Lohstroh, "Static-noise margin analysis of MOS SRAM Cells", *IEEE Journal of Solid-State Circuits*, SC-22 (Oct. 1987), 748–754.
22. Singh, Sapna, Neha Arora, Meenkshi Suthar and Neha Gupta, "Performance evaluation of different SRAM cell structures at different technologies", *International Journal of VLSI Design & Communication Systems (VLSICS)* 3, no.1, February 2012. doi:10.5121/vlsic.2012.3108
23. Rabaey, J., *Digital Integrated Circuits, a Design Perspective* (Prentice Hall, Upper Saddle River, NJ), 1996.
24. Pavlov, Andrei and Manoj Sachdev, *CMOS SRAM Circuit Design and Parametric Test in Nano-Scaled Technologies, Process Aware SRAM Design and Test* (Springer, Cham, Switzerland), 2008.
25. Adiseshaiah, M., D. sharathBabuRao and V. Venkateswara Reddy, "Implementation and design of 6T SRAM with read and write assist circuits", *International Journal of Research in Engineering and Applied Sciences*, 2, no. 5 (May 2012), 2249–3905.
26. Kumar, Ravindra and Dr. Gurjit Kaur, "A novel approach to design of 6T (8 × 8) SRAM cell low power dissipation using MCML technique on 45NM", *International Journal of Engineering Research and applications (IJERA)*, 2, no. 4, (July–August 2012), 093–097.
27. Chuang, C.-T., S. Mukhopadhyay, J.-J. Kim, K. Kim and R. Rao, "High performance SRAM in nanoscale CMOS: Design challenges and techniques," *IEEE*

International Workshop on Memory Technology, Design and Testing, Taipei, Taiwan, 3–5 December 2007.
28. Yamauchi, H., “A discussion on SRAM circuit design trend in deeper nanometer-scale technologies”, *IEEE Transactions on Very Large Scale Integration Systems*, 18, no. 5 (May 2010), 763–774.
29. Saun, Shikha, and Hemant Kumar, “Design and performance analysis of 6T SRAM cell on different CMOS technologies with stability characterization”, *IOP Conference Series: Materials Science and Engineering*, 2019.
30. Banga, H. and D. Agarwal, “Single bit-line 10T SRAM cell for low power and high SNM”, *International Conference on Recent Innovation in Signal Processing and Embedded Systems*, pp. 433–438, 2017.

9 Nanotechnology in the Agriculture Industry

Sujit Kumar, and N D Spandanagowda
Department of Electrical and Electronics Engineering, Jain (Deemed-To-Be-University), Bengaluru, Karnataka, India

Ritesh Tirole
Department of Electrical Engineering, Sir Padampat Singhania University, Udaipur, Rajasthan, India

Vikramaditya Dave
Department of Electrical Engineering, College of Technology and Engineering, Udaipur, Rajasthan, India

9.1 INTRODUCTION

Nanotechnology is used in our culture and influences society as a whole. In September 2003, the U.S. Department of Agriculture released their roadmap [1]. The speed of this research has significantly increased in the last decade. That speed is most extensive in the food and agriculture industries, such as agriculture, water/purification and packaging, animal care and feeding, and food processing [2–6]. There is a lot of money to be made in the food and beverage business, with \$8 trillion (about USD 8 trillion) spent annually on agriculture, \$1 trillion on food production and \$1 trillion (approximately USD 1.4 trillion) on preparation and delivery, and another \$4 trillion (about USD 4 billion) spent on marketing and distribution [7]. The most current projections forecast nanotechnology's global economic effect will be at least \$3 trillion and perhaps as many as 6 million jobs by 2020 [8].

This content highly appeals to food producers seeking to produce innovative goods, food characteristics, protection and flavor enhancements. Over the last decade, hundreds of goods have already been developed. Most of these items are used in the food sector, but consumers never eat them, such as wrapping materials and utensils. The only nanomaterials that have been inserted into food items that have been consumed are titanium dioxide and iron oxide. Policy and regulations for nanomaterials are only in their infancy, since there are very few cases that are comprehensively covered by code and case-specific procedures [9–11].

Deeper reasons for restricted control include novel nanomaterials' lack of toxicity and risk [5, 12–14]. Nothing is known about toxicity in animals in vivo

DOI: 10.1201/9781003220350-9

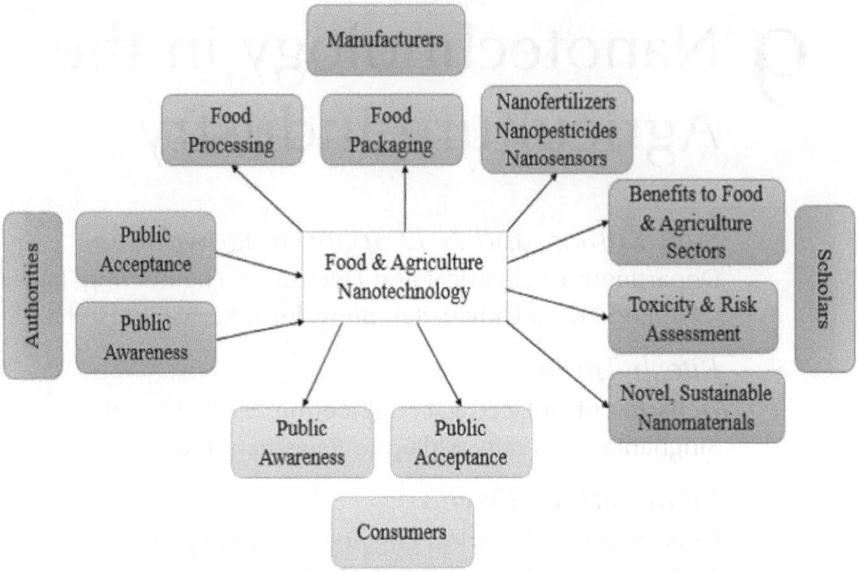

FIGURE 9.1 Advancement of food nanotechnology [71].

(animal research), although various studies have explored the toxicity of nanomaterials in vitro [15]. Many holes must be filled: nanoparticle (NP) toxicity to mammal cells, tissues and organs; movement of NPs to food; breakdown in the environment; the process of bioaccumulation.

Publicity and public recognition are often overlooked by scholars and government agencies [formal 16, 17]. Ultimately, consumers decide whether or not nanotechnology should be used. Applied nanomaterials have been used extensively in the agricultural sector, from food production, packaging, refining and transportation. Regardless of how many individuals want to consume these novel items, the pollution would negatively affect the climate, vegetation and wildlife.

A significant amount of knowledge and commitment is needed to resolve food nanotoxicity, as seen in Figure 9.1.

In each segment, reports and/reviews are available in great abundance. The chapter gives an overview of the latest risk management advancements and nanomaterials for policy and public awareness. Under these conditions, biosynthesized nanomaterials (or "green synthesized nanomaterials") can present a novel way to smear innovative nanomaterials in foodstuff manufacturing through comparatively insignificant influence.

9.2 RECENT EMINENCE ON FOODSTUFF AND FARMING

Nanotechnology focuses on products ranging from 1 nm to 100 nm. The use of nanomaterials like silver NPs is widely processed because all NPs provide an antimicrobial function, while gold is often researched for sensors. "Titanium dioxide" NPs are extensively examined for their utility as a disinfectant, coloring and sweetener. Non-food-nip commodity NPs are commonly applied in food production. For instance, shown in Figure 9.1, numerous nanomaterials provided a promising ability in any field of foodstuff manufacturing, from farming to dishes, though a great deal of knowledge remains undisclosed. The recent change is addressed in the following sections, utilizing delegates from the last five years.

9.2.1 Existing Standing on Foodstuff Nanotechnology

Food nanotechnology has permeated several facets of consumer goods, including food processing, flavors and foodstuff protection. Adoption of this original expertise has improved foodstuff production and packing in terms of maintaining foodstuff protection. Many typical chemicals used as food additives or wrapping components were discovered to occur in part at the nanometer scale. Foodstuff rating "TiO2 NPs", for example, was detected up to roughly 40% in the small variety [18,19]. While materials such as "TiO2 NPs" are commonly regarded as nonhazardous at ambient temperatures, long-term acquaintance to such materials can result in contrary effects [20].

Usage of original foodstuff technology and the existence of microscopic scale substances have piqued the populace's imagination in terms of possible threats. In this portion, we carefully discuss recent progress in the implementation of food nanotechnology. Table 9.1 lists several of the nanomaterials used in food items.

Any authorizations issued by these two agencies, shown in Table 9.1, are focused on the peril evaluation of a substance's standard particle size; thus, designed NPs need approval on a circumstance-by-circumstance basis by experts.

9.2.1.1 Foodstuff Dispensation

When used as pigment or essence extracts or as foodstuff preservatives, nanomaterials are better than micro encapsulation or micro emulsion. Planned nanomaterial agreement has unique compensations for food or dispensation purposes. Additionally, the US FDA allows the use of inorganic oxide chemicals, such as SiO2 (anti-caking agent), MgO (food additive), and TiO (food additive) (see Table 9.1). Additionally, TiO2 is an additive used in white sauces, candy, icing and pudding. All current authorizations of Table 9.1 for foodstuff dispensation are based on conformist element size, excluding carbon (authorized by EC 10/2011 but previously unreachable by the FDA). Nevertheless, it is commonly found in food at submicroscales.

Other than food, animal feed also ensures the production of various agricultural and animal products and food products worldwide. Copper, iron, zinc and zinc oxide have been determined safe as a dietetic increment by the FDA.

9.2.1.2 Foodstuff Packing

Food contact materials come into direct contact with food products during the engineering, conveyance and storage processes. Nanotechnology has been extensively considered and established as a novel solution for foodstuff packing in foodstuff production [33]. When compared to traditional packaging materials, nanomaterials aimed at foodstuff packing have numerous benefits. Because of its

TABLE 9.1
Existing Position of Nanotechnology – Permitted Foodstuff Goods [21]

Segment	Use	Nano-Materials	Existing Status	Highlights	References
Foodstuff dispensation	Color additives	TiO2	Exempt from accreditation	Less than 1% by food weight	[21]
		"Synthetic iron oxide"	Exempted from accreditation	<0.25% (for canines and kittens) and 0.1 (for human) % by weight of the finished food	[21,22]
	Preservatives	Silver-silica	Food Contact Substance Inventory	"silver as an antimicrobial agent merged into polymers"	[23]
	Essence hauler	"Silicon dioxide"	Sanctioned by "EC"	Less than 10,000 mg/kg, without foodstuffs for infants and young children	[24]
	Marking fruit and vegetables	Silicon dioxide	Exempt from certification	<2% of the ink solids	[21]
	Nutritional dietary supplement	Copper oxide		Permitted for animal feedstuff	[25]
Foodstuff exchange packing	Pesticides recognition	ZnO2 QDs	R&D		[26]
	Pathogens detection	Magnetic nanosensors	R&D		[27,28]
	Toxins detection	Fluorescent nanosensors	R&D		[29]
	Edible film/coating	Nanoemulsion with lemongrass essential oil	R&D	Verified on new cut "Fuji apples"	[30]
		Bentonite (Al2O34SiO2nH2O)	Generally recognized as safe	U.S. FDA	[31]
	Flare obstruction spices, etc. Avoid abrasive wear	Montmorillonite	Food Contact Substance Inventory	FCN No. 1163.	[23]
	Montmorillonite Chromium (III) oxide	Toyo Seikan Kaisha Limited and Nanocor Incorporated		FCN No. 932.	[32]
	Prevent abrasive wear Heating enhancer in polyethylene terephthalate (PET) polymers	Titanium aluminum nitride	Normally accepted as harmless	Supreme width of the external covering shouldn't surpass 5 mm.	[31]

TABLE 9.2
Existing Position of Technologies – Enabled Food Products [71]

Use	Viable Terms	Existing Position	Structures	References
Nano-fertilizer	Ag-Nano	In use	Unknown nanomaterials	[a]Note
	NanoPro™, NanoRise™, Nan-Gro™, Nan-Phos™, Nan-K™, Nan-Pack™, Nan-Stress™, Nan-Zn™	In use	Unknown nanomaterials	[a]Note
	NovaLand-Nano	In use	Micro-elements NPs	[a]Note
Nanopesticides	–	Research & Development	"Cu(OH)2 NPs"	[34]
	NANOCU®	In use	Copper NPs	[a]Note
Nano-herbicides	N/A	Research & Development	"Poly (epsilon-caprolactone) (PCL)" nanocapsules	[35]
Nano-sensors	–	Research & Development	"Copper doped Montmorillonite"	[36]
	–	Research & Development	Graphene	[37]

[a] Material may be found online

mechanical, thermal and barrier properties, as well as its low cost, "nanoclay" is one of the utmost extensively utilized and researched novel nanomaterials for foodstuff wrapping (shown in Table 9.2).

They interact to influence new foodstuff exhalation in addition to infectious activity in the goods. Reedy coating of fusion nano-edible structures (typically 100 μm [38]) is used as a vapor and dampness buffer to strengthen mechanical belongings and sensual senses, avoid bacterial decay and extend the shelf lifespan of new foodstuff items. In contrast to the metal-containing nanomaterials described in Section 9.2.1, the majority of comestible coverings are built on biological compounds derived from natural extracts (shown in Table 9.2). "Pectin" from apples [32], "lemongrass" vital lubricant [39] and "quinoa protein/chitosan" [40], for example, were developed as palatable coatings to extend the shelf lifespan of unpreserved goods.

9.3 RECENT EMINENCE ON FARMING NANOTECHNOLOGY

Food development through nanotechnology is increased for similar or even better nutritional value. Using a good fertilizer, pesticide, herbicide and plant supplement is the best way to increase crop yields – nanocarriers' use for skillful relief of insecticides, herbicides and vegetable development controllers. For instance, poly ("epsilon-caprolactone") is currently being used as an herbicide with atrazine. The treatment of "Brassica juice" with poly ("epsilon-caprolactone") caps showed a dramatic decrease in net photosynthesis and development relative to industrial atrazine. Still, it significantly increased the herbicidal activity of the plants. Other NP formulations, such as silica NPs and polymeric NPs, unleash pesticides in a regulated manner. Applying the nanoscale carriers allows for the prolonged release of these organisms.

9.4 TOXIC ESSENTIALS WITH HAZARD VALUATION

9.4.1 Introduction Directions and Relations

Over the last decade, the growing use of "nanotechnology" in foodstuff and farming got the public's attention. Intentional contamination is widespread in food and agriculture materials; whether deliberately or accidentally, NPs are most commonly applied to them [41]. The spread of nanoproducts then increases the public's fear of the atmosphere and human health due to the unusual chemical properties of nanomaterials. As the adverse health effects of nanomaterials, nanopesticides and nanoherbicides are exacerbated, the impact of nanofertilizers and immobilized nanobots cannot be excluded [42]. At a nano level, the nanomaterials' action and destiny are extremely reliant on the chemical and physical belongings of the system [43]. The wide range of environmental conditions under which nanomaterials may be expected to operate restricts their predictability in their actions and destiny. The distribution of NPs is complex as a consequence of bio-nano interactions [44].

At the nano-bio-ecosystem level, systemic approaches have been endorsed, although case-specific experiments are needed to prove the environment's effects on NPs' impact. Although "aluminum NPs" in consumption serve as a sample to show difficulty, trial-run-to-failure assessment of nano substituents can affect digestion (shown in Figure 9.2 [45]).

Four parts of the digestion scheme show how Al particles agglomerate. No measurable chemical changes take place in artificial saliva at the outset of digestion. Aluminum NPs start to melt and release small amounts of aluminum ions in the fluid of the stomach. There can be particle agglomeration while a reaction is happening in the intestine, although it is more severe in the fluid phase.

9.4.2 Nano-Toxicity Mechanisms

It is a big issue for human health and safety whether or not food and other agricultural items come into contact with nanomaterials as ingredients. Direct food additives that use nanomaterials can have health risks. Reactive oxygen species (ROS) are poisonous due to their production [46]. Increased levels of ROS can cause autophagy, neuronal injury, cancer and aging-related diseases in humans. Furthermore, metal allergy may occur with exposure to food nanocompounds.

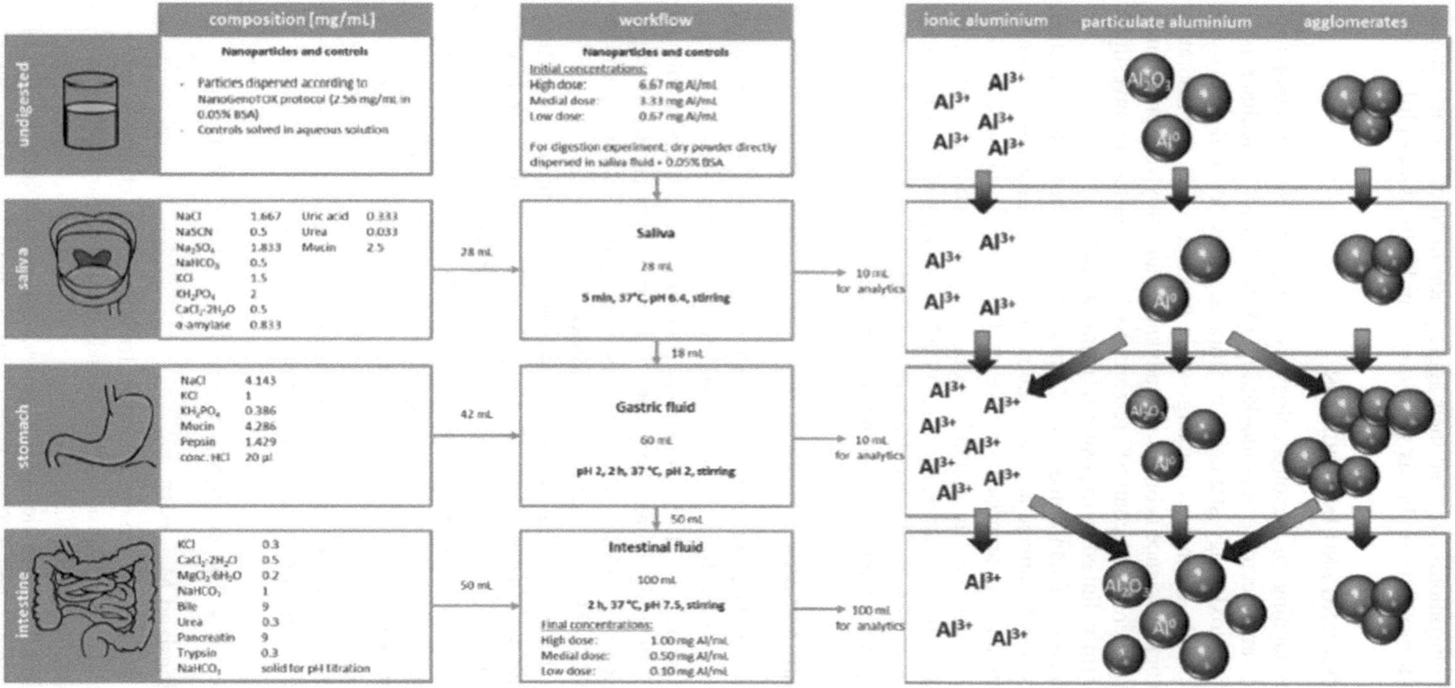

FIGURE 9.2 Destiny of Al particles in the digestive scheme serves as a clear indication of the level of sophistication and experimental procedure needed to evaluate potential nanotoxicity [71].

The amount of nanometalloids [46] and their aggregation in edible tissue (human) sections could result in severe problems with long-term health issues [46].

9.4.3 Data Producing and Investigation

An in-depth understanding of nanotoxicology is critical to secure nanomanufacturing, handling and accurate nanomarketing of food and agriculture [47]. Furthermore, existing toxicology strategies traditionally provide no data that can help develop sustainable techniques for large-scale production. In the laboratory and people, toxicological science is only so far advanced, and only very preliminary evidence can be gathered. In contrast, researchers have started to investigate whether cellular harm is linked to exposure to nanomaterials.

Toward complication of nano-biochemical connections, it has been explained that ROS synthesis, DNA damage and immune responses are easier to predict for conventional in vitro and situ approaches; however, the cost is far higher and not as expected [48]. It is a common understanding that the use of cellular models, including Escherichia coli [49] and human "A549 adenocarcinoma" [50] for omics data generation, would become the standard in the field of nanotoxicity. Machines must adjust to cope with more and delve through increasing amounts of data.

9.5 FRONTLINE AREAS

Much earlier research indicated that nanomaterials pose little to no threat to people and may be included in the food supply; nevertheless, toxicities may be changed due to continuous exposure [51]. We are unsure concerning the bioavailability and biodistribution of NPs in the human body, but taking them conservatively is prudent. Over the past year, a global wave has arisen of citizens in France who are reconsidering TiO2 (E171) as a natural food additive.

Along with advances in general knowledge and government policy, some other important considerations prevent nanotechnology from flourishing in the food industry. People are debating whether the possible dangers of conventional NPs have been thoroughly researched. Risk management is long overdue. Various strategies have been employed to minimize the toxicity of designed nanomaterials, while at the same time improving their choice and conducting their tasks more accurately and reliably. For example, uniform tailoring and shape formation are useful ways to help these nanotubes last longer and be less harmful. Use of bio-synthesizers or "green synthesized" materials is reflected for imminent usage.

9.5.1 Viewpoints for "Biosynthesized and Bio-Inspired" Materials

9.5.1.1 Biosynthesized Materials

In the quest for a safe and environmentally friendly green chemical process, "biosynthesis" became a sizzling subject for the scheme and construction of several nanomaterials (shown in Figure 9.3).

As seen in Table 9.3, several biological systems, such as microbes, "fungi, yeast, actinomycetes, enzymes," and different herbal sections, have revealed encouraging

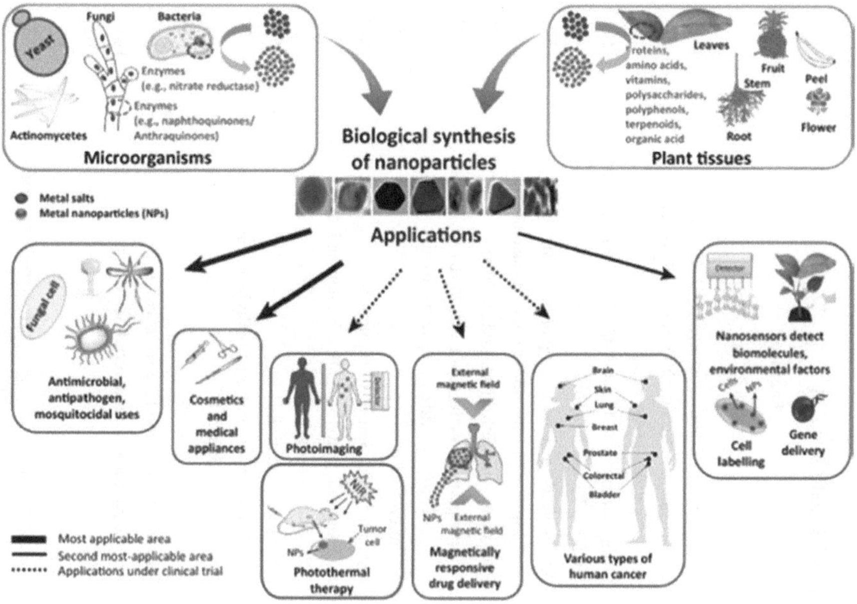

FIGURE 9.3 Applications of nanoparticles and biological synthesis [71].

appropriateness for biosynthetic nanomaterials. Many scholars have identified three significant benefits, which are as follows:

1. The organic structure as an industrial mass may serve as a covering, steadying and decreasing agent, resulting in fewer dangerous materials throughout the manufacturing procedures.
2. Biosynthetic frequently receipts in atmospheric temperature and pressure, as well as neutral pH, can further minimize the usage of energy assets and perilous substances.
3. Most biosynthetic materials are biocompatible with little poison.

In this section, we will include a summary of the three facets for biosynthetic usage in developing and manufacturing materials.

In He et al. [52], it has been demonstrated that "SDS-PAGE" outlines of extracellular proteins are accountable for the stabilization of Ag NPs. Furthermore, capping agents as a function of biomolecules have been discovered through numerous studies in nanomaterials' synthesis (shown in Table 9.3). Again, typical conditions for biological systems include ambient temperature and strain, as well as neutral pH.

9.5.1.2 Nanomaterials Approached for Bio Inspiration

Biological approaches and biosynthesis are often conflated [64] mismanagement of a bio-enthused method [65]. The inherent essence of a bio-inspired way has been commonly implemented in biomedical studies and other disciplines (Figure 9.4) via biological processes.

TABLE 9.3
Illustrations of Selected Biosynthesized Nanomaterials in Current 5 years (2015–2020) [71]

	Biological System	Biogenic Nanoparticles (NPs)	Features	Reference
Microorganisms	"Pichia fermentans JA2"	Ag and Zn oxide NPs	Silver NPs inhibited most of the G^- Clinical pathogens; zinc oxide NPs reserved solitary Pseudomonas aeruginosa.	[52]
	"Bacillus cereus strain"	Magnetic Fe(OH) NPs	Little cytotoxicity: IC50, MCF-7 > 5 mg/ml and IC50, 3T3 > 7.5 mg/ml	[52]
	Serratia sp. BHU - S4	Silver NPs	As fungicide against phytopathogen Bipolaris sorokiniana causing spot blotch disease in wheat	[52]
Fungi	Saccharomyces cerevisiae	Silver NPs	Photo catalytic filth of methylene blue	[53]
	Aspergillus flavus TFR 7	TiO2 NPs	Arouse plant growth: shoot length (þ17%), root length (Promote rhizospheric microbes	[54]
Yeast	Candida lusitaniae	Ag/AgCl NPs	Antimicrobial activity	[55]
	Magnusiomyces ingens LH-F1	Au NPs	Catalytic reduction of nitrophenols	[56]
Actinomycetes	Isolate VITBN4	CuO NPs	Antibacterial activity against human and fish bacterial pathogens	[57]
	Streptomyces sp. strain NH21	Silver and gold NPs	Antibacterial activity	[58]
Enzymes	α-amylase	TiO2 NPs	"MIC of 62.50 mg/ml on Staphylococcus aureus and Escherichia coli"	[59]
Plant extracts	"Red ginseng root"	Ag and Au NPs	Antimicrobial activity	[60]
	Aloe vera plant	Nanoscale zero-valent iron	Eliminating (As) and Se from H2O	[61]
	Leaf of Cassia Tora	Ag NPs	Antioxidant and antibacterial activities	[61]
	Pineapples and oranges fruits	Silver NPs	N/A	[62]
Marine algae	Macroalga Sargassum muticum	ZnO NPs	N/A	[63]
	Brown alga Cystoseira trinodis	CuO NPs	Catalytic, antioxidant and antibacterial properties	[63]

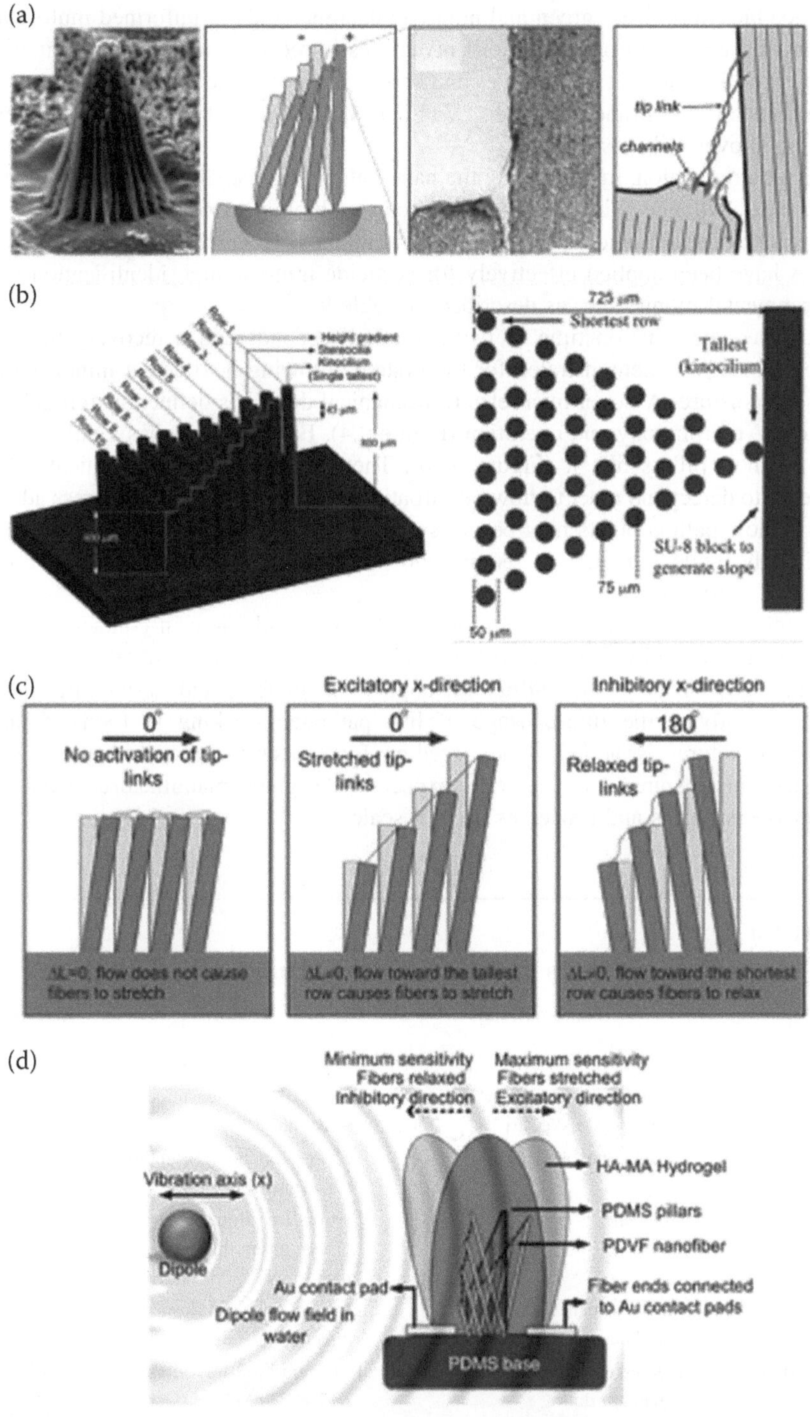

FIGURE 9.4 a) Morphology of real sensors and the mathematical model of how hair bundles function, b) cells resembling hair, c) various designs for pillars, d) nanofiber sensors [71].

Once biosynthesized, green and nontoxic nanomaterials in unformed multiscale systems have since informed various inventions in recent human history that contain gecko-like adhesives, false eyes and compound eyes. The graded bond method exploits nanoscale strands to create "Van der Waals," and adhesion and detaching developed over millions of years.

At high resolution, manipulating the nanomaterials is a challenge. It has become a complex challenge to imitate naturally occurring biological processes. The field of biomimetics in food and agriculture is in its infancy. However, bio-inspired techniques have been applied effectively for pesticide transmission, identification and environmental monitoring, as described in Table 9.4.

An adequate reaction time of two-and-a-half seconds and a recovery time of three seconds was demonstrated for the system, allowing it to detect minor variations in moisture. A novel microelectromechanical device is defined as mimicking hair cells' morphology and operation (Figure 9.4). Each package has a taller pillar and 54 short pillars like it (Figure 9.4b). These transducers are specifically engineered to detect and react to flow perturbations (Figure 9.4c). Tensile stress added to both flow-path contacts may be measured and corrected (Figure 9.4d). It is exceptionally biodegradable and requires no power source, exhibiting strong flow/direction altering sensitivity.

Graphene was fabricated onto bioresorbable silk and came in contact with an electromagnetic coil on the top (Figure 9.5).

Although it has no straightforward application in food and agriculture, these findings motivate the future usage of live pathogen tracking in foodstuff and farming structures. Nonetheless, advanced bio functions remain to be identified, and advanced bio-inspired nanomaterials expect to designate manufactured from inexpensive practices and processes on vast scales.

TABLE 9.4
Bio-Inspired Nanocatalytes in Grain and Agricultural Industries [71]

Bio-inspired Template	Nanomaterials	Features	Reference
Mussel	Mussel avermectin NPs [P (SteMAA)eAveCat]	Advanced poisonousness in the direction of "aphids"	[66]
Cactus	ZnSnO3 nanostructures	High-performance Humidity nanosensors	[67]
Silk	Graphene	Sensors free from battery for distant observing of pathogens	[68]
Biological cilia	Polyvinylidenefluoride piezoelectric nanofiber	Flow velocity and flow direction	[69]
Zwitterion	Fluorescent biomimetic carbon quantum dots	Detection limit for "vitamin B12 at 81 nM"	[70]

B

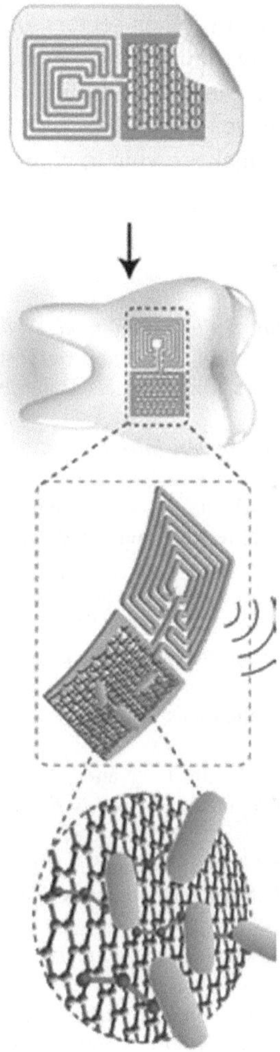

FIGURE 9.5 Wireless pathogen attracted to teeth that is dependent on silk [71].

9.6 CONCLUSION

Nanotechnology has a solid potential to be commonly used in the food industry. This is based on data that came from laboratories. Until they can be controlled and tested at the nanoscale, it is unknown to what extent products using nanotechnology can benefit or damage the ecosystem. This slows down the advancement of regulations and laws that block the use of new technology. The public has been slow to embrace food nanotechnology, but attitudes toward it can be tuned. The battle appears to be what the community needs to know about foodstuff nanotechnology (particularly new goods), but foodstuff producers want the reverse. For both consumers and food producers, adequate details should be documented, and logistic help is often essential.

Expecting bio-friendly practices has become essential for biotechnological endeavors; such procedures are becoming common in many other sectors of the economy. However, compared to biomedical fieldwork, research and production in food and agriculture, bio-engineered NPs are very conservative for a billion-dollar agricultural sector. A lot of new nanotechnologies have found their way into packaged food goods. This would be a battlefield for the maker owing to regulatory oversight of protection codes.

REFERENCES

1. US DOA. *Nanoscale science and engineering for agriculture and food systems: A report submitted to cooperative state research, education and extension service.* Washington, DC: Department of Agriculture; 2003. The United States Department of Agriculture: National Planning Workshop.
2. Dasgupta N., and Ranjan S. *Nanotechnology in food sector. An introduction to food grade nanoemulsions.* Springer, 1–18, 2018.
3. Peters R.J., Bouwmeester H., Gottardo S., Amenta V., Arena M., and Brandhoff P. "Nanomaterials for products and application in agriculture, feed and food." *Trends Food. Sci. Technol.* 54 (2016): 155–164.
4. Finglas P.M., Yada R.Y., and Toldr A.F. "Nanotechnology in foods: Science behind and future perspectives." *Trends. Food. Sci. Technol.* 40 (2014): 125–126.
5. Bryksa B.C., and Yada R.Y. "Challenges in food nanoscale science and technology." *J. Food. Drug. Anal.* 20 (2012): 418–421.
6. Sozer N., and Kokini J.L. "Nanotechnology and its applications in the food sector." *Trends. Biotechnol.* 27 (2009): 82e9.
7. Cushen M., Kerry J., Morris M., Cruz-Romero M., and Cummins E. "Nanotechnologies in the food industry: Recent developments, risks and regulation." *Trends Food. Sci. Technol.* 24 (2012): 30–46.
8. Roco M.C., Mirkin C.A., and Hersam M.C. *Nanotechnology research directions for societal needs in 2020: Retrospective and outlook.* Springer Science & Business Media, 2020.
9. Kavitha P., Manjunath M., and Huey-Min H. *Nanotechnology applications for environmental industry.* Elsevier, 2018.
10. Xiaojia H., Hua D., AW G., and Huey-Min H. *Regulation and safety of nanotechnology in the food and agriculture industry.* CRC Press: Taylor & Francis Group; 2018, 2018.
11. Marrani D. "Nanotechnologies and novel foods in European law." *NanoEthics* 7 (2013): 177–188.
12. Dasari T., Deng H., McShan D., and Yu H. *Nanosilver-based antibacterial agents for food safety.* Science Publishers, 2014.
13. Deng H., Zhang Y., and Yu H. "Nanoparticles considered as mixtures for toxicological research." *J. Environ. Sci. Health. C. Environ. Carcinog. Ecotoxicol. Rev.* 36 (2018): 1–20.
14. Senjen R. "Nanotechnology and patents – How can potential risks be assessed?" *Recent. Pat. Food, Nutr. Agric.* 4 (2012): 245–249.
15. Duncan T.V. "Applications of nanotechnology in food packaging and food safety: Barrier materials, antimicrobials and sensors." *J. Colloid Interface Sci.* 363 (2011): 1–24.
16. Cormick C. "Why do we need to know what the public thinks about nanotechnology?" *NanoEthics* 3 (2009): 167–173.

17. Arnaldi S., and Muratorio A. "Nanotechnology, uncertainty and regulation. A guest editorial." *NanoEthics* 7 (2013): 173–175.
18. Dudefoi W., Terrisse H., Richard-Plouet M., Gautron E., Popa F., and Humbert B. "Criteria to define a more relevant reference sample of titanium dioxide in the context of food: A multiscale approach." *Food Addit. Contam. A* 34 (2017): 653–665.
19. Weir A., Westerhoff P., Fabricius L., Hristovski K., and Von Goetz N. "Titanium dioxide nanoparticles in food and personal care products." *Environ. Sci. Technol.* 46 (2012): 2242–2250.
20. Dorier M., Beal D., Marie-Desvergne C., Dubosson M., Barreau F., and Houdeau E. "Continuous in vitro exposure of intestinal epithelial cells to E171 food additive causes oxidative stress, inducing oxidation of DNA bases but no endoplasmic reticulum stress." *Nanotoxicology* 11 (2017): 751–761.
21. Code of Federal Regulations (CFR). *Electronic code of federal regulations. Title 21: Food and drugs. PART 73 – LISTING OF COLOR ADDITIVES EXEMPT FROM CERTIFICATION.* The United States office of the federal register (OFR) and the United States. Government Publishing Office, 2018.
22. U.S. FDA. *Color additive status list.* United States Food & Drug Administration, 2018.
23. U.S. FDA., 2018. Inventory of effective food contact substance (FCS) notifications, https://www.fda.gov/food/packaging-food-contact-substances-fcs/inventory-effective-food-contactsubstance-fcs-notifications
24. European Commission. *Regulation (EC) No. 1333/2008 of the European Parliament and of the Council of 16 December 2008 on Food Additives.* The European Parliament and the Council of the European Union, 2008.
25. U.S. FDA. *Food additive status list.* US FDA/CFSAN Office of Food Additive Safety, 2018.
26. Sun A., Chai J., Xiao T., Shi X., Li X., and Zhao Q. "Development of a selective fluorescence nanosensor based on molecularly imprinted-quantum dot optosensing materials for saxitoxin detection in shellfish samples." *Sens. Actuators, B* 258 (2018): 408–414.
27. Shi S., Wang W., Liu L., Wu S., Wei Y., and Li W. "Effect of on the physicochemical characteristics of longan fruit under ambient temperature." *J. Food. Eng.* 118 (2013): 125–131.
28. Code of Federal Regulations (CFR). *Electronic code of federal regulations. Title 21: Food and drugs. Part 184-direct food substances affirmed as generally recognized as safe. Subpart b-listing of specific substances affirmed as GRAS.* The United States office of the federal register (OFR) and the United States. Government Publishing Office, 2018.
29. Gorrasi G., and Bugatti V. "Edible bio-nano-hybrid coatings for food protection based on pectins and LDH-salicylate: Preparation and analysis of physical properties." *LWT-Food Sci. Technol.* 69 (2016): 139–145.
30. Pereda M., Marcovich N.E., and Ansorena M.R. *Nanotechnology in food packaging applications: Barrier materials, antimicrobial agents, sensors, and safety assessment.* Springer, 2018.
31. Zhao L., Ortiz C., Adeleye A.S., Hu Q., Zhou H., and Huang Y. "Metabolomics to detect response of lettuce (Lactuca sativa) to Cu(OH)2 nanopesticides: Oxidative stress response and detoxification mechanisms." *Environ. Sci. Technol.* 50 (2016): 9697–9707.
32. Oliveira H.C., Stolf-Moreira R., Martinez C.B.R., Grillo R., de Jesus M.B., and Fraceto L.F. "Nanoencapsulation enhances the post-emergence herbicidal activity of atrazine against mustard plants." *PLoS One* 10 (2015): 013297–81.

33. Abbacia A., Azzouz N., and Bouznit Y. "A new copper doped montmorillonite modified carbon paste electrode for propineb detection." *Appl. Clay Sci.* 90 (2014): 130–134.
34. Khalifa N.S., and Hasaneen M.N. "The effect of chitosanePMAAeNPK nanofertilizer on Pisum sativum plants." *Biotech* 8 (2018): 193–205.
35. Flores-Lopez M.L., Cerqueira M.A., de Rodrı´guez D.J., and Vicente A.A. "Perspectives on utilization of edible coatings and nano-laminate coatings for extension of postharvest storage of fruits and vegetables." *Food. Eng. Rev.* 8 (2016): 292–305.
36. Salvia-Trujillo L., Rojas-Grau¨ M.A., Soliva-Fortuny R., and Martı´n- Belloso O. "Use of antimicrobial nanoemulsions as edible coatings: Impact on safety and quality attributes of fresh cut Fuji apples." *Postharvest Biol. Technol.* 105 (2015): 8–16.
37. Robledo N., Lopez L., Bunger A., Tapia C., and Abugoch L. "Effects of antimicrobial edible coating of thymol nanoemulsion/quinoa protein/chitosan on the safety, sensorial properties, and quality of refrigerated strawberries (Fragaria ananassa) under commercial storage environment." *Food Bioprocess Technol.* 11 (2018): 1566–1574.
38. Perc in I., Idil N., Bakhshpour M., Yılmaz E., Mattiasson B., and Denizli A. "Microcontact imprinted plasmonic nanosensors: Powerful tools in the detection of Salmonella paratyphi." *Sens. Actuators, B* 17 (2017): 1375–1386.
39. Banerjee T., Sulthana S., Shelby T., Heckert B., Jewell J., and Woody K. "Multiparametric magneto-fluorescent nanosensors for the ultrasensitive detection of Escherichia coli O157: H7." *ACS Infect. Dis.* 2 (2016): 667–673.
40. Zhang C.H., Liu L.W., Liang P., Tang L.J., Yu R.Q., and Jiang J.H. "Plasmon coupling enhanced Raman scattering nanobeacon for single-step, ultrasensitive detection of cholera toxin." *Anal. Chem.* 88 (2016): 447–452.
41. Zhang W., Han Y., Chen X., Luo X., Wang J., and Yue T. "Surface molecularly imprinted polymer capped Mn-doped ZnS quantum dots as a phosphorescent nanosensors for detecting patulin in apple juice." *Food Chem.* 232 (2017): 145–154.
42. Kumar S., Kumar D., and Dilbaghi N. "Preparation, characterization, and bio-efficacy evaluation of controlled release carbendazim-loaded polymeric nanoparticles." *Environ. Sci. Pollut. Res.* 24 (2017): 926–937.
43. Duhan J.S., Kumar R., Kumar N., Kaur P., Nehra K., and Duhan S. "Nanotechnology: The new perspective in precision agriculture." *Biotechnol. Rep.* 15 (2017): 11–23.
44. Dimkpa C.O., McLean J.E., Britt D.W., and Anderson A.J. "Antifungal activity of ZnO nanoparticles and their interactive effect with a biocontrol bacterium on growth antagonism of the plant pathogen Fusarium graminearum." *Biometals* 26 (2013): 913–924.
45. Rajiv P., Rajeshwari S., and Venckatesh R. "Bio-Fabrication of zinc oxide nanoparticles using leaf extract of Parthenium hysterophorus L. and its size-dependent antifungal activity against plant fungal pathogens." *Spectrochim. Acta A* 112 (2013): 384–387.
46. Lin Y.W., Huang C.C., and Chang H.T. "Gold nanoparticle probes for the detection of mercury, lead and copper ions." *Analyst* 136 (2011): 863–871.
47. Jokar M., Safaralizadeh M.H., Hadizadeh F., Rahmani F., and Kalani M.R. "Design and evaluation of an apta-nano-sensor to detect Acetamiprid in vitro and in silico." *J. Biomol. Struct. Dyn.* 34 (2016): 2505–2517.
48. Graham J.H., Johnson E.G., Myers M.E., Young M., Rajasekaran P., and Das S. "Potential of nano-formulated zinc oxide for control of citrus canker on grapefruit trees." *Plant Dis.* 100 (2016): 2442–2447.
49. Hannon J.C., Kerry J.P., Cruz-Romero M., Azlin-Hasim S., Morris M., and Cummins E. "Assessment of the migration potential of nanosilver from nanoparticle-coated lowdensity polyethylene food packaging into food simulants." *Food Addit. Contam. A* 33 (2016): 167–178.

50. Hwang H.M., Ray P.C., Yu H., and He X. *Toxicology of designer/engineered metallic nanoparticles*. Cambridge, Royal Society of Chemistry, 2012.
51. He X., Aker W.G., Leszczynski J., and Hwang H.-M. "Using a holistic approach to assess the impact of engineered nanomaterials inducing toxicity in aquatic systems." *J. Food Drug Anal.* 22 (2014): 128–146.
52. He X., Aker W.G., Fu P.P., and Hwang H.-M. "Toxicity of engineered metal oxide nanomaterials mediated by nanoebioeecoe interactions: A review and perspective." *Environ. Sci.: Nano.* 2 (2015): 564–582.
53. He X., Aker W.G., Huang M.-J., Watts D.J., and Hwang H.-M. "Metal oxide nanomaterials in nanomedicine: Applications in photodynamic therapy and potential toxicity." *Curr. Top. Med. Chem.* 15 (2015): 1887–1900.
54. He X., and Hwang H.-M. "Nanotechnology in food science: Functionality, applicability, and safety assessment." *J. Food. Drug. Anal.* 24 (2016): 671–681.
55. He X., Fu P., Aker W.G., and Hwang H.-M. "Toxicity of engineered nanomaterials mediated by nano-bio-eco interactions." *J. Environ. Sci. Health. C Environ. Carcinog. Ecotoxicol. Rev.* 36 (2018): 21–42.
56. Deng H., and Yu H. "A mini review on controlling the size of Ag nanoclusters by changing the stabilizer to Ag ratio and by changing DNA sequence." *Adv. Nat. Sci.* 8 (2015): 1–9.
57. McShan D., Zhang Y., Deng H., Ray P.C., and Yu H. "Synergistic antibacterial effect of silver nanoparticles combined with ineffective antibiotics on drug resistant Salmonella typhimurium DT104." *J. Environ. Sci. Health. C* 33 (2015): 369–384.
58. Deng H., McShan D., Zhang Y., Sinha S.S., Arslan Z., and Ray P.C. "Mechanistic study of the synergistic antibacterial activity of combined silver nanoparticles and common antibiotics." *Environ. Sci. Technol.* 50 (2016): 8840–8848.
59. Dasari T., Deng H., McShan D., and Yu H. "Nanosilver-based antibacterial agents for food safety." InRay C. Paresh, *Food poisoning: Outbreaks, bacterial sources and adverse health effects*. Nova Science Publisher, 35–62, 2014.
60. Zhang Y., Dasari T.P.S., Deng H., and Yu H. "Antimicrobial activity of gold nanoparticles and ionic gold." *J. Environ. Sci. Health C Environ. Carcinog. Ecotoxicol. Rev.* 33 (2015): 286–327.
61. Chen X.X., Cheng B., Yang Y.X., Cao A., Liu J.H., and Du L.J. "Characterization and preliminary toxicity assay of nano-titanium dioxide additive in sugar-coated chewing gum." *Small* 9 (2013): 1765–1774.
62. Maertens A., and Plugge H. "Better metrics for "sustainable by design": Toward an in silico green toxicology for green(er) chemistry." *ACS Sustainable Chem. Eng.* 6 (2020): 1999–20010.
63. Gou N., Onnis-Hayden A., and Gu A.Z. "Mechanistic toxicity assessment of nanomaterials by whole-cell-array stress genes expression analysis." *Environ. Sci. Technol.* 44 (2010): 5964–5970.
64. Li X., Zhang C., Bian Q., Gao N., Zhang X., and Meng Q. "Integrative functional transcriptomic analyses implicate specific molecular pathways in pulmonary toxicity from exposure to aluminum oxide nanoparticles." *Nanotoxicology* 10 (2016): 957–969.
65. Liu R., Zhang H., and Lal R. "Effects of stabilized nanoparticles of copper, zinc, manganese, and iron oxides in low concentrations on lettuce (Lactuca sativa) seed germination: Nanotoxicants or nanonutrients?" *Water, Air Soil Pollut.* 227 (2016): 42–61.
66. Zhang Y., Leu Y.-R., Aitken R.J., and Riediker M. "Inventory of engineered nanoparticle-containing consumer products available in the Singapore retail market and likelihood of release into the aquatic environment." *Int. J. Environ. Res. Publ. Health.* 12 (2015): 8717–8743.

67. Quadros M.E., Pierson R., Tulve N.S., Willis R., Rogers K., and Thomas T.A. "Release of silver from nanotechnologybased consumer products for children." *Environ. Sci. Technol.* 47 (2013): 8894–8901.
68. Benn T., Cavanagh B., Hristovski K., Posner J.D., and Westerhoff P. "The release of nanosilver from consumer products used in the home." *J. Environ. Qual.* 39 (2010): 1875–1882.
69. Azamat A., and Kunal S. "Risks of nanotechnology in the food industry: A review of current regulation." *Nanotechnol. Percept.* 11 (2015): 27–30.
70. Jain A., Ranjan S., Dasgupta N., and Ramalingam C. "Nanomaterials in food and agriculture: An overview on their safety concerns and regulatory issues." *Crit. Rev. Food Sci. Nutr.* 58 (2018): 297–317.
71. He X., Deng H., and Hwang H. "The current application of nanotechnology in food and agriculture." *J. Food and Drug Analy.* 27 (2019): 1–21.

10 Recent Advancements in the Applications of ZnO

A Versatile Material

Chandra Prakash Gupta, and Amit Kumar Singh
Department of Electronics and Communication Engineering, Manipal University Jaipur, Rajasthan, India

10.1 INTRODUCTION

In this chapter, the historical background along with the properties and the applications of ZnO-based devices are reviewed. A general history of ZnO and the growth of ZnO and related devices are listed chronologically. Further, the material properties of ZnO (structural, physical, electrical, and optical) are summarized. ZnO, an inorganic compound, is obtained from the smelting process. The first use of ZnO in electronics was reported back in the 1920s [1]. Research on ZnO took pace in 1935, when its properties were studied in detail [2]. In one of the first works by Hubert M. James and Vivian A. Johnson, the crystal structure of ZnO was reported [3]. Bell Laboratories, which was a pioneer in the semiconductor transistor, brought about a significant growth in the semiconductor industry in the 1940s [4]. Research in the area of ZnO devices has recently leaped, mainly due to the advancement in fabrication techniques for crystals. However, the *p*-type ZnO fabrication remains a challenge to date, but the development of viable devices based on ZnO does not look very far off.

Recently, metal oxide-based nanostructures have been an extensive choice for sensing and memory applications due to their tunable structural and electrical properties. These nanostructures offer a high surface-to-volume ratio, and the transport mechanism of electrons in them improves the performance of these devices. Further improvement in performance is due to the possible quantum confinement of the metal oxide-based nanostructures. NiO, CuO, TiO_2 and ZnO are some of the metal oxides that have been studied as a candidate for the above-mentioned applications [5], and ZnO has emerged as an effective choice, not only in terms of cost but in performance, too. The other reasons for the success of ZnO are that it is physically and chemically stable, and the conducting electrons possess high mobility. Also, the operating temperature for ZnO is low. ZnO/Si heterojunction-based photodetectors can detect ultraviolet light, which has many potential applications, such as pollution monitoring, UV detection in the ozone layer, water purification, flame detection and space communication, etc. [6, 7].

DOI: 10.1201/9781003220350-10

Many reports are available in literature where the n-ZnO thin film-based heterojunctions with p-type substrates, such as AlGaN, GaN and Si, have been investigated [5], and n-ZnO/p-Si heterojunction has surfaced as a structure of interest because it offers flexibility, is compatible with silicon IC (integrated circuit) fabrication technology [6] and is very cost effective.

In continuation, in the area of technology and semiconductor science, a remarkable development is seen in thin-film metal oxide-based devices. In the semiconductor industry, a revolutionized impact was made by the ZnO nanostructure-based thin films.

In the last decade, because of ZnO's excellent properties as a wide bandgap semiconductor material, it has drawn major research interest from researchers. High mobility of electrons, large exciton binding energy, direct and wide bandgap, high thermal conductivity, good transparency and ease of growing many nanostructures with several low-cost deposition methods make it a potential candidate for a variety of nano-electronic and optoelectronic device applications [8]. Figure 10.1 shows the wide variety applications of ZnO in electronics. ZnO is used in lasers because of an intense near-band-edge excitonic emission, which is produced due to its significant exciton binding energy [9]. Moreover, it is used in solar cell applications and UV detection due to its high optical transparency. In the current review, ZnO is found to be an upcoming material of choice in the fields of spintronics, flat panel displays, thin-film transistors, piezo electronic nanogenerators and gas-sensing applications. In literature [10], there are several reports on p-doped ZnO films that use various elements of group-V as dopants. For example, phosphorus (P), nitrogen (N), arsenic (As) and antimony (Sb), etc. but until now, high quality and stable p-type conductivity of ZnO films has not been attained. To produce high-quality p-type ZnO films, numerous research efforts by various research groups across the globe are taking place. On the other hand, to explore the potentiality of ZnO in many

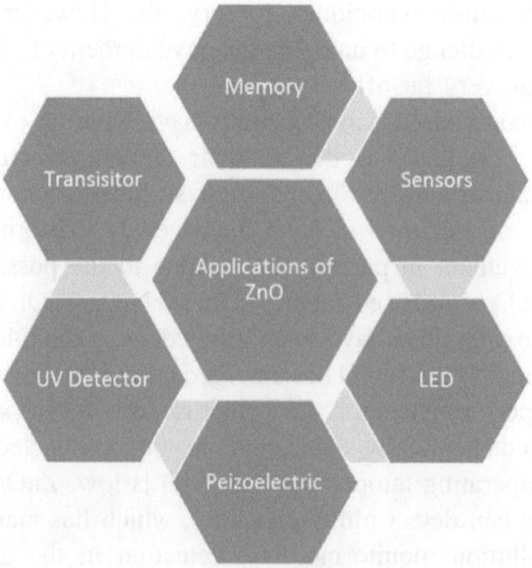

FIGURE 10.1 Different applications of ZnO in electronics.

advanced device applications, some efforts to deposit n-type ZnO over various p-type substrates have been done. Studies are available in the literature where n-ZnO has been deposited over various p-type substrates, namely, CuO, ZnTe, CdTe [11], NiO, AlGaN, GaN, Si and SiC [12].

10.2 MATERIAL PROPERTIES OF ZNO

This section presents the important properties of ZnO. In the next section, applications of ZnO that have originated from its material properties are presented.

10.2.1 Structural Properties

ZnO is from the class of II-VI group compound semiconductors. In ZnO crystals, each anion is enclosed by cations that reside at each corner of a tetrahedron. A substantial ionic character is also exhibited by ZnO apart from the covalent character due to the tetrahedral arrangement [13], which results in the unexpectedly high bandgap of ZnO. Further, ZnO is found in either zinc blende, rock salt or wurtzite crystal forms, and wurtzite is found to be the most thermodynamically stable arrangement under normal atmospheric conditions.

The wurtzite structure is reported to be in a hexagonal unit cell structure in 3D along with its three lattice constants, where a and b are equal, and a and c are in the ratio of $a/c=\sqrt{3/8} = 0.6123$. For the ideal wurtzite structure, the density is equal to 5.6 g cm^{-3}. The lattice parameters for the wurtzite ZnO cell structure generally lies from 5.204 to 5.207 Å for the c- lattice constant and from 3.2475 to 3.2501 Å for the a-lattice constant and a/c ratio ranges between 0.6123 and 0.6277. For the ideal wurtzite structure, the deviance in the value of lattice constants is mainly due to changes in ionicity and lattice stability. The large difference of the electronegativity of the two atoms of ZnO (O = 3.44 and Zn = 1.65) is the cause for the strong ionic bonding between the two atoms [14].

10.2.2 Optical Properties

The direct and large bandgap of ZnO offers advantages of low electronic noise, high breakdown voltage and the ability to sustain high electric fields, temperature and power. The exciton binding energy of ZnO is established to be 60 meV, which is much higher than that of GaN with the binding energy of 25 meV [15, 16]. This high exciton energy is responsible for bright emissions and stability against thermal disassociation of excitons. The excitons are electron-hole pairs held together by Coulomb forces and are the main reason for the emission of light at low temperatures. Young-Sung Kim et al. obtained the optical bandgap of 3.22 eV, and ultraviolet emission was obtained at a green-yellow wavelength of 490–620 nm [17]. C. Gumo et al. conveyed the bandgap energy of ZnO thin films as 3.27 eV; also, a higher transmittance value above 90% was reported [18]. Xue-Yong Li reported a transmittance greater than 86% on their work on aluminium-doped ZnO thin films [19]. Jaehyeong Lee et al. have a resistivity of 9.7×10^{-4} Ωcm and transmittance value greater than 90%

for their work on aluminium-doped ZnO thin films [20]. Optical properties such as reflectance, absorbance, responsivity, quantum efficiency, detectivity, transmittance, dielectric constant, refractive index, cathodoluminescence and photoluminescence can be obtained for ZnO nanostructures. In nanoplates, the higher surface-to-volume ratio of ZnO nanoplates enhances the number of adsorption sites over the ZnO surface which can significantly contribute to improving the optical device sensitivity. Further, the carrier lifetime is enhanced due to the large surface-to-volume ratio of ZnO nanoplates. The shorter transit time and longer lifetime of the charge carriers are jointly accountable for the higher sensitivity of a fabricated optical device under UV illumination [21].

10.2.3 Electrical Properties

For understanding the potential of ZnO thin films for nanoelectronics device applications, several recent studies have been done. H. Nanto et al. have stated a resistivity of 10^{-4} Ωcm and hall mobility of 120 $cm^2/V.sec$ at the maximum conductivity in zinc oxide thin films [22]. Jin Hong Lee observed hall mobility of 15–26 $cm^2/V.sec$ in zinc oxide thin films [23].

The semiconductor devices work either under low electric field or high electric field. In the low electric field, the operation is dependent on the thermal energy, and the devices follow Ohm's law. When the device is operating under a high electric field, the device operation is dependent on the electric field only. Device performance can be measured using V-I characteristics and C-V characteristics. V-I characteristics are used in characterizing solar cells, heterostructures, sensors, memories, etc. The properties, such as the barrier height, ideality factor, threshold voltage, forming a current, switch process, etc., are measured in Özgür et al. [24]. In Table 10.1, the properties of ZnO are summarized.

10.3 FABRICATION TECHNIQUES

With the advancements in technology, many controlled fabrication methods have emerged for semiconductor devices. Many methods exist for controlled deposition, and the choice of method depends on various factors like material for deposition, cost and deposition rate, as shown in Figure 10.2. The main steps for the deposition method are (i) generation of specimen to be deposited, (ii) transportation of the generated specimen to the target and (iii) deposition on the target. Gas or vapor phase deposition methods are most suitable for thin film deposition.

These are categorized as chemical vapor deposition (CVD) and physical vapor deposition (PVD). Vapors of depositing species undergo chemical reaction at the target surface for thin film deposition by CVD, whereas PVD deposits the thin film by condensing the atoms and molecules in the vapor. CVD encompasses the thermal CVD, atomic layer deposition (ALD), metal-organic CVD (MOCVD), plasma-enhanced CVD (PECVD), low-pressure CVD (LPCVD), laser CVD (LCVD) and molecular-beam epitaxy (MBE).

PVD may be carried out as thermal evaporation, sputtering and pulse laser deposition (PLD).

TABLE 10.1
Properties of ZnO

	Physical Properties of Zinc Oxide (ZnO)
Appearance	Pure microcrystalline—white; when heated—lemon yellow; when cooled back down—white; single-crystal—colorless
Molar mass	81.37 g/mol
Heat capacity (Cp)	9.62 cal/deg/mole at 25°C
Melting point	1975 °C
Solubility in water	0.16 mg/100 mL
Boiling point	2360 °C
Density	5.606 g/cm^3
Coefficient of Thermal Expansion	4×10^{-6}/°C
	Electrical Properties of ZnO
Intrinsic carrier concentration (per cm^3)	10^{16} to 10^{20}
E_{me} (Effective mass- electrons)	0.24m_o
E_{mh} (Effective mass- holes)	0.59m_o
Electron mobility (at room temperature)	200 cm^2/V.sec.
Hole mobility (at room temperature)	5–50 cm^2/V.sec
Refractive index	2.0041
Relative dielectric constant	8.66

A number of deposition methods, such as the hydrothermal deposition, atomic layer deposition thermal evaporation, spin coating, PLD and RF sputtering, have been utilized in recent times for fabricating devices using ZnO thin films. RF magnetron sputtering technique and thermal vaporizing techniques have been used as they were found suitable for ZnO thin films [25]. Sputtering belongs to the class of physical vapor deposition method, which bombards energetic ions to erode the targeted surface physically.

10.4 UTILITIES OF THE ZNO THIN FILMS

10.4.1 Ultraviolet Light Detection by ZnO/Si Heterojunction Diodes

ZnO nanostructures have attracted a momentous interest of researchers in the area of photonics and nano-electronics. Remarkable properties of the ZnO, such as large bandgap, high exciton energy, low growth cost, large saturation velocity, high radio resistance and high mechanical and thermal stability have made it a suitable choice for many devices and instruments, such as laser diodes, light-emitting diodes and ultraviolet detectors [26]. In view of difficulty to produce stable and high-quality

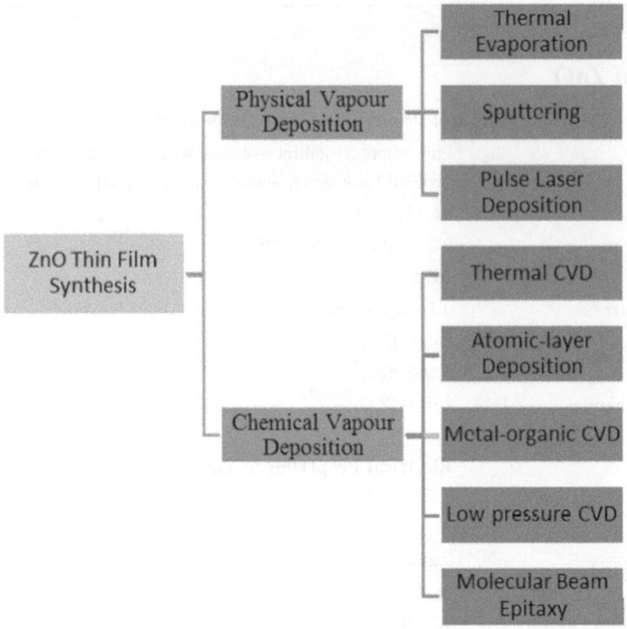

FIGURE 10.2 Fabrication techniques.

p-type ZnO, considerable attention has been given for n-ZnO based p-n heterojunctions, mainly due to its compatibility with silicon IC fabrication technology, which makes it further cost-effective.

In thin-film technology, the growth direction of a grown nanostructure plays a key role in device performance. In comparison to other orientations, (002) orientation of the ZnO thin films exhibits better thermal stability. Many deposition techniques, such as pulsed laser deposition, thermal evaporation, RF sputtering and spin coating, have been used by many researchers to demonstrate the c-axis (002) growth of the ZnO thin films [27]. RF sputtering can be a preferential choice among the different options due to its good controllability over the film growth. Various deposition parameters, such as RF power, deposition temperature and argon flow rate in RF sputtering, can help to achieve very good quality films with dominating (002) crystallographic orientation [28].

ZnO/Si heterojunction-based photodetectors can detect ultraviolet light that can have many potential applications such as pollution monitoring, UV detection in the ozone layer, water purification, flame detection and space communication, etc. Despite many potential advantages, the performance of a ZnO thin film-based devices is affected due to mismatch of lattice and thermal expansion coefficient between the layers of ZnO and silicon [29], which can be of the order of 15% and 56%, respectively. To eliminate these mismatches, a very thin layer of a suitable material is deposited as a buffer over the silicon substrate before depositing ZnO thin films. The insertion of ultra-thin films can help to passivate defects and improves the quality of the interface, which results in better device performance [30].

In the recent times, ZnO nanostructures, namely nanorods (NRs), nanowires (NWs), and nanoplates (NPs), have been extensively examined and investigated [31–33]. These ZnO structures hold exclusive properties and are being utilized in a wide variety of applications, such as bio-sensors, UV detectors, photo-detectors, light emitters, chemical sensors and memories, etc. A diverse range of bulk and nanoscale heterojunction devices, namely p-NiO/n-ZnO [34] and n-ZnO/p-Si [35], have been examined in the previous decades, but currently p-Si/n-ZnO is more appropriate due to the affordability of silicon-based devices.

Some reports are available in literature, where improvement in device performances by introducing buffers layers have been demonstrated. Outstanding chemical and physical properties of AlN (aluminum nitride), such as high thermal conductivity, high insulating resistance and good stability in severe settings, can be very helpful to improve the performance of ZnO/Si-based devices [36]. Wang et al. have conveyed that the AlN buffer layer in n-ZnO/p-Si heterojunction changed the crystallographic orientation of the ZnO thin films from (100) to (002). They also observed that the crystalline quality of a ZnO thin film has also enhanced by introducing the buffer layer of the AlN [37]. Chen et al. investigated that the use of the AlN as a buffer layer has dramatically improved the luminescence performance of the ZnO/Si heterojunction-based light-emitting diodes. Insulating properties of the AlN as a buffer layer in ZnO/Si heterojunctions has decreased the reverse leakage current significantly [38,39]. All these reports available in the literature suggest that the AlN can be a promising buffer layer that can improve both crystalline qualities as well as the electrical performance of the ZnO/Si heterojunctions.

10.4.2 Gas Sensing by the ZnO/Si Heterojunctions

The ability of ZnO to adsorb oxygen molecules on the surface is the main reason for its use in vapor and gas sensing as metal oxide semiconductor heterojunction sensors. The adsorbed oxygen molecules react with the sensed gas molecules and change the conductivity of the material by either increasing or decreasing it.

Among the various methods available, RF sputtering has been widely used for the growth of commercially viable ZnO thin films on various substrates for gas sensing. In the case of gas sensing, repeatability and stability of the deposited films is the foremost challenge being faced [40–42]. Such issues can be readily sorted in the RF sputtering technique with accurately controlling the different deposition metrics, namely gas flow rate, sputtering power, deposition pressure and the substrate temperature. A distinct benefit of the RF technique is the c-axis favored growth of ZnO with greater controllability. For sensing gases such as NH_3, NO_2, CO and ethanol and methanol, ZnO is found to be extremely appropriate [43]. Ethanol sensing is preferred in a large variety of applications related to the food production, medical and clinical fields [44, 45]. Past a certain concentration, ethanol may lead to health hazards and skin and eye irritation. Quantity of ethanol in human blood, if excessive, may lead to severe health problems, such as pain, headaches and vomiting [46–49].

10.4.3 Semiconductor Memories by using ZnO

Due to the recent advancements in complementary metal-oxide semiconductor (CMOS) and other semiconductor devices, semiconductor memories are in huge demand. Memory occupies a significant portion in an IC, so it should be smaller in size, power efficient and stable. Due to the scaling in CMOS technology, the size of these conventional memories is reducing and has now reached a saturation point. The scaling has also increased the leakage power in CMOS circuits. To get rid of the limits of conventional memories, and in order to bring progress in contemporary technologies, such as big data and Internet of Things (IOT) applications, the current day memories must be robust, highly compact and consume less power [50–52]. Current nonvolatile (NV) memories, such as flash memories, are charge storing memories and are now reaching their physical limits [53]. Hence, nanosized memories, which do not operate by charge storing, like FRAM (ferroelectric RAM), MRAM (magnetoresistive RAM), PCRAM (phasechange RAM), and RRAM (resistive RAM) have attracted substantial curiosity of researchers for forthcoming NV memories [54–57]. RRAM is a probable applicant for forthcoming memories due to its modest structures, lower power consumption, astonishing compactness and extraordinary scalability [58]. The device structure of the RRAM is a capacitor-like configuration with a metal-insulator-metal (M-I-M) structure. It is observed that the resistive switching that occurs in the M-I-M structure can be changed by an electrical signal applied to it [59]. Recent reports on memory arrays are focused on the metal oxide-based RRAM due to the ease of the materials and exceptional compatibility with the fabrication procedure of the CMOS.

The working principle of the RRAM is established on the reversible resistive switching (RS) mechanism among two stable resistance states, which are low resistance state and high resistance state. This reversible switching happens in transition metal oxides with the M-I-M configuration. There are two types of the switching memories related to electrical polarity, i.e. unipolar and bipolar [60]. The process that brings variation in the resistance states of the device, i.e. beginning LRS upto HRS, is known as the SET process, while the variation from LRS to HRS is known as RESET process. An explicit resistive state (HRS or LRS) can be reserved after the cancellation of the electric stress that specifies the NV nature of the RRAM. Generally, in the initial resistance state of a fresh sample, a higher voltage (more than the set voltage) is required in order to initiate the resistive switching behaviour. This process is known as forming/electroforming process.

The mechanism of switching in the unipolar RRAM device is described as the formation of the conductive filament when voltage is applied, which sets the device into a low resistance state. The Joule heating produced is responsible for the rupture back to HRS. The polarity of the applied current does not affect the Joule heating effect, but its amplitude does. This type of device shows unipolar switching behavior [61]. Since switching direction is dependent on the applied voltage polarity in a bipolar RRAM, different polarity is used for erasing and writing the data. To circumvent the dielectric breakdown in every switching mode, we need to implement set compliance, which is being supplied by the semiconductor parameter analyzer or series resistor or memory cell transistor. A small voltage is required for data reading from the memory cell, and it does not affect the memory cell to sense whether the cell is in HRS or LRS. The effect of varying annealing temperatures ranging from 100°C to 300°C on the switching characteristics of ZnO metal oxide-based RRAM has been reported [62].

Zinc oxide (ZnO) has an exciton-binding energy of 60 meV with a wide direct bandgap 3.37 eV at 300 K [63, 64], which aids in the fabrication of photonics and optoelectronic devices. On a variety of the substrates, these ZnO nanostructures can be grown, which includes ZnO, Si, GaN, polymer, sapphire and even glass. These properties make the nanostructured ZnO as a material of choice for the fabrication of transducers/nano-sensors [65], nano-photonic [66], piezoelectric nano-generator [67], transparent and spin electronics devices [68]. Usability of ZnO for gas sensing and UV light detection applications have mainly been explored in several recent works. Heterojunctions have been utilized to construct ZnO-based UV photodetectors due to the absence of stable and well-regulated *p*-type ZnO films with a different *p*-type semiconductor, such as SiC [12], Si [27], NiO [32], GaN [25] and so on. Certainly, the commercial silicon has obtained ample consideration for ZnO-based *p-n* heterojunctions within all of these p-type semiconductor materials because of its usability and low cost. A wide bandgap material in UV photodetectors is utilized as a window for transmission of the lower energy photons that are collected by a material having a narrow bandgap.

Due to the higher transparency of about 80% in the range of visible wavelength, ZnO is a very promising versatile material with a wide bandgap of 3.3 eV as a II-VI semiconductor. Therefore, in the optoelectronics and photonics applications, the n-ZnO/p-Si heterojunction diode has materialized as a replacement device. In recent times, significant attention has been drawn by the metal oxide (e.g. ZnO) thin film-based gas sensors mainly due to their several advantages, such as easy fabrication, low cost, good compatibility with micro-fabrication and high response [69]. When compared to bulk ZnO the nanostructured ZnO have a higher surface to volume ratio, and they possess a wide energy gap, which makes them impending applicants for gas sensing and other applications [70].

10.5 CONCLUSION

ZnO is an essential material, and based on its performance and applications, it has penetrated all aspects of life. This has encouraged research work in various aspects of ZnO heterojunction-based devices. ZnO-based photodetectors have many potential applications, such as pollution monitoring, UV detection in the ozone layer, water purification, flame detection and space communication, etc. It is the need of the hour to design and develop low-cost, easy to fabricate sensor devices for detecting various classes of hazardous vapors and gases at fairly lower concentrations. Selectivity and sensitivity enhancement in ZnO based gas sensors is being undertaken aggressively. Several studies on gas-sensing abilities of the ZnO/Si-based gas sensors for various gases such as methanol, ethanol, benzene, etc. have been reviewed. RRAM is a strong applicant for the forthcoming memories due to its modest structure, higher density, lower power consumption and extraordinary scalability. The switching characteristics of various ZnO-based devices have been outlined. The fabrication, modeling, and characterization of ZnO-based nanosized devices are being taken for some efficient and useful applications that can contribute in household, military, irrigation and medical applications.

REFERENCES

1. Jagadish, C., and Pearton, S. (Eds.). *Zinc Oxide Bulk, Thin Films and Nanostructures: Processing, Properties, and Applications*. Amsterdam, Elsevier, 2006.
2. Yearian, H.J. "Intensity of diffraction of electrons by ZnO." *Physical Review* 48 (1935): 631
3. James, H.M., and Johnson, V.A. "Electron distribution in ZnO crystals." *Physical Review*, 56, no.1 (1939): 119.
4. Ross, I.M. "The invention of the transistor." *Proceedings of the IEEE*, 86, no. 1 (1998): 7–28.
5. Ozgur, U., Hofstetter, D., and Morkoc, H. "ZnO devices and applications: a review of current status and future prospects." *Proceedings of the IEEE*, 98, no. 7 (2010): 1255–1268.
6. Jagadish, C., and Pearton, S. "Basic properties and applications of ZnO." In *Zinc Oxide Bulk, Thin Films and Nanostructures: Processing, Properties, and Applications*, 1st Ed., pp. 1–20. Elsevier New York, 2006.
7. Fan, Z., and Lu, J.G. "Zinc oxide nanostructures: synthesis and properties." *Journal of Nanoscience and Nanotechnology*, 5, no. 10 (2005): 1561–1573.
8. Lee, Ching-Ting. "Fabrication methods and luminescent properties of ZnO materials for light-emitting diodes." *Materials* 3, no. 4 (2010): 2218–2259.
9. Patrinoiu, G., Calderon-Moreno, J.M., Birjega, R., Culita, D.C., Somacescu, S., Musuc, A.M., and Carp, O. "Sustainable one-pot integration of ZnO nanoparticles into carbon spheres: manipulation of the morphological, optical and electrochemical properties." *Physical Chemistry Chemical Physics*, 18, no. 44 (2016): 30794–30807.
10. Piyadasa, A., Wang, S., and Gao, P.X. "Band structure engineering strategies of metal oxide semiconductor nanowires and related nanostructures: A review." *Semiconductor Science and Technology*, 32, no. 7 (2017): 073001.
11. Olgar, M.A., Atasoy, Y.A.V.U.Z., Bacaksız, E., and Aydoğan, Ş. "Synthesis and characterization of ZnO micro-rods and temperature-dependent characterizations of heterojunction of ZnO microrods/CdTe and ZnO microrods/ZnTe structures." *Sensors and Actuators A: Physical*, 261 (2017): 56–65.
12. Kwietniewski, N., Masłyk, M., Werbowy, A., Taube, A., Gierałtowska, S., Wachnicki, Ł., and Sochacki, M. "Electrical characterization of ZnO/4H-SiC n–p heterojunction diode." *Physica Status Solidi (a)*, 213, no. 5 (2016): 1120–1124
13. Xin-Yu, Z., Zhou-Wen, C., Yan-Peng, Q., Yan, F., Liang, Z., Li, Q., Ming-Zhen, M., Ri-Ping, L., and Wen-Kui, W. "Ab initio comparative study of zincblende and wurtzite ZnO." *Chinese Physics Letters*, 24, no. 4 (2006): 1032–1034.
14. Morkoc, H., and Ozgur, U. "General properties of ZnO." In *Zinc Oxide: Fundamentals, Materials and Device Technology*. pp. 1–76. Wiley, NJ, 2009.
15. Lee, J.S., Park, K., Kang, M.I., Park, I.W., Kim, S.W., Cho, W.K., … and Kim, S. "ZnO nanomaterials synthesized from thermal evaporation of ball-milled ZnO powders." *Journal of Crystal Growth*, 254, no. 3–4 (2003): 423–431.
16. Riyaj, Md, Singh, A.K., Alvi, P.A., and Rathi, A. "Wavefunctions and optical gain in In 0.24 Ga 0.76 N/GaN Type-I nano-heterostructure under external uniaxial strain." In *Intelligent Computing Techniques for Smart Energy Systems*, pp. 341–349. Springer, Singapore, 2020.

17. Kim, Y.S., Tai, W.P., and Shu, S.J. "Effect of preheating temperature on structural and optical properties of ZnO thin films by sol–gel process." *Thin Solid Films*, 491, no. 1–2 (2005): 153–160.
18. Gumus, C., Ozkendir, O.M., Kavak, H., and Ufuktepe, Y. "Structural and optical properties of zinc oxide thin films prepared by spray pyrolysis method." *Journal of Optoelectronics and Advanced Materials*, 8, no. 1 (2006): 299.
19. Li, X.Y., Li, H.J., Wang, Z.J., Xia, H., Xiong, Z.Y., Wang, J.X., and Yang, B.C. "Effect of substrate temperature on the structural and optical properties of ZnO and Al-doped ZnO thin films prepared by dc magnetron sputtering." *Optics Communications*, 282, no. 2 (2009): 247–252.
20. Lee, J., Lee, D., Lim, D., and Yang, K. "Structural, electrical and optical properties of ZnO: Al films deposited on flexible organic substrates for solar cell applications." *Thin Solid Films*, 515, no. 15 (2007): 6094–6098.
21. Sharma, S., Bayer, B.C., Skakalova, V., Singh, G., and Periasamy, C. "Structural, electrical, and UV detection properties of ZnO/Si heterojunction diodes." *IEEE Transactions on Electron Devices*, 63, no. 5 (2016): 1949–1956.
22. Nanto, H., Minami, T., Shooji, S., and Takata, S. "Electrical and optical properties of zinc oxide thin films prepared by rf magnetron sputtering for transparent electrode applications." *Journal of Applied Physics*, 55, no. 4 (1984): 1029–1034.
23. Lee, J.H., Ko, K.H., and Park, B.O. "Electrical and optical properties of ZnO transparent conducting films by the sol–gel method." *Journal of Crystal Growth*, 247, no. 1-2 (2003): 119–125.
24. Özgür, Ü., Alivov, Y.I., Liu, C., Teke, A., Reshchikov, M., Doğan, S., ... and Morkoç, A.H. "A comprehensive review of ZnO materials and devices." *Journal of Applied Physics*, 98, no. 4 (2005): 11.
25. Martin, P.M. (Ed). "Deposition technologies: An overview." In *Handbook of Deposition Technologies for Films and Coatings* (Third Edition). William Andrew Publishing, pp. 1–31, ISBN 9780815520313, 2010.
26. Mozharov, A., Bolshakov, A., Cirlin, G., and Mukhin, I. "Numerical modeling of photovoltaic efficiency of n-type GaN nanowires on p-type Si heterojunction." *Physica status solidi (RRL)–Rapid Research Letters*, 9, no. 9 (2015): 507–510.
27. Sun, H., Priante, D., Min, J.W., Subedi, R.C., Shakfa, M.K., Ren, Z., and Ryou, J.H. "Graded-index separate confinement heterostructure AlGaN nanowires: toward ultraviolet laser diodes implementation." *ACS Photonics*, 5, no. 8 (2018): 3305–3314.
28. Pietruszka, R., Luka, G., Kopalko, K., Zielony, E., Bieganski, P., Placzek-Popko, E., and Godlewski, M. "Photovoltaic and photoelectrical response of n-ZnO/p-Si heterostructures with ZnO films grown by an atomic layer deposition method." *Materials Science in Semiconductor Processing*, 25 (2014): 190–196.
29. Tsay, C.Y., and Hsu, W.T. "Comparative studies on ultraviolet-light-derived photoresponse properties of ZnO, AZO, and GZO transparent semiconductor thin films." *Materials*, 10, no. 12 (2017): 1379.
30. Eames, C., Frost, J.M., Barnes, P.R., O'Regan, B.C., Walsh, A., and Islam, M.S. "Ionic transport in hybrid lead iodide perovskite solar cells." *Nature Communications*, 6, no. 1 (2015): 1–8.
31. Dutta, K., Bhowmik, B., and Bhattacharyya, P. "Resonant frequency tuning technique for selective detection of alcohols by TiO2 nanorod-based capacitive device." *IEEE Transactions on Nanotechnology*, 16, no. 5 (2017): 820–825.
32. Dasgupta, N.P., Sun, J., Liu, C., Brittman, S., Andrews, S.C., Lim, J., and Yang, P. "Semiconductor nanowires–synthesis, characterization, and applications." *Advanced Materials*, 26, no. 14 (2014): 2137–2184.

33. Sirelkhatim, A., Mahmud, S., Seeni, A., Kaus, N.H.M., Ann, L.C., Bakhori, S.K.M., and Mohamad, D. "Review on zinc oxide nanoparticles: antibacterial activity and toxicity mechanism." *Nano-micro Letters*, 7, no. 3 (2015): 219–242.
34. Hasan, M.R., Xie, T., Barron, S.C., Liu, G., Nguyen, N.V., Motayed, A., and Debnath, R. "Self-powered p-NiO/n-ZnO heterojunction ultraviolet photodetectors fabricated on plastic substrates." *APL Materials*, 3, no. 10, pp 106101, 2015.
35. Sharma, S., and Periasamy, C. "Simulation study and performance analysis of n-ZnO/p-Si heterojunction photodetector." *Journal of Electron Devices*, 19 (2014): 1633–1636.
36. Gupta, Chandra Prakash, Singh, A.K., Jain, P.K., Sharma, S.K., Birla, S., and Sancheti, S. "Electrical transport properties of thermally stable n-ZnO/AlN/p-Si diode grown using RF sputtering." *Materials Science in Semiconductor Processing* 128 (2021): 105734.
37. Wang, W., Chen, C., Zhang, G., Wang, T., Wu, H., Liu, Y., and Liu, C. "The function of a 60-nm-thick AlN buffer layer in n-ZnO/AlN/p-Si (111)." *Nanoscale Research Letters*, 10, no. 1 (2015): 91.
39. Sharma, S. Bayer, B.C., Skakalova, V., Singh, G., and Periasamy, C. "Structural, electrical and uv detection properties of ZnO/Si heterojunction diodes" *IEEE Transaction on Electron Devices*, 63, no. 05 (2016): 1949–1956.
38. Cheng, B., Choi, S., Northrup, J.E., Yang, Z., Knollenberg, C., Teepe, M., and Johnson, N.M. "Enhanced vertical and lateral hole transport in high aluminum-containing AlGaN for deep ultraviolet light emitters." *Applied Physics Letters*, 102, no. 23 (2013): 231106.
40. Sharma, S.K., Bhowmick, B., Pal, V., and Periasamy, C. "Electrical and low temperature methanol sensing characteristics of RF sputtered n-ZnO/p-Si heterojunction diodes." *IEEE Sensors Journal*, 17, no. 22 (2017): 7332–7339.
41. Bhowmik, B., Hazra, A., Dutta, K., and Bhattacharyya, P. "Repeatability and stability of room-temperature acetone sensor based on TiO2 nanotubes: Influence of stoichiometry variation." *IEEE Transactions on Device and Materials Reliability*, 14, no. 4 (2014): 961–967.
42. Chinh, N.D., Quang, N.D., Lee, H., Hien, T.T., Hieu, N.M., Kim, D., Kim, C., and Kim, D. "NO gas sensing kinetics at room temperature under UV light irradiation of In_2O_3 nanostructures." *Scientific Reports*, 6, no. 35066 (2016): 1–11.
43. Hjiri, M., Mir, L.E., Leonardi, S.G., Donato, N., and Neri, G. "CO and NO2 selective monitoring by ZnO-based sensors." *Nanomaterials*, 3 (2013): 357–369.
44. Zhang, Y., Jia, Q., Ji, H.M., and Yu, J. "Semiconducting nano-structured $SmFeO_3$-based thin films prepared by novel sol-gel method for acetone gas sensors." *Integrated Ferroelectrics*, 152, no.1 (2014): 29–35.
45. Gupta, C.P., Sharma, S.K., Bhowmik, B., Sampath, K.T., Periasamy, C., and Sancheti, S. "Development of highly sensitive and selective ethanol sensors based on RF sputtered ZnO nanoplates." *Journal of Electronic Materials* 48, no. 6 (2019): 3686–3691.
46. Kumar, I.M., Bhatt, V., Abhyankar, A.C., Kim, J., Kumar, A., Patil, S.H., and Hyung, J. "New Insights towards Strikingly Improved Room Temperature Ethanol Sensing Properties of p-type Ce-doped SnO2 Sensors." *Scientific Reports*, 8, no. 8079 (2014): 1–12.
47. Jahan, K., Mahmood, D., and Fahim, M. "Effects of methanol in blood pressure and heart rate in the rat." *Journal of Pharmacy Bioallied Sciences*, 7, no. 1 (2015): 60–64.
48. Panconesi, A. "Alcohol-induced Headaches: Evidence for a Central Mechanism." *Journal of Neuroscience in Rural Practice*, 7, no. 2 (2016): 269–275.

49. Bhowmik, B., Dutta, K., Hazra, A., and Bhattacharyya, P. "Low temperature acetone detection by p-type nanocrystalline TiO_2 thin film: Equivalent circuit model and sensing mechanism." *Solid State Electronics*, 99 (2014): 84–92.
50. Hwang, C.S. "Prospective of semiconductor memory devices: from memory system to materials." *Advanced Electronic Materials*, 1, no. 6 (2015): 1400056.
51. Duari, C., Birla, S., and Singh, A.K. "A dual port 8T SRAM cell using FinFET & CMOS logic for leakage reduction and enhanced read & write stability." *Journal of Integrated Circuits and Systems* 15, no. 2 (2020): 1–7.
52. Singhal, S., Singh, R., and Singh, A.K. "Design of a sub-0.4 V reference circuit in 0.18 μm CMOS technology." In *Advanced Materials Research*, 816 (2013): 882–886. Trans Tech Publications Ltd.
53. Meena, J.S., Sze, S.M., Chand, U., and Tseng, T.-Y. "Overview of emerging non-volatile memory technologies." *Nanoscale Research Letters*, 9, no. 526 (2014): 1–34.
54. Sarwat, S.G. "Materials science, and engineering of phase change random access memory." *Materials Science and Technology*, 33, no. 16 (2017): 1890–1906.
55. Wang, Lei, Yang, C., Wen, J., and Gai, S. "Emerging non-volatile memories to go beyond scaling limits of conventional CMOS nanodevices." *Journal of Nanomaterials*, 2014, Article ID 927696: 10.
56. Kumar, D., Aluguri, R., Chand, U., and Tseng, T.Y. "Metal oxide resistive switching memory: Materials, properties, and switching mechanisms." *Ceramics International*, 43 (2017): S547–S556.
57. Fujisaki, Y. "Review of emerging new solid-state non-volatile memories." *Japanese Journal of Applied Physics*, 52, no. 4R (2013): 040001.
58. Zhoua, Linggang, Zhu, J., and Sunab, Z.G.Z. "An overview of materials issues in resistive random access memory." *Journal of Materiomics* 1, no. 4 (2015): 285–295.
59. Gorshkov, O.N., Antonov, I.N., Belov, A.I., Kasatkin, A.P., and Mikhaylov, A.N. "Resistive switching in metal–insulator–metal structures based on germanium oxide and stabilized zirconia." *Technical Physics Letters*, 40, no. 2 (2014): 101–103.
60. Lelmini, D. "Resistive switching memories based on metal oxides: mechanisms, reliability, and scaling." *Semiconductor Science and Technology*, 31, no. 6 (2016): 063002.
61. Gupta, C.P., Jain, P.K., Chand, U., Sharma, S.K., Birla, S., and Sancheti, S. "Effect of top electrode materials on switching characteristics and endurance properties of zinc oxide based RRAM device." Journal of Nano- and Electronic Physics (2020): 01007(1-8).
62. Gupta, C.P., Jain, P.K., Chand, U., Sharma, S.K., Birla, S., and Sancheti, S. "Effect of annealing temperature on switching characteristics of zinc oxide based RRAM Device." In, pp. 112–115. IEEE, 2019.2019 International Conference on Innovative Trends and Advances in Engineering and Technology (ICITAET)
63. Chang, T.-C., Chang, K.-C., Tsai, T.-M., Chu, T.-J., and Sze, S.M. "Resistance random access memory." *Materials Today*, 19, no. 5 (2016): 254–264.
64. Moors, M., Adepalli, K.K., Lu, Q., Wedig, A., Bäumer, C., Skaja, K., and Valov, I. "Resistive switching mechanisms on TaO_x and $SrRuO_3$ thin-film surface probed by scanning tunneling microscopy." *ACS Nano*, 10, no. 1 (2016): 1481–1492.
65. Zhu, C., Yang, G., Li, H., Du, D., and Lin, Y. "Electrochemical sensors and biosensors based on nanomaterials and nanostructures." *Analytical Chemistry*, 87, no. 1 (2015): 230–249.
66. Perevedentsev, A., Sonnefraud, Y., Belton, C.R., Sharma, S., Cass, A.E., Maier, S.A., and Bradley, D.D. "Dip-pen patterning of poly (9, 9-dioctylfluorene) chain-conformation-based nano-photonic elements." *Nature Communications*, 6, no. 1 (2015): 1–9.

67. Nam, S., Song, M., Kim, D.H., Cho, B., Lee, H.M., Kwon, J.D., and Park, Y.C. "Ultrasmooth, extremely deformable and shape recoverable Ag nanowire embedded transparent electrode." *Scientific Reports*, 4, no. 1 (2014): 1–7.
68. Shin, S.H., Kim, Y.H., Lee, M.H., Jung, J.Y., and Nah, J. "Hemispherically aggregated BaTiO3 nanoparticle composite thin film for high-performance flexible piezoelectric nanogenerator." *ACS Nano*, 8, no. 3 (2014): 2766–2773.
69. Willander, M., Khun, K., and Ibupoto, Z.H. "ZnO based potentiometric and amperometric nanosensors." *Journal of Nanoscience and Nanotechnology*, 14, no. 9 (2014): 6497–6508.
70. Basyooni, M.A., Shaban, M., and El Sayed, A.M. "Enhanced gas sensing properties of spin-coated Na-doped ZnO nanostructured films." *Scientific Reports*, 7 (2017): 41716.

11 SRAM Designing with Comparative Analysis using Planer and Non-Planer Nanodevice

Neha Mathur, Deepika Sharma, and Shilpi Birla
Manipal University Jaipur, Jaipur, India

11.1 INTRODUCTION

The innovation of new technology node limits the MOSFET scaling as far as SCE and new operating principle based devices are taken into consideration as an alternative proposed approach.

In [1] Authors have revealed that FINFET devices have a predominant capacity to limit and control the leakage current and short channel effects by providing an enhanced drive current. More extensive non-planer device FINFET is manufactured with multi fins which are arranged in parallel consequently, independent controlling of the FINFET's multi gates permits remarkable leakage current reduction.

The author revealed in [2] a fin-shaped structure with a novel bottom spacer, which is more promising for system-on-chip requirements for logic applications. Thus the proposed approach for non-planer devices accomplished less power delay with improved short-channel effect and self-heating performance.

The author has investigated in [3], another FINFET structure of STI. Results from the simulation show that FINFET structure with STI-type reduces the short-channel effect and threshold voltage (V_{th}) with well-controlled DIBL. Thus, this STI-type FINFET likewise shows a higher trans-conductance.

The author demonstrated a non-planer device optimization method to make it applicable for high-performance applications of SoC. In this method, high-k/ metal-gate EOT scaling has been carried out, for drive current (I_{on}) improvement, and leakage current (I_{off}) reduction with carrier mobility improvement [4].

In [5] author investigated drain-extended novel FINFET structure for the high value of voltage and speedy applications. This structure provides improved breakdown voltage without influencing the MOS behavior, by adding additional longitudinal fin. This structure is supposed to perform with improved gate-oxide consistency and electrostatic discharge hardness.

The author of [6] studied gate-induced drain loss (GIDL) in double-doped FINFET devices which can be suppressed by forming steep junctions underlapping for doped and non-doped channel devices specific to the application for designing the memory cell.

DOI: 10.1201/9781003220350-11

In [7], it is analyzed that the array multipliers utilize models of FINFET for 7 nm, 10 nm, 14 nm, 16 nm, and 20 nm feature size for speedy applications. To increase static power, the feature size is decreased, hence delay is decreased. FINFET based arithmetic circuits can be designed for speedy applications.

In [8], the author observed that due to extraordinary properties, FINFET becomes an encouraging VLSI technology for the recent future. 7nm advanced technology-based FINFET has been developed with standard cell libraries which are operated at various supply voltages, provides voltage scaling with multi-threshold technologies. Simulation result with this technology node demonstrated that the 7nm FINFET circuits burn 5 times more power, and require 600 times low energy, with normal V_{th}, while with high V_{th} it requires 10 times more power and 1000 times low energy when it is compared with high technology node. Thus with a 7nm node, energy efficiency can be enhanced by 7 times and 16 times with normal and high V_{th} against the conventional technology with 14-nanometer bulk, respectively.

FINFET technology with Layout dependent effect has been set on in [9] to overcome many issues such as Length of Oxidation, Metal Boundary Effect, Gate Line End Effect, Well Proximity Effect, and Neighboring Diffusion Effect [10].

FINFET when compared to iFINFET, provides excellent integration with electrostatics and more suitable to operate at low power. This is demonstrated by the author in [11], that the evolutionary performance of FINFET (iFINFET) is enhanced by the 3D device simulations.

FINFET faces lithographic and work-function engineering challenges. In this article [12], a multi-core PVT is introduced to integrate the framework for power simulation, delay, and PVT variants to take control of FINFET challenges.

In [13], the author revealed an idea for the performance improvements in RF, analog and digital field that allows vertical tunneling or area-scaling in conventional technologies beyond the scaling limit. The gate length which is 10-nm, followed by the proposed device when this is compared with conventional FET, Provides ON state current improvement by 100%, output resistance is improved by 30% with a frequency whose gain is unity which is improved by 55%. Besides this, it offers to scale down in the I_{off} current by 15 times, trans-conductance is increased by 3 times, and the most important footprint area is reduced by 6 times for a drive given capability. The average sub-threshold slopes down to 40 mV/decade and minimum sub-threshold slopes down to 11 mV/decade at a gate length of 10-nm.

In [14], the author analyzed negative capacitance FINFET's (NC-FINFET). The author revealed a calibrated model by putting V_{dd} to 0.25 V, which shows reduced energy to use at extraordinary low power, having high performance.

In [15], the author developed 10nm Si-FINFET logic technology. The simulation result when compared to the 14nm provides excellent improvement with the evolution of the next generation of Fin technology, source/drain with heavy doping, and optimizing contact resistance.

To boost the ratio of I_{on} to I_{off} current, Hyper-FETs were proposed by the author by employing phase transition material in [16] for improvement in performance over conventional FET. Phase transition thresholds were also established to maintain the purpose of the Hyper-FETs. Simulation results of Hyper-FETs shown I_{on} yields larger over FINFET.

The author described in [17] that the non-planer device demonstrates a record peak F_t (314/285GHz) and significantly achieves higher F_{max} (180/140 GHz) over planer devices for 14nm and 28nm technology, respectively with larger L_g and DGCs.

11.2 GENERAL ISSUES WITH CMOS

11.2.1 Small Dimension or Channel Effect

In long channel devices, the applied gate voltage is used to control generated transverse electric field that exists everywhere. But gate voltage is not a contributor to control the devices having short channels, because of generated longitudinal electric field [18,19]. CMOS showed several unwanted effects with the short channel, resulting in the potential distribution in two-dimension and generation of effective high electric fields [20,21].

11.2.2 Carrier Velocity Saturation

SCE introduces an effective longitudinal electric field. For the low value of the electric field, Vd is proportional to the applied Electric field while for a high value of the electric field, drift velocity becomes saturate because of which V-I characteristics of the planer device are directly affected [22].

11.2.3 Mobility Degradation

A higher value of the electric field which is vertical in the channel provides dispersion of carriers outside the oxide interface. Thus mobility of charge carrier degrades, hence depletion current is decreased [4,23].

11.2.4 DIBL

The inversion channel formation is possible by applying efficient gate voltage. If the voltage applied at the gate terminal is not efficient to alter the channel surface, then the channel current is blocked by a potential barrier [24]. The elimination of this barrier is possible when the applied gate voltage is increased.

V_{ds} along with V_{gs} plays an equal contribution for the controlling of the potential barrier. Now by increasing the value of V_{ds}, the depletion region width of the drain body can be improved [25]. Therefore, DIBL comes into the picture because, channel potential barrier is decreased, causing the carriers to flow even at V_{gs} below V_{th} [6]. This is referred to as roll-off, of threshold voltage and the corresponding current is said to be sub-threshold current or off-state current [23].

11.2.5 Hot Carrier Effects

Decreasing channel length is responsible for the increased electric field near the drain terminal. Consequently, charge carriers retrieve energy in a considerable amount and become a hot carrier, and then some charge carriers get almost enough energy close to the drain region that results in impact ionization consequently, more pairs of electron-hole are generated [26]. This leads to an increase in the drain-to-body current (I_{db}). Additionally, hot electrons which are small in number can traverse the oxide and collect through a gate terminal. While oxide is damaged by some hot carriers with subsequent degradation of the device [27].

11.2.6 Leakage Current and Dynamic Current

The persistent scaling of planar devices provides improved performance. However, the continuation of this trend is very difficult in the nanometer range only because of the dramatic enhancement in the sub-threshold off-state current (Ioff) [28,29]. Thus, the gate cannot completely switch off the channel in the switched-off state, which tends to increase the gap in the channel [4,6,20].

11.2.7 Power Dissipation

The continuous growth of dynamic as well as static power dissipation in each CMOS process technology having higher cost and efficiency is increased of the system [2]. The switching activity of the charge and discharge load capacities are the main source of dynamic energy consumption when the output switches between high and low logic while the switched off current is responsible for the power dissipation of static nature [7].

11.2.8 Device Performance

For the new technology node, when the channel length shrinks, more leakage current is introduced because of which device performance is degraded [16].

11.3 ALTERNATE SOLUTION OVER PLANER DEVICE: NON-PLANER DEVICE (FINFET)

With the reduced technology node, the gate terminal losses control over the channel, in the conventional planer structures. Consequently, a greater loss of sub-threshold leakage in the channel [30,31]. In the conventional planer device, the gate terminal is not able to control the leakage path that is far away. These effects can be improved or overcome by using different structures that allow a transistor to resize beyond the conventional planer scaling limit [32].

FINFET has become a leading technology as feature size shifts to 20 nm or less and offers improved performance over conventional planer devices [7].

11.3.1 FINFET Devices

FINFET is a non-planar field-effect transistor that may build on a bulk substrate or SOI substrate. The gate terminal offers dominant control over the channel because it is confined around the channel [6].

11.3.1.1 *Drive Current Equation of FINFET*

The I-V Characteristics of FINFET Varies as per the respective region. If mobility is denoted by μ, gate oxide capacitance is represented by C_{ox}, a ratio of width to length is represented by W/L, the threshold voltage is V_{th}, X_d is the thickness of depletion layer and t_{ox} is the oxide thickness, then the respective current equation will be:

Linear Region: When $V_g > V_{th}$, I_D is directly proportional to drain voltage and increases linearly with V_{ds} [33].

$$I_D = 2 * \mu * C_{ox} * (W/L) * (V_g - V_{th} - V_{ds}/2) * V_{ds}$$

Saturation Region: I_D is independent of drain voltage and does not increase as V_{ds} increases. For a saturation, region I_D is given by:

$$I_D = \mu * C_{ox} * (W/L) * (V_g - V_{th})^2/2m$$

Where, $m = 1 + (3 * t_{ox}/X_d)$

Cut-Off Region: when V_g is less than V_{th}, no channel is induced between drain and source terminal, consequentially $I_D = O$.

11.3.1.2 *FINFET Scaling Parameters*

Channel width can be considered as a perfect scaling parameter for FINFET and can be represented as:

$$\text{Channel width} = 2 * \text{Fin Height} + \text{Fin Width}$$

Thus for FINFET, the performance-dependent output current can be improved by dealing with the width of the channel [34]. If multiple fins are connected parallel, then it also increases the device drive current. So any arbitrary channel width cannot be considered for FINFET [35].

11.3.1.3 *Advantages of FINFET*

- Excellent control of short channel effects by suppressing short channel effects in the submicron regime.
- Low cost
- Non-planer devices having high maturity in technology than planar DG.
- Higher Drive Current

TABLE 11.1
Literature Review of SRAM using CMOS & FINFET

S. No.	No. of T	Tech.	Author	Year & Publication	Device	Advantages	Limitation
1	10T	180 nm	1. V. Rukkumani 2. M. Saravanakumar 3. K. Srinivasan [36]	2016 IEEE:TENCON [36]	CMOS	Power efficiency is improved under various temperature conditions.	Area and delay are increased because of 10T.
2		90 nm	1. Gande Bhargav 2. Govind Prasad [37]	2016 AEEICB-16 [37]	CMOS	98% reduction in required power and 55% reduction in static power.	
3		16 nm	1. Navneet 2. Neha Gupta 3. Hitesh Pahuja 4. Balvinder Singh 5. Sudhakar Pandey [38]	2016 IEEE: CIPECH-16[38]	FINFET	This provides reduced PDP and Leakage current with improved read stability over conventional SRAM cells from 6T.	
4		32 nm	1. Arundhati Bhattacharya 2. Soumitra Pal 3. Aminul Islam [39]	2014 IEEE: ICACCCT [39]	FINFET over CMOS	SRAM cell based on FINFET device offers 2.33× improvement in Read Access Time (TRA) and 1.29× improvement in Write Access Time (TWA) when compared with conventional device-based 10T SRAM cell.	
5		45 nm	1. Singh 2. Pattanaik 3. Shukla 4. Birla 5. Nagpal [40]	2012 EEC-12 [40]	CMOS	When V_{DD}=0.8V, static leakage is reduced by 74% and when it is 0.7V then reduced by 77%.	

6	9T	32 nm	1. Nidhi Sharma 2. Uday Panwar 3. Virender Singh [41]	2016 IEEE: CSBE [41]	FINFET	Proposed circuits show a maximum saving of dynamic power up to 76.57% in 4T, maximum leakage power saving up to 53.21% in 6T.	It improves write stability only.
7		22nm	1. Juhyun 2. Younghwi Yang 3. Seong Ookjung [42]	2016 IEEE Transaction on VLSI Systems [42]	FINFET	The proposed SRAM cell uses a bit interleaving approach.	
8		22nm	1. Juhyun 2. YounghwiYang 3. Seong Ook jung [43]	2015 IEEE Transaction on VLSI Systems [43]	FINFET	This approach provides one transistor read path. To improve the read performance of the cell.	
9		45nm	1. R. K. Singh 2. Shilpi Birla 3. Manisha Pattanaik [44]	2011 IACSIT [44]	CMOS	When V_{dd}= 0.8V and temperature is 50 then, 1. SNM = 0.33V, 2. Write Margin = 0.35 V 3. Read current = 14μA	
10	8T	60nm	1. Pooran Singh 2. Santosh Kumar [45]	2017 IEEE Transaction on VLSI Systems [45]	CMOS	Leakage power is reduced to 82 times for the conventional 6T SRAM and 75 times for RD 8T SRAM.	Read Disturbances
11		180nm	1. P. Raikwal 2. V. Neema 3. A. Verma [46]	2017 IEEE: ICECA [46]	CMOS	67.37% improvement in RSNM when compare with the traditional structure of SRAM cell.	
12		180nm	1. P. Raikwal 2. V. Neema 3. A. Verma [47]	2016 IEEE: SCOPES [47]	CMOS	59.12% power is saved by 8T SRAM array during write operation. While 80.49%	

(Continued)

TABLE 11.1 (Continued)
Literature Review of SRAM using CMOS & FINFET

S. No.	No. of T	Tech.	Author	Year & Publication	Device	Advantages	Limitation
						power is saved during read operation.	
13		90nm	1. Ghasem Pasandi 2. Sied Mehdi Fakhraie [48]	2013 IEEE [48]	FINFET	This technology reduces write and read delay over conventional 6T SRAM cell.	
14		32 nm	1. Aminul Islam 2. Mohd. Hasan [49]	2012 IEEE [49]	FINFET	This design implies approx. 5 times and approx. 2 times improvement in access time for both read and write operation respectively.	
15	7T	45 nm	1. G. Sneha 2. B. Hari Krishna 3. Ashok Kumar [50]	2017 IEEE: ICISC [50]	FINFET	This provides maximum of 60.8% reduction in required power.	poor read stability and write ability
16		14 nm	1. Younghwi Yang 2. Hanwool Jeong [51]	2016 IEEE Transactions on Circuits & System [51]	FINFET	Read access time is improved by 13%, energy is enhanced by 42% and standby power by 23%.	
17		90 nm	1. Majid Moghaddam 2. Mohammad Hossein 3. Mohammad Eshghi [52]	2015 IEEE:ICEE [52]	CMOS	DIBL effect is controlled to provide efficient read operation structure in the hold '1' state.	
18	6T	32 nm	1. Sudarshan Patil 2. V S Kanchana [53]	2017 IEEE International Conference on Next gen Electronic Technologies [53]	FINFET	Offers minimum power and energy dissipation.	Conventional SRAM cell using 6T is the better than 7T, 8T and 9T in terms of read delay, write

19	45 nm	1. Apoorva Pathak 2. Divyesh Sachan 3. Harish Peta 4. Manish Goswami [54]	2016 IEEE :29th International Conference on VLSI Design[54]	CMOS	Hold power dissipation of 4.74154pW which is much less as compared to the standard 6T SRAM cell.	delay, power dissipation but data retention is major issue for 6T SRAM.
20	32 nm	1. Saurabh 2. P. Srivastava [55]	2012 IEEE[55]	CMOS	Write operation shows approx. 8% improvement in Power consumption while hold operation shows approx. 25% improvement.	

11.4 CONCLUSION

The comparative study between planer and non-planer devices has been completed. Non planer device FinFET can be considered as an efficient alternative over planar device CMOS with technology scaling beyond 32nm and it is more dominant in terms of superior performance for low power applications.

REFERENCES

1. Brian Sawn and Soha Hassoun. "Gate Sizing: FinFETs vs. 32nm Bulk MOSFETs." *DAC 2006*, (July 24–28, 2006): 528-531.
2. Mayank Shrivastava and Maryam Shojaei Baghini. "A Novel Bottom Spacer FinFET Structure for Improved Short-Channel, Power-Delay, and Thermal Performance." *IEEE Transactions on Electron Devices* 57, No. 6 (June 2010): 1287–1293.
3. Jyi-Tsong Lin, Po-Hsieh Lin and Yi-Chuen Eng. "A New STI-type FinFET Device Structure for High-Performance Applications." *IEEE*, (2010): 25–27.
4. Chih-Chieh Yeh, Chih-Sheng Chang, Hong-Nien Lin and Wei-Hsiung Tseng. "A Low Operating Power FinFET Transistor Module Featuring Scaled Gate Stack and Strain Engineering for 32/28nm SoC Technology." *2010 International Electron Devices Meeting* (2010) pp. 772–775, doi: 10.1109/IEDM.2010.5703473.
5. Mayank Shrivastava, Harald Gossner and V. Ramgopal Rao. "A Novel Drain-Extended FinFET Device for High-Voltage High-Speed Applications." *IEEE Electron Device Letters* 33, no. 10 (October 2012): 1432–1434.
6. Pranita Kerber, Qintao Zhang, Siyuranga Koswatta, and Andres Bryant. "GIDL in Doped and undoped FinFET Devices for Low-Leakage Applications." *IEEE Electron Device Letters* 34, no.1 (January 2013): 6–8.
7. Joseph Whitehouse and Eugene John. "Leakage and Delay Analysis in FinFET Array Multiplier Circuits." *IEEE*, (2014): 909–912.
8. Qing Xie and Massoud Pedram. "Performance Comparisons Between 7-nm FinFET and Conventional Bulk CMOS Standard Cell Libraries." *IEEE* 62, no. 8 (August 2015): 761–765.
9. David C. Chen, Guan Shyan Lin, Tien Hua Lee, Ryan Lee, Y.C. Liu, Meng Fan Wang, Yi Ching Cheng and D.Y. Wu. "Compact Modeling Solution of Layout Dependent Effect for FinFET Technology." *IEEE* , (2015): 110–115.
10. Pouya Hashemi, Karthik Balakrishnan, Amlan Majumdar and Ali Khakifirooz. "Strained Si1-xGex-on-Insulator PMOS FinFETs with Excellent Sub-Threshold Leakage, Extremely High Short-Channel Performance and Source Injection Velocity for 10nm Node and Beyond." *IEEE Symposium on VLSI Technology Digest of Technical Papers*, (2014).
11. Peng Zheng, Daniel Connelly, Fei Ding, and Tsu-Jae King Liu. "Simulation-Based Study of the Inserted-Oxide FinFET for Future Low-Power System-on-Chip Applications." *IEEE Electron Device Letters* 36, no. 8 (August 2015): 742–744.
12. Aoxiang Tang, Yang Yang, Chun-Yi Lee and Niraj K. Jha. "McPAT-PVT: Delay and Power Modeling Framework for FinFET Processor Architectures under PVT Variations." *IEEE Transactions on Very Large Scale Integration (VLSI) Systems* 23, no. 9 (September 2015): 1616–1627.
13. Kuruva Hemanjaneyulu and Mayank Shrivastava. "Fin Enabled Area Scaled Tunnel FET." *IEEE Transactions on Electron Devices* 62, no. 10 (October 2015): 3184–3191.
14. S. Khandelwal, A.I. Khan, J.P. Duarte, A.B. Sachid, S. Salahuddin, and C. Hu. "Circuit Performance Analysis of Negative Capacitance FinFETs." *IEEE Symposium on VLSI Technology Digest of Technical Papers*, (2016).

15. H.J. Cho and K.H. Yeo. "Si FinFET based 10nm Technology with Multi Vt Gate Stack for Low Power and High Performance Applications." *IEEE Symposium on VLSI Technology Digest of Technical Papers*, (2016).
16. Ahmedullah Aziz, Nikhil Shukla, Suman Datta and Sumeet Kumar Gupta. "Steep Switching Hybrid Phase Transition FETs (Hyper-FET) for Low Power Applications: A Device-Circuit Co-design Perspective," *IEEE Transactions on Electron Devices* 64, no. 3 (March 2017): 1350–1357.
17. Jagar Singh, J. Ciavatti, K. Sundaram, J.S. Wong, A. Bandyopadhyay, X. Zhang, S. Li, A. Bellaouar, J. Watts, J.G. Lee, and S.B. Samavedam. "14-nm FinFET Technology for Analog and RF Applications." *IEEE Transactions on Electron Devices* 65, no. 1 (January 2018): 31–37.
18. K.I. Seo, B. Haran, D. Gupta, D. Guo, T. Standaert and R. Xie4. "A 10nm Platform Technology for Low Power and High Performance Application Featuring FINFET Devices with Multi Work function Gate Stack on Bulk and SOI." *IEEE Symposium on VLSI Technology Digest of Technical Papers*, (2014).
19. M.D. Giles and N. Arkali Radhakrishna. "High sigma measurement of random threshold voltage variation in 14nm Logic FinFET technology." *Symposium on VLSI Technology Digest of Technical Papers*, (2015): 150–151.
20. Bhattacharya and Niraj, K.. "FinFETs: From Devices to Architectures." *Hindawi Publishing* 2014, Article ID 365689, 21 pages.
21. S.K. Mohapatra, P. Pradhan, D. Singh, and P.K. Sahu. "The Role of Geometry Parameters and Fin Aspect Ratio of Sub-20nm SOI-FinFET: An Analysis towards Analog and RF Circuit Design." *IEEE Transactions on Nanotechnology* 14, no. 3 (May 2015): 546–554.
22. Peng Zheng, Daniel Connelly, Fei Ding, and Tsu-Jae King Liu. "FinFET Evolution toward Stacked-Nanowires FET for CMOS Technology Scaling." *IEEE Transactions on Electron Devices* 62, no. 12 (December 2015): 3945–3950.
23. Pavan H Vora, and Ronak Lad. "A Review Paper on CMOS, SOI and FinFET Technology," [online]. Available: https://www.design-reuse.com/articles/41330/cmos-soi-finfet-technology-review-paper.html
24. R. Xie, P. Montanini, K. Akarvardar, N. Tripathi and B. Haran. "A 7nm FinFET Technology Featuring EUV Patterning and Dual Strained High Mobility Channels." *2016 IEEE International Electron Devices Meeting (IEDM)*, (2016), pp. 47–50, doi: 10.1109/IEDM.2016.7838334.
25. Mohamed T. Ghoneim, Nasir Alfaraj, Galo A. Torres-Sevilla, Hossain M. Fahad and Muhammad M. Hussain. "Out-of-Plane Strain Effects on Physically Flexible FinFET CMOS." *IEEE Transactions on Electron Devices* 63, no. 7 (July 2016): 2657–2664.
26. Xiaoliang Dai and Niraj K. Jha. "Improving Convergence and Simulation Time of Quantum Hydrodynamic Simulation: Application to Extraction of Best 10-nm FinFET Parameter Values." *IEEE Transactions on Very Large Scale Integration (VLSI) Systems* 25, no. 1 (January 2017): 319–329.
27. Hassan Ghasemzadeh Mohammadi, Pierre-Emmanuel Gaillardon and Giovanni De Micheli. "Efficient Statistical Parameter Selection for Nonlinear Modeling of Process/Performance Variation." *IEEE Transactions on Computer-Aided Design of Integrated Circuits and Systems* 35, no. 12 (December 2016): 1995–2007.
28. Juan Nunez and Maria J. Avedillo. "Comparative Analysis of Projected Tunnel and CMOS Transistors for Different Logic Application Areas." *IEEE Transactions on Electron Devices*, 63, no. 12 (December 2016): 5012–5020.
29. Kumar Prasannajit Pradhan, Samar K. Saha, Prasanna Kumar Sahu and Priyanka. "Impact of Fin Height and Fin Angle Variation on the Performance Matrix of Hybrid FinFET's." *IEEE Transactions on Electron Devices* 64, no. 1 (January 2017): 52–57.

30. Paula Ghedini Der Agopian, Joao Antonio Martino, Rita Rooyackers, Anne Vandooren, Eddy Simoen, and Cor Claeys. "Experimental Comparison Between Trigate p-TFET and p-FinFET Analog Performance as a Function of Temperature." *IEEE Transactions on Electron Devices* 60, no. 8 (August 2013): 2493–2497.
31. Tomasz Brozek, Stephen Lam, Shia Yu and Mike Pak. "Gated Contact Chains for Process Characterization in FinFET Technologies." *IEEE Conference on Microelectronic Test Structures*, (March 24–27, 2014): 64–69.
32. Huichu Liu, Matthew Cotter, Suman Datta and Vijay krishnan Narayanan. "Soft-Error Performance Evaluation on Emerging Low Power Devices." *IEEE Transactions on Devices* and Materials Reliability 14, no. 2 (JUNE 2014): 732–741.
33. Maria Glória Cano de Andrade, Joao Antonio Martino, Eddy Simoen, and CorClaeys, "Comparison of the Low-Frequency Noise of Bulk Triple-Gate FinFETs With and Without Dynamic Threshold Operation." *IEEE: Electron Device Letters* 32, no. 11 (November 2011): 1597–1599.
34. Prasad, M., and Dr. U.B. Mahadevaswamy. "Comparative Study of MOSFET, CMOS and FINFET." *Proc. of Int. Conf. on Current Trends in Eng., Science and Technology, ICCTEST*, (2017): 355–362.
35. Venkata P. Yanambaka, Saraju P. Mohanty, Elias Kougianos, Dhruva Ghai and Garima Ghai, Member. "Process Variation Analysis and Optimization of a FinFET-Based VCO." *IEEE Transactions on Semiconductor Manufacturing* 30, no. 2 (May 2017): 126–134.
36. Dr. V. Rukkumani, Dr. M. Saravana kumar and Dr. K. Srinivasan. "Design and Analysis of SRAM Cells for Power Reduction Using Low Power Techniques." *IEEE Region 10 Conference (TENCON): Proceedings of the International Conference*, (2016): 3058–3062.
37. Govind Prasad and Gande Bhargav. "Novel Low Power 10T SRAM Cell on 90nm CMOS." *2016: International Conference on Advances in Electrical, Electronics, Information, Communication and Bio-Informatics (AEEICB16)*, (2016): 978–982.
38. Navneet Kaur, Neha Gupta, Hitesh Pahuja, Balvinder Singh and Sudhakar Pandey. "Low Power FinFET Based 10T SRAM Cell." *IEEE:CIPECH-16*, (2016): 227–233.
39. Soumitra Pal, Arundhati Bhattacharya and Aminul Islam. "Comparative Study of CMOS- and FinFET-based 10T SRAM Cell in Sub threshold regime." *IEEE: ICACCCT*, (2014): 507–511.
40. R.K. Singh, Manisha Pattanaik, Neeraj Kr. Shukla, S. Birla and Sveen Nagpal. "Analysis and Simulation of a Low-Leakage 10T SRAM Bit-Cell using Dual-Vth Scheme at Deep Sub-Micron CMOS Technology." *2012: Proc. of the Intl. Conf. on Advances in Electronics, Electrical and Computer Science Engineering—EEC 2012*, (2012): 84–87.
41. Nidhi Sharma, Uday Panwar and Virendra Singh. "A Novel Technique of Leakage Power Reduction in 9T SRAM Design in FINFET Technology." *2016 IEEE: 6th International Conference – Cloud System and Big Data Engineering*, (2016): 737–743.
42. Tae Woo Oh, Hanwool Jeong, Kyoman Kang, Juhyun Park, Younghwi Yang and Seong-Ook Jung. "Power-Gated 9T SRAM Cell for Low-Energy Operation." *IEEE: IEEE Transactions on Very Large Scale Integration (VLSI) Systems*, (2016): 1–5.
43. Younghwi Yang, Juhyun Park, and Seung Chul. "Single-Ended 9T SRAM Cell for Near-Threshold Voltage Operation With Enhanced Read Performance in 22-nm FinFET Technology." *IEEE: IEEE Transactions on Very Large Scale Integration (VLSI) Systems* 23, no. 11 (November 2015): 2748–2752.
44. R.K. Singh, Shilpi Birla and Manisha Pattanaik. "Characterization of 9T SRAM Cell at Various Process Corners at Deep Sub-micron Technology for Multimedia Applications." *International Journal of Engineering and Technology(IACSIT)*, 3, no. 6 (December 2011): 696–700.

45. Pooran Singh and Santosh Kumar Vishvakarma. "Ultra-Low Power High Stability 8T SRAM for Application in Object Tracking System." *IEEE Transactions on Very Large Scale Integration (VLSI) Systems* 6 (December 2017): 2279–2290.
46. P. Raikwal, V. Neema and A. Verma. "High Speed 8T SRAM Cell Design with Improved Read Stability at 180nm Technology." *IEEE: International Conference on Electronics, Communication and Aerospace Technology, ICECA*, (2017): 563–568.
47. P. Raikwal, V. Neema and A. Verma. "Design and Analysis of Low Power Single Ended 8T SRAM ARRAY (4x4) at 180nm Technology." *2016 IEEE: International conference on Signal Processing, Communication, Power and Embedded System (SCOPES)-2016*, (2016): 312–316.
48. Ghasem Pasandi and Sied Mehdi Fakhraie. "A New Sub-300mV 8T SRAM Cell Design in 90nm CMOS." *2013 IEEE*, (2017): 39–44.
49. Aminul Islam and Mohd. Hasan. "FinFET-based Variation Resilient 8T SRAM Cell." *2012 IEEE*, (2012): 121–125.
50. G. Sneha, Dr. B. Hari Krishna and C. Ashok Kumar. "Design of 7T FinFET Based SRAM Cell Design for Nanometer Regime." *2017 IEEE: International Conference on Inventive Systems and Control (ICISC-2017)*, (2017): 1–4.
51. Younghwi Yang and Hanwool Jeong. "Single Bit-Line 7T SRAM Cell for Near-Threshold Voltage Operation With Enhanced Performance and Energy in 14 nm FinFET Technology." *2016 IEEE Transactions on Circuits and Systems—I: Regular Papers* 63, no. 7 (July 2016).
52. Majid Moghaddam. "Ultra Low-Power 7T SRAM Cell Design Based on CMOS." *2015 IEEE: 23rd Iranian Conference on Electrical Engineering (ICEE)*, 1357–1361, July.
53. Sudarshan Patil and V. S. Kanchana Bhaaskaran. "Optimization of Power and Energy in FinFET Based SRAM Cell Using Adiabatic Logic." *2017 IEEE: International Conference on Nextgen Electronic Technologies*, (2017): 394–402.
54. Apoorva Pathak, Divyesh Sachan, Harish Peta and Manish Goswami. "A Modified SRAM Based Low Power Memory Design." *2016 IEEE: 29th International Conference on VLSI Design and 15th International Conference on Embedded Systems*, (2016): 1354–1359.
55. Saurabh and P. Srivastava. "Low Power 6T -SRAM Comparative Study of Different Architecture in 32 nm Technology using Microwind." *IEEE*, (2012): 978–982.

12 A Study of Leakage and Noise Tolerant Wide Fan-in OR Logic Domino Circuits

Ankur Kumar
Department of Electronics and Communication Engineering, Institute of Engineering and Technology, Lucknow

Sajal Agarwal
Department of Electronics Engineering, Rajeev Gandhi Institute of Petroleum Technology, Jias, India

Vikrant Varshney
Department of Electronics and Communication Engineering, Institute of Engineering and Technology, Lucknow

Abhilasha Jain
Department of Electronics and Communication Engineering, Institute of Engineering and Technology, Lucknow

R. K. Nagaria
Department of Electronics and Communication Engineering, Meerut Institute of Engineering and Technology, Meerut

12.1 INTRODUCTION

Wide fan-in OR domino circuits are widely used to perform many operations in signal processors and microprocessors [1–10]. These operations must be performed within a few nanoseconds with no leakage. Hence, low-power and high-speed domino circuits are required to implement these operations and their applications [11–67]. In order to achieve dynamic power at a satisfactory level, the supply as well as the size of transistors must be downsized as per the technology rule [1–6]. Moreover, the threshold voltage of the devices must be reduced to maintain the performance of the circuits [6–10, 68]. The reduced

DOI: 10.1201/9781003220350-12

threshold voltage causes a large leakage current in the device and reduces the noise immunity (NI) of the circuit [6–29, 68]. Therefore, a static inverter and week keeper transistor is connected to the dynamic node (DN) to preserve the high logic at the output node [30–43]. The ratio of the current drive capacity of the keeper transistor and that of the evaluation transistor is termed keeper ratio (KR) [44–57]. The NI of the transistor is boosted by changing the size of the keeper transistor. However, the various performance parameters, such as delay, power, variability and so on, are increased due to the increase in KR value [48–67]. Hence, this process is not a consistent solution because of the negative effect on the various performance parameters [11–50].

The operation of the domino circuit is to interpretate by stored charge in output node capacitor [11], which is controlled by DN, as given in Figure 12.1. This node should ideally be at high in the evaluation phase if there is no discharging path through PDN. Eventually, the charge of DN starts to leak because of the leakage current, which degrades the signal integrity. There are various types of leakage currents that exist in the MOS [4–10]. These leakage currents degrade the performance of the circuits. It can be observed through literature [11–67] that subthreshold leakage is the main contributor among all components of leakage currents in the PDN. It is also found that subthreshold leakage increases rapidly with the temperature and scaling of technology node [14–25]. The charge leakage of DN occurs significantly because of subthreshold current and partially by reverse diode leakage [29–43].

The subthreshold leakage is a current that flows from drain to source in weak inversion operation of MOS transistor [6, 7]. The subthreshold current is at its peak when the voltage difference between source and drain is equal to the power supply in a turned OFF transistor [8, 9]. In the present scenario, the subthreshold leakage is the predominant source of leakage current [36, 57], which is expressed by eq. (12.1)

$$I_{sub} = I_0\left(1 - e^{\left(\frac{-V_{ds}}{V_t}\right)}\right)\left(e^{\frac{-V_{gs}-V_{th0}-\eta V_{ds}}{\eta V_t}}\right) \tag{12.1}$$

As the subthreshold current in the PDN increases, the fan-in increases; this results in an increase in the power dissipation and charge sharing. Therefore, the probability of logic failure in domino circuit increases with reduction in NI [47–61]. Thus, it is very essential to design said domino circuit. Hence, the various circuit-level techniques [11–66] are designed in the literature to drop the leakage current and enhance the NI of circuit at low PDP.

In the second section, a detailed review of various existing leakage-tolerant domino circuits is given with the advancements and limitations of their structure. In the third section, various reported wide fan-in domino circuits are compared and discussed based on their advanced performances. In the last section, the chapter is concluded based on the arguments discussed in the whole chapter.

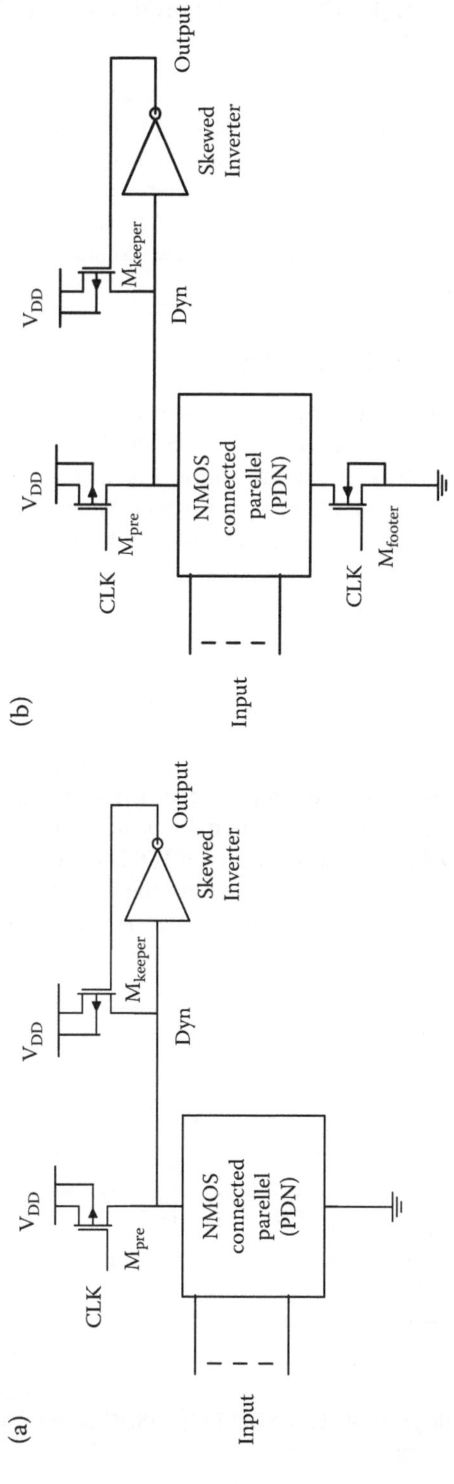

FIGURE 12.1 The circuit diagram of conventional domino circuit (CDC) (a) without footer and (b) with footer [11]. It has a keeper network to recharge the DN and an evaluation network to verify the output logic.

12.2 REVIEW OF LEAKAGE-TOLERANT WIDE FAN-IN OR LOGIC DOMINO CIRCUITS

In the past few decades, many techniques have been designed and modified to overcome leakage and improve the NI for wide fan-in OR logic domino circuits [11–67]. Thus, a detailed review of some existing techniques is studied in a systematic manner. In this section, most of the existing circuits are explained in detail for better understanding of the readers. Therefore, the researchers/VLSI designers can acquire complete knowledge to propose a new domino circuit further by overcoming the said performance limitations.

12.2.1 High-Speed Domino Circuit (HSDC)

The HSDC was designed by Anis et al. [13], as given in Figure 12.2, which minimizes the PD and leakage current. Here, NMOS and PMOS are integrated to effectively turn OFF and ON the keeper transistor in good manner. This integration of transistors enables one to save the PD to a great extent. Moreover, dual-threshold transistors are also included to minimize the effect of leakage current on the DN. Thus, it improves the NI of the designed circuit. This arrangement of transistors is not sufficient to achieve the NI matric in case of wide fan-in.

CKDC added two keeper (WK and SK) to recharge the DN.

12.2.2 Conditional Keeper Domino Circuit (CKDC)

Alvandpour et al. presented CKDC [14], as represented in Figure 12.3, in which dual keeper, i.e week keeper (WK) and strong keeper (SK), are added to improve the NI of the circuit. WK turns ON unconditionally at the start of the evaluation phase. Whereas SK turns ON after a certain delay, if DN is not discharging through PDN. Moreover, a NAND gate is included to control the switching activity of SK, resulting in reduction in PD of overall performance. However, NI is enhanced because of both keepers at the cost of PC and area overhead.

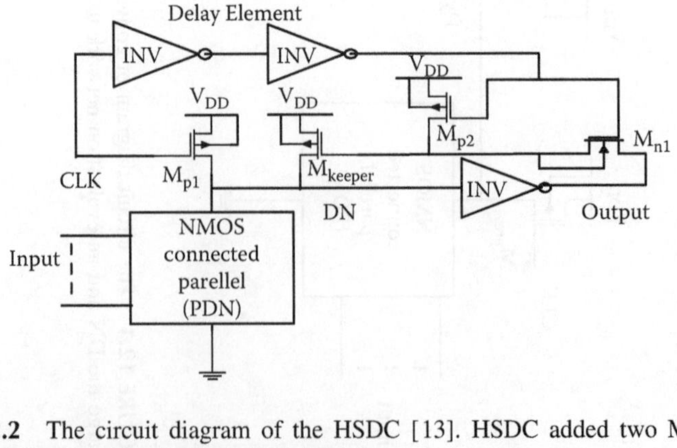

FIGURE 12.2 The circuit diagram of the HSDC [13]. HSDC added two M_{n1} and M_{n2} transistors to control the switching of the keeper.

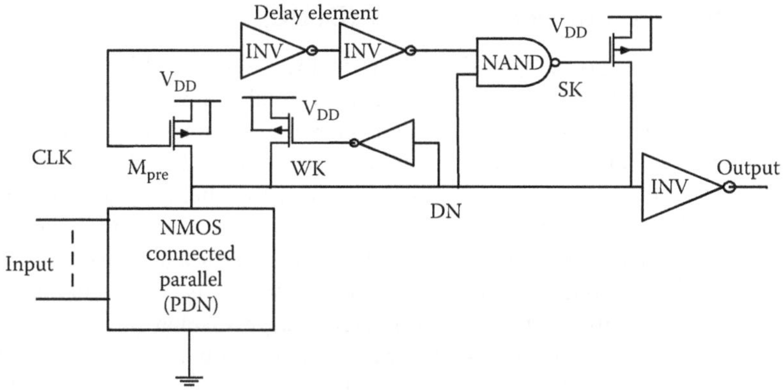

FIGURE 12.3 The circuit diagram of the CKDC [14].

12.2.3 Current Mirror-Footed Domino Circuit (CMFDC)

The CMFDC are presented by Moradi et al. [16], as represented in Figure 12.4, which overcome the effect of leakage current by redesigning the evaluation network. In this, three NMOS transistors are added at the bottom of the footer transistor. The transistor M_{n1} is connected in diode-configuration to develop the body and stacking effect. Further, M_{n2} is included in the feedback loop to minimize the leakage current. Furthermore, M_{n3} transistor is included as a mirror to the M_{n1} for quick discharging of the DN when any input is applied in the evaluation phase. This domino is incapable of overcoming the area, delay and NI in the case of large fan-in.

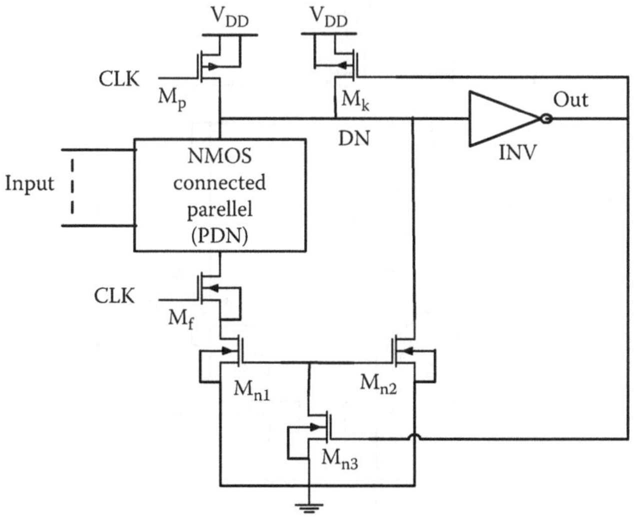

FIGURE 12.4 The circuit diagram of CMFDC [16]. CMFDC added three M_{n1}, M_{n2} and M_{n3} transistors to decrease leakage current.

12.2.4 Diode-Footed Domino Circuit (DFDC)

Mahmoodi et al. proposed DFDC [17], as in Figure 12.5, which is specially designed to minimize the leakage current to a notable level. For this, footer transistors are introduced in a diode-configuration scheme, which develop a negative potential at the source terminal of the PDN when DN is not discharging in the evaluation phase. Thus, this potential behaves as a negative V_{GS} (voltage between gate to source), resulting in enhancement of threshold voltage in PDN. Therefore, subthreshold leakage current and noise margin of the gate is decreased and increased, respectively. Mirror is used to balance the performance degradation of the circuit. Here, large discharging path is developed due to incorporation of many transistor in evaluation network. Thus, these transistors should be of large size, which leads to large area and delay overhead. Hence, DFDC is not suitable to implement the VLSI circuits at low technology node.

12.2.5 Leakage Current Replica Domino Circuit (LCRDC)

The LCRDC [19] is the new challenging domino circuit, which is presented by Lih et al., as in Figure 12.6. An extra PMOS transistor is cascaded with keeper in this domino, which is driven by an analog current mirror. This arrangement of transistors guides the circuit to track the corners of process, voltage and temperature. Circuit works as rate controller to truncate the subthreshold leakage during the standby mode. Moreover, switching activity of keeper is improved to boost the NI because of analog current mirror. LCRDC is not crucial in the case of the large fan-in and low technology node.

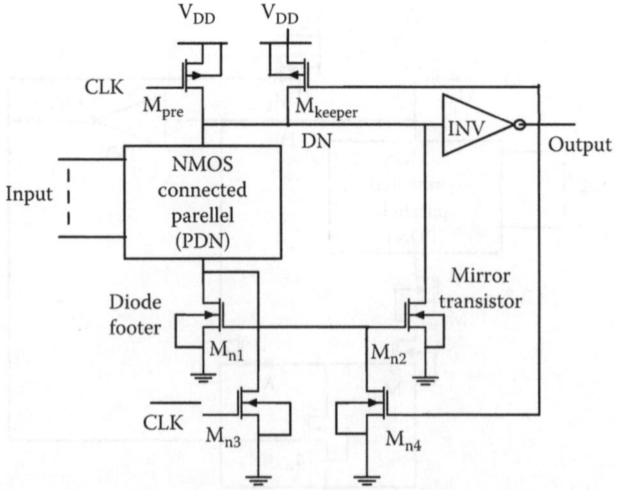

FIGURE 12.5 The circuit diagram of the DFDC [17]. DFDC added four M_{n1}, M_{n2}, M_{n3}, and M_{n4} transistors to minimizes leakage current.

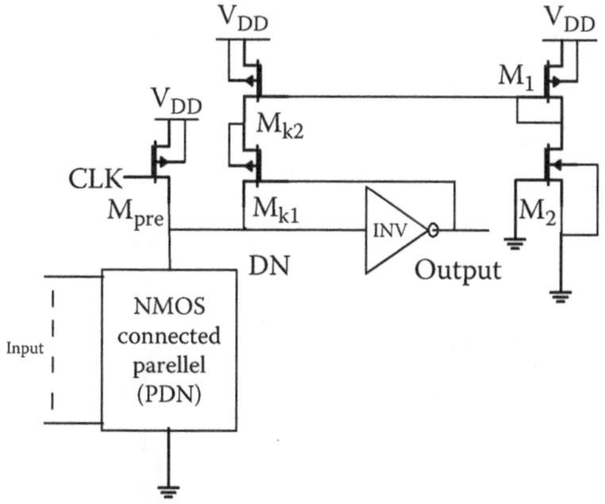

FIGURE 12.6 The circuit diagram of the LCRDC [19], LCRDC added analog current mirror (M_1 and M_2 transistors) to track variation.

12.2.6 Single-Phase Domino Circuit (SPDC)

Akl et al. introduced SPD [22], as shown in Figure 12.7, in which dynamic PD is reduced by limiting the switching activity. Here, CLK_1 and CLK_2 (delayed version of CLK_1) pulse are used to charge the DN upto V_{DD} by the keeper at starting of evaluation phase, i.e. DN in precharge phase is not charged due to which switching activity of the DN is limited. This circuit is limited to 8- or 16-bit fan-in.

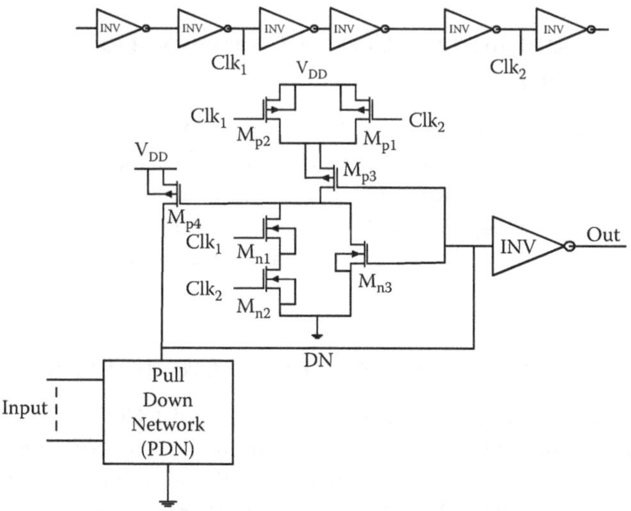

FIGURE 12.7 The circuit diagram of the SPDC [22], SPDC redesigned the keeper network for better controlling of DN.

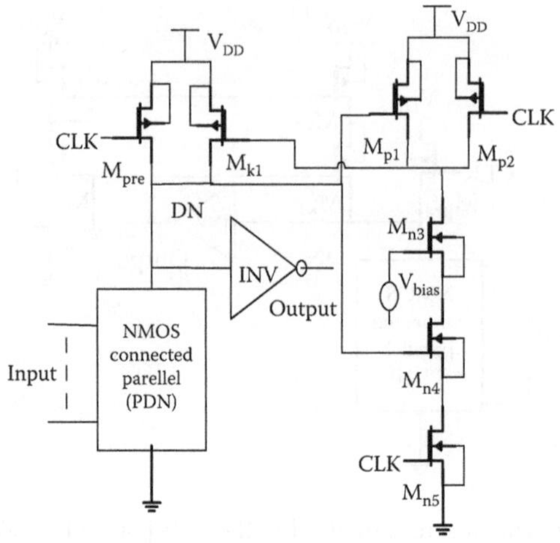

FIGURE 12.8 The circuit diagram of the RSKDC [25]. It has rate controller (M_{n3}, M_{n4}, M_{n5}, M_{p1} and M_{p2}) to upgrade the noise immunity.

12.2.7 Rate Sensing Keeper Domino Circuit (RSKDC)

The RSKDC [25] was introduced by Jeyasingh et al. to upgrade the NI and track the process corner, as shown in Figure 12.8. The rate controller is included to make effective controlling of the keeper transistor. Also, this rate sensed keeper is useful to track the different process corners along with faster switching of circuit. Moreover, NI of domino circuit is enhanced. RSKDC is less effective because of high digital load at DN.

12.2.8 Controlled Keeper by Current Comparison Domino Circuit (CKCCDC)

Peiravi et al. designed a new robust design CKCCDC [28] for low leakage, as shown in Figure 12.9. In this, keeper turns ON and OFF on the difference of leakage current between PMOS transistor (M_{p2}) and PDN. This efficient controlling of keeper improves the noise margin. Further, stacking effect is developed in evaluation network to truncate the leakage current. However, area is increased with a large factor.

12.2.9 Current Comparison-Based Domino Circuit (CCDC)

Peiravi et al. introduced another domino circuit called CCDC [29], as shown in Figure 12.10, in which leakage current is dropped by comparing the worst case leakage to the current of PUN (Pull Up Network). The most influenced factor is use of PUN at the place of PDN. Therefore, unwanted charge sharing will not occur as in the PDN due to which NI of domino circuit improves at the cost of area. A complex reference circuit is incorporated to produce the reference voltage, which leads to complexity and process variations.

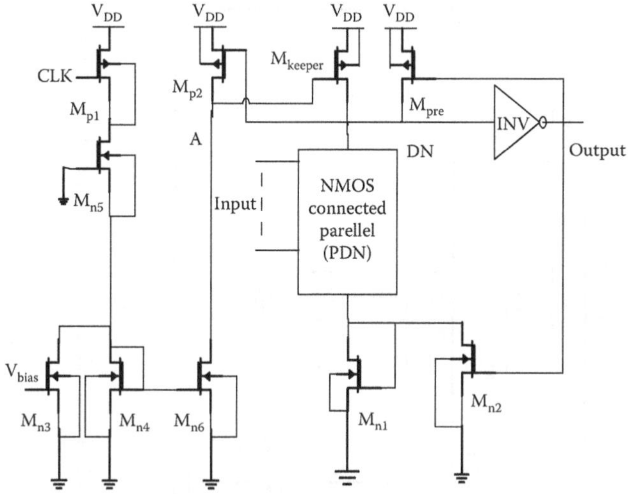

FIGURE 12.9 The circuit diagram of the CKCCDC [28]. I added a referenced circuit at left side and (M_{n1} and M_{n2}) transistors to decrease the leakage current.

12.2.10 Voltage Comparison Domino Circuit (VCDC)

Asyaei et al. proposed VCDC [36], as shown in Figure 12.11, in which DN is separated to the PDN so that the influence of the leakage current can be overcome on the DN. A sense amplifier is also introduced to get the correct output logic by measuring the difference of top and bottom node voltages of PDN. This arrangement reduces the leakage current in a large amount. Moreover, switching PC can be overcome be reducing the swing of the DN in this circuit. Sense amplifier is crucial to determine the true logic at output node. However, this amplifier increases the area and process variation of VCDC. Hence, the performance starts to degrade for wide fan-in.

12.2.11 Foot-Driven Stack Transistor Domino Circuit (FDSTDC)

Garg et al. introduced FDSTDC [47] to minimize the leakage current by developing the body and stacking effect, as shown in Figure 12.12. Here, two NMOS transistors are added to avoid the unnecessary discharging of DN in the evaluation network. Both transistors will be turned ON and OFF alternatively to limit the leakage. This inclusion of transistors in the footer is capable of overcoming the leakage current. However, it is limited in terms of delay and area.

12.2.12 Low Power Domino Circuit (LPDC)

The LPDC is designed by Asyaei et al. [49] to reduce the PD in the case of bit lines, as shown in Figure 12.13, in which PD is minimized by decreasing the voltage swing of the PDN. Also, a reference inverter is included to turn ON and OFF the footer transistor, which minimizes the PD and leakage current. It is limited because

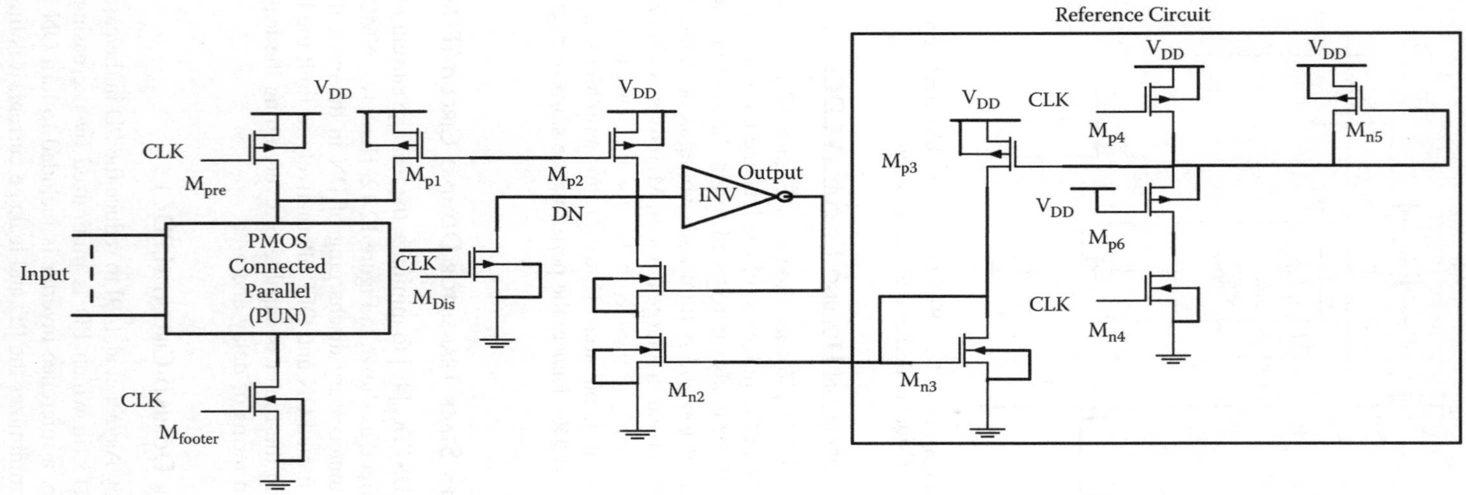

FIGURE 12.10 The circuit diagram of the CCDC [29]. Here, a virtual DN is created, which is controlled by a reference circuit to provide a better noise immunity.

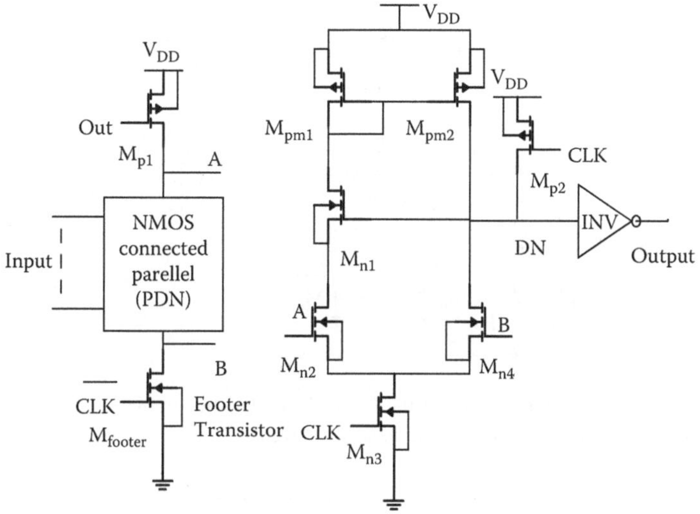

FIGURE 12.11 The circuit diagram of the VCDC [36]. Here, the voltage comparator is designed to control the isolated DN.

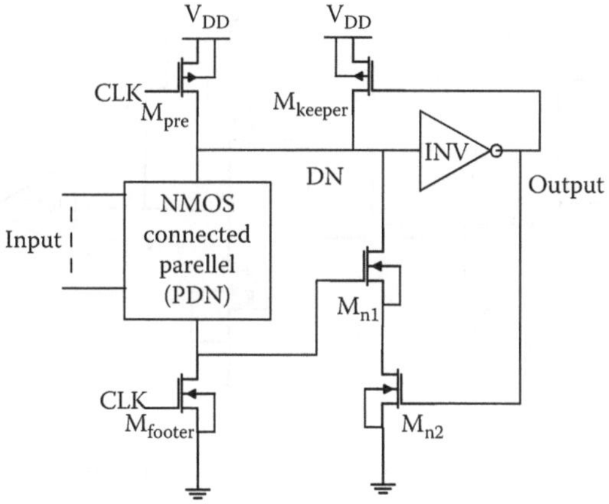

FIGURE 12.12 The circuit diagram of the FDSTDC [47]. The FDSTDC added three (M_{n1}, M_{n2} and M_{footer}) transistors to overcome the charge sharing.

output logic starts to degrade the performance of the circuit in the case of high fan-in.

12.2.13 New Low-Power Domino Circuit (NLPDC)

The NLPDC [50] was designed by Asyaei et al., as given in Figure 12.14. Here, output is isolated from PDN to mitigate the effect of leakage current. The inverter is added in different ways to scale down the voltage swing at DN. Thus, the PD is reduced by a large quantity of the domino circuit. Moreover, body effect

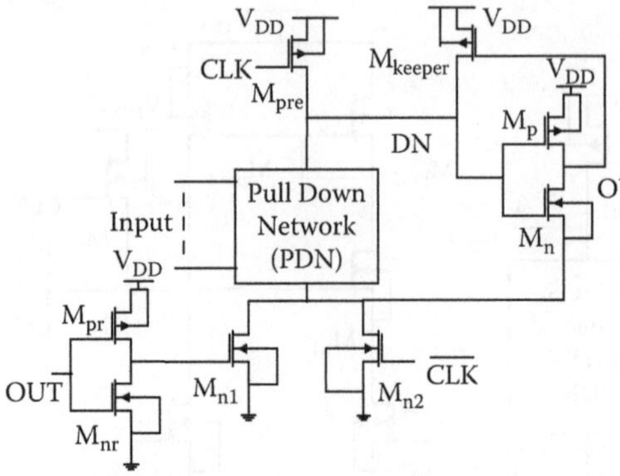

FIGURE 12.13 The circuit diagram of the LPDC [49]. The two transistors (M_p and M_n) are inserted in footer of PDN to truncate the leakage current.

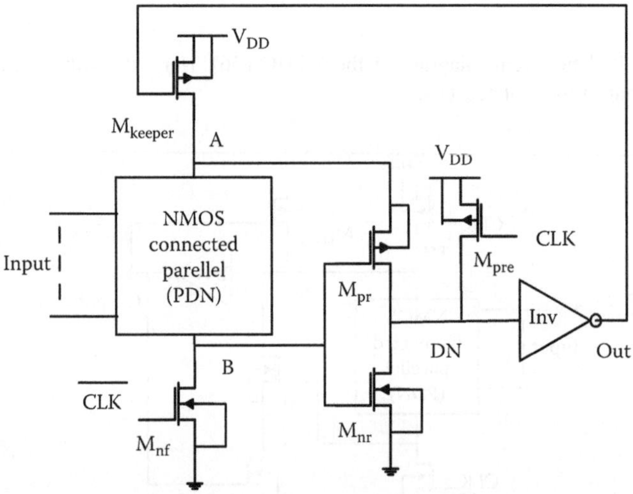

FIGURE 12.14 The circuit diagram of NLPDC [50]. An inverter (Mpv and Mnv) is inserted to decrease the voltage swing.

phenomena is developed to truncate the subthreshold leakage current. This circuit is not suitable for high-speed application because of the increased delay.

12.2.14 Controlled-Current Comparison-Based Domino Circuit (C3DC)

Asyaei et al. proposed C3DC [56] to upgrade the NI and reduction of PD for large fan-in, as shown in Figure 12.15. In this, output is isolated from PDN to overcome the effect of leakage current. Here, output is decided by comparing the two currents, which are controlled by the voltages across the PDN. Moreover, a diode-connected

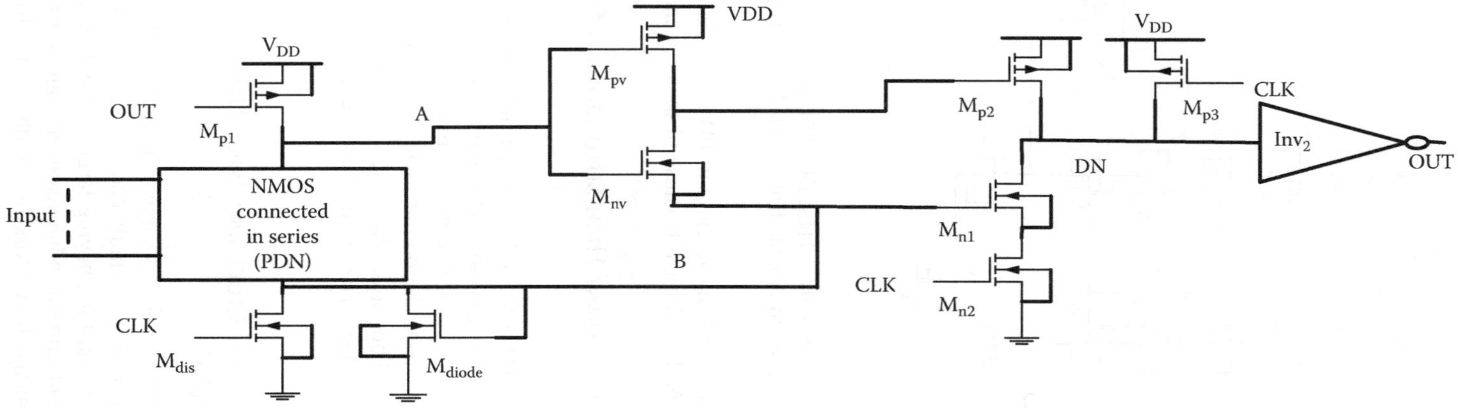

FIGURE 12.15 The circuit diagram of the C3DC [56]. An inverter (Mpv and Mnv) is added to control the virtual DN and reduce the voltage swing.

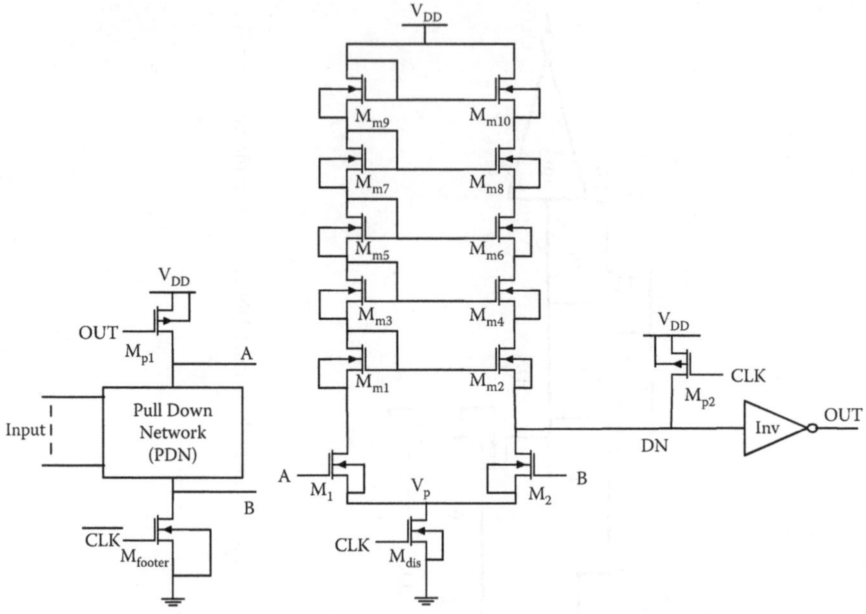

FIGURE 12.16 The circuit diagram of the LTHSCDC [57]. In this, a voltage comparator, having 5 mirror branches, is designed to provide better noise immunity and power saving.

transistor is included to lessen the leakage current. PD is also reduced by lowering the voltage swing of the PDN. It is limited because of the area and delay overhead.

12.2.15 A New Leakage-Tolerant High-Speed Domino Circuit (LTHSDC)

Ankur et al. introduced the LTHSDC [57], as given in Figure 12.16, which is the advance of the VCDC. Here, output is isolated to the PDN to mitigate the effect of the leakage current. Further, a novel is designed to boost the NI and overcome the PD. This voltage comparator elects the output logic on the base of voltage difference across the PDN (at top and bottom). The switching power is also minimized by dropping the voltage swing of DN. The circuit lacks because of area and process variations overhead.

12.2.16 Domino Logic with Clock and Input Dependent Transistors Circuit (DOINDC)

Shah et al. proposed DOINDC [58], as shown in Figure 12.17, in which dual threshold techniques are incorporated simultaneously to decrease the leakage current in a large amount. The transistors, having large threshold voltage, are inserted to overcome the subthreshold current, whereas sleep transistors are used to turn off these high threshold voltage transistors in standby. Therefore, overall PC is

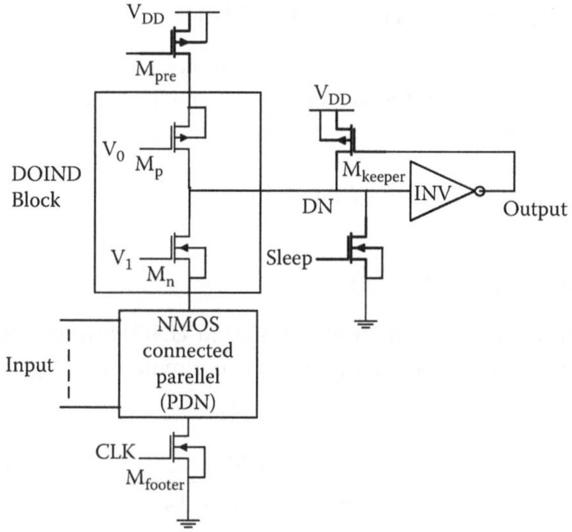

FIGURE 12.17 The circuit diagram of the DOINDC [58]. DOING block (Mp and M_n) are inserted between precharge and evaluation network to truncate leakage current.

decreased in the designed circuit at improved robustness. DOINDC has complications to decide both voltages for the DOIND block.

Moreover, authors are using different techniques, e.g. FINFET, CNFET etc., to implement the wide fan-in OR logic domino circuit [54, 59, 62, 64, 65]. In these circuits, MOS transistors are simply replaced by the FINFET, CNFET etc. to improve the performance. It has been observed through the literature review that leakage current could be reduced in two possible way. In the first way, output is isolated to the PDN to decease the effect of leakage current on DN [29, 36, 50, 56, 57]. This technique also reduces the PD of the domino circuits. In the second way, the evaluation network is redesigned by incorporating a transistor at the bottom of the PDN in a different fashion [16, 17, 28, 28, 47, 49, 58]. This incorporation increases the threshold voltage of transistors in the PDN to reduce the subthreshold leakage current. Further, it is also found out through the literature review that NI of the domino circuit is improved by modifying the keeper network [13, 14, 19, 19, 22, 25]. This modification effectively controls the switching of the keeper transistor, which decreases the switching PD of domino circuits.

12.3 PERFORMANCE PARAMETER OF DOMINO CIRCUITS

12.3.1 Power Dissipation

The average PD in domino circuit is the algebraic sum of switching PD (P_{sw}), short circuit PD (P_{sh}) and leakage PD (P_{le}), which is defined as [2, 21]:

$$P_{avg} = P_{sw} + P_{sh} + P_{le} \tag{12.2}$$

12.3.2 Unity Noise Gain Margin (UNGM)

UNGM is the most used matric to determine the noise margin of digital circuits. The value of input signal that may developed the equal amount of amplitude at the output is termed (UNGM) [28, 36, 56].

$$UNGM = \{Vnoise: Vnoise = Vout\} \tag{12.3}$$

12.3.3 Figure of Merit (FOM)

The FOM can be obtained by ration of normalized UNGM to the product of mean normalized of PD (PD_{nm}), delay factor (t_{nm}) and circuit area (A_{nm}) in accordance with (CDC) [25, 28, 36, 57].

$$FOM = \frac{UNGM_{nm}}{PD_{nm} \times t_{nm} \times A_{nm}} \tag{12.4}$$

12.4 SIMULATION RESULTS AND DISCUSSION OF REPORTED DOMINO CIRCUITS

The performance evaluation of the reviewed work is done to verify the outcomes. The various existing domino circuits are structured and simulated by "SPECTRE" simulator under the environment of "CADENCE VIRTUOSO" at 45 nm technology node. The supply voltage, temperature and capacitive load are considered as 1 V, room temperature (27°C), and 5 fF, respectively for the simulation of all the circuits. Most of the performance parameters, e.g. FOM, NI, delay, area and PD etc., are obtained under similar delay and compared in a similar simulation environment to make an honest and true comparison. The performance parameters at 8, 16, 32 and 64 fan-in are found by simulating some reported domino circuits for the OR logic.

The performance of the CDC [11], HSDC [13], CKDC [14], DFDC [17], LCRDC [19], RSKDC [25], CKCCDC [28], VCDC [36], LPDC [49], C3DC [56] and DOINDC [58] is evaluated at different fan-in. The above-mentioned circuits are selected based on different design techniques/methods as discussed in the review section. The difference among reported OR logic domino circuits of PD and normalized PD at several fan-in is given in Table 12.1. It has been found through the results that PD increases as the fan-in increases in each domino circuits. The DFDC [17] and C3DC [56] have large reduction of PD among all the reported domino circuits. The difference among reported OR logic domino circuits of UNGM and normalized UNGM at several fan-in is given in Table 12.2. It has been detected through the results that NI decreases as the fan-in increases in each domino circuits. The VCDC [36] and C3DC [56] have large improvement of NI among all the reported domino circuits.

The observations among reported domino circuits of normalized leakage current for 32-bit OR logic are given in Figure 12.18. The results show that DFDC [17],

TABLE 12.1
Difference among reported OR logic domino circuits of PD at several fan-in

Fan-in	Gates / Parameters	CDC [11]	HSDC [13]	CKDC [14]	DFDC [17]	LCRDC [19]	RSKDC [25]	CKCCDC [28]	VCDC [36]	LPDC [49]	C3DC [56]	DOINDC [58]
8-bit	PD (μW)	9.48	9.36	9.41	9.25	9.13	9.35	9.32	9.18	9.35	9.21	9.44
	Normalized PD	1.00	0.99	0.99	0.98	0.96	0.99	0.98	0.97	0.99	0.97	1.00
16-bit	PD (μW)	12.33	12.11	12.65	11.75	11.84	11.89	11.83	11.82	11.75	12.01	12.01
	Normalized PD	1.00	0.98	1.03	0.95	0.96	0.96	0.96	0.96	0.95	0.97	0.97
32-bit	PD (μW)	18.01	16.85	17.85	16.35	17.48	16.74	17.29	16.83	16.54	16.89	17.29
	Normalized PD	1.00	0.94	0.99	0.91	0.97	0.93	0.96	0.93	0.92	0.94	0.96
64-bit	PD (μW)	28.32	24.57	27.12	23.38	26.12	25.62	25.83	25.32	25.36	25.16	25.46
	Normalized PD	1.00	0.87	0.96	0.83	0.92	0.90	0.91	0.89	0.90	0.89	0.90

TABLE 12.2
Difference among reported OR logic domino circuits of UNGM at several fan-in

Fan-in	Gates / Parameters	CDC [11]	HSDC [13]	CKDC [14]	DFDC [17]	LCRDC [19]	RSKDC [25]	CKCCDC [28]	VCDC [36]	LPDC [49]	C3DC [56]	DOINDC [58]
8-bit	UNGM (V)	0.33	0.36	0.35	0.33	0.34	0.34	0.37	0.38	0.36	0.37	0.32
	Normalized UNGM	1.00	1.09	1.06	1.00	1.03	1.03	1.12	1.15	1.09	1.12	0.97
16-bit	UNGM (V)	0.29	0.32	0.32	0.30	0.30	0.31	0.33	0.35	0.33	0.33	0.29
	Normalized UNGM	1.00	1.10	1.10	1.03	1.03	1.07	1.14	1.21	1.14	1.14	1.00
32-bit	UNGM (V)	0.24	0.27	0.28	0.25	0.26	0.25	0.28	0.30	0.29	0.29	0.25
	Normalized UNGM	1.00	1.13	1.17	1.04	1.08	1.04	1.17	1.25	1.21	1.21	1.04
64-bit	UNGM (V)	0.19	0.22	0.23	0.21	0.20	0.19	0.22	0.25	0.21	0.24	0.20
	Normalized UNGM	1.00	1.16	1.21	1.11	1.05	1.00	1.16	1.32	1.11	1.26	1.05

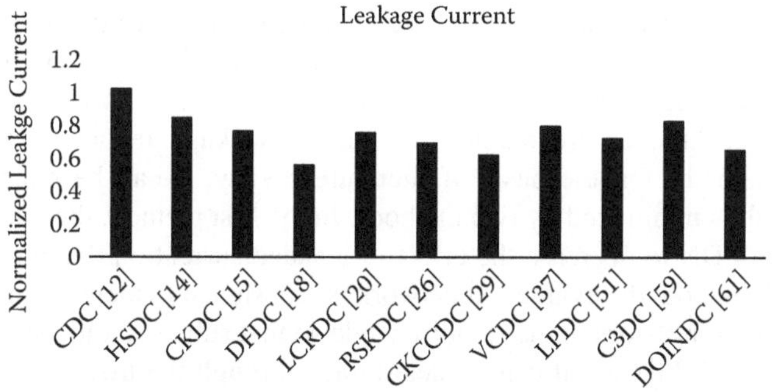

FIGURE 12.18 The observations among reported domino circuits of normalized leakage current for 32-bit OR logic

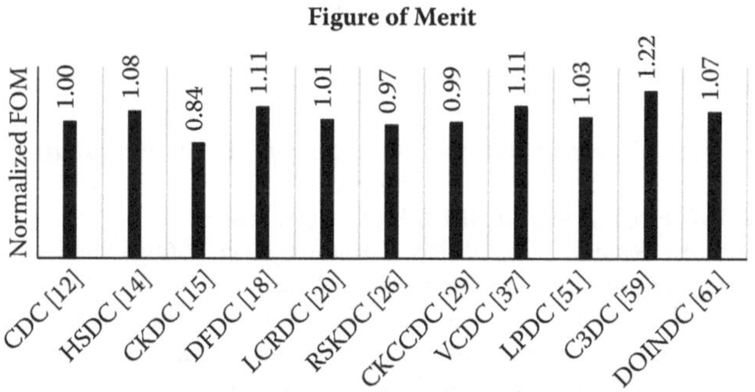

FIGURE 12.19 The observations among reported domino circuits of normalized FOM for 32-bit OR logic

CKCCDC [25] and DOINDC [58] have large reduction in the leakage current among reported domino circuits. Further, the observations among reported domino circuits of normalized FOM for 32-bit OR logic is given in Figure 12.19. This figure depicts that DFDC [17], VCDC [36] and C3DC [56] have large improvement in FOM among reported domino circuits.

From this section, it is determined that PD is minimizing maximum in those domino circuits in which evaluation network is redesigned. However, leakage current is minimizing maximum in those domino circuits in which output is isolated from the PDN. Moreover, NI is enhanced by modifying the keeper network. Therefore, the researcher/VLSI designer can select any reported domino circuits to implement the VLSI circuits as per requirements.

12.5 CONCLUSION

In this chapter, an overview of leakage and noise-tolerant wide fan-in OR logic domino circuits is explained in a detailed manner. The design procedures, analysis and simulation results of domino circuits have been studied and verified. On the basis of literature, various observations are identified to enhance the performance of the domino circuit for wide fan-in OR logic. The standard domino circuit has high subthreshold leakage current, low NI and high PDP. Thus, various leakage and noise-tolerant techniques are studied to find out leakage reduction and NI enhancement methods. On the basis of literature review, it can be concluded that leakage current is minimized by two methods. In the first method, the output node is isolated to the PDN to decrease the effect of leakage current on the dynamic node. In the second method, the evaluation network is redesigned to increase the threshold voltage of PDN transistors in the standby mode so that subthreshold leakage current can be minimized. Further, it can be determined through the literature review that the NI of the domino circuit is improved by modifying the keeper network, which effectively controls switching of the keeper transistor.

Finally, the comparative performance analysis of the reported domino circuits has been done at 45 nm CMOS technology node. All the outcomes of literature are verified through the simulation in CADENCE VIRTUSO.

REFERENCES

1. P. Gronowski. "Issues in dynamic logic design," In *Design of High-Performance Microprocessor Gates*, A. Chandrakasan, W.J. Bowhill, and F. Fox, Eds., pp. 140–157. Piscataway, NJ: IEEE Press, 2001, ch. 8.
2. Krishnamurthy Ram K., et al. "A 130-nm 6-GHz 256/spl times/32 bit leakage-tolerant register file." *IEEE Journal of Solid-State Gates* 37.5 (2002): 624–632.
3. M. Nasserian, M. Kafi-Kangi, M. Maymandi-Nejad, and F. Moradi. "A low-power fast tag comparator by modifying charging scheme of wide fan-in dynamic or gates." *INTEGRATION, the VLSI Journal* 52 (2016): 129–141.
4. A.P. Chandrakasan, S. Sheng, and R.W. Brodersen. "Low-power CMOS digital design." *IEICE Transactions on Electronics* 75, no. 4 (1992): 371–382.
5. J.M. Rabaey, A.P. Chandrakasan, and B. Nikolic. *Digital Integrated Circuits*, vol. 2. Englewood Cliffs: Prentice Hall, 2002.
6. S.-M. Kang and Y. Leblebici. *CMOS Digital Integrated Circuits*. Tata McGraw-Hill Education, 2003.
7. M.C. Johnson, D. Somasekhar, L.Y. Chiou, and K. Roy. "Leakage control with efficient use of transistor stacks in single threshold CMOS." *IEEE Transactions on Very Large Scale Integration (VLSI) Systems* 10, no. 1 (2002): 1–5.
8. Kaushik Roy, Saibal Mukhopadhyay, and Hamid Mahmoodi-Meimand. "Leakage current mechanisms and leakage reduction techniques in deep-submicrometer CMOS circuits." *Proceedings of the IEEE* 91, no. 2 (2003): 305–327.
9. N. Hanchate and N. Ranganathan. "A new technique for leakage reduction in CMOS circuits using self-controlled stacked transistors." In 17th International Conference on VLSI Design. Proceedings., IEEE, (2004): 228–233.
10. P. Butzen and R. Ribas. "Leakage current in sub-micrometer CMOS gates." *Universidade Federal do Rio Grande do Sul* (2006): 1–28.
11. M. Lee. "High-speed compact circuits with CMOS." *IEEE Journal of Solid-State Circuits* 17, no. 3 (1982): 614–619.
12. Alvandpour A., Larsson-Edefors, P. and Svensson, C.. "A leakage-tolerant multi-phase keeper for wide domino circuits". In ICECS'99. Proceedings of ICECS'99. 6th IEEE International Conference on Electronics, Circuits and Systems (Cat. No. 99EX357), IEEE. 1 (1999, September): 209–212.

13. M.H. Anis, M.W. Allam, and M.I. Elmasry. "Energy-efficient noise-tolerant dynamic styles for scaled-down CMOS and MTCMOS technologies." *IEEE Transactions on Very Large Scale Integration (VLSI) Systems* 10, no. 2 (2002): 71–78.
14. A. Alvandpour, R.K. Krishnamurthy, K. Soumyanath, and S.Y. Borkar. "A sub-130-nm conditional keeper technique." *IEEE journal of solid-state circuits* 37, no. 5 (2002): 633–638.
15. Z. Liu and V. Kursun. "Sleep switch dual threshold voltage domino logic with reduced subthreshold and gate oxide leakage current." *IEEE Transactions on Very Large Scale Integration (VLSI) Systems* 12, no. 5 (2004): 485–496.
16. Moradi Farshad , Ali Peiravi, and Hamid Mahmoodi. "A new leakage-tolerant design for high fan-in domino circuits." In Proceedings. The 16th International Conference on Microelectronics, ICM 2004, IEEE, (2004): 493–496.
17. H. Mahmoodi-Meimand and K. Roy. "Diode-footed domino: A leakage-tolerant high fan-in dynamic circuit design style." *IEEE Transactions on Circuits and Systems I: Regular Papers* 51, no. 3 (2004): 495–503.
18. B. Chatterjee, M. Sachdev, and R. Krishnamurthy. "Designing leakage tolerant, low power wide-OR dominos for sub-130 nm CMOS technologies." *Microelectronics Journal* 36, no. 9 (2005): 801–809.
19. Y. Lih, N. Tzartzanis, and W.W. Walker. "A leakage current replica keeper for dynamic circuits yolin." *IEEE Journal of Solid-State Circuits* 42, no. 1 (2006): 48–55.
20. A. Amirabadi, A. Afzali-Kusha, Y. Mortazavi, and M. Nourani. "Clock delayed domino logic with efficient variable threshold voltage keeper." *IEEE Transactions on Very Large Scale Integration (VLSI) Systems* 15, no. 2 (2007): 125–134.
21. Liu Zhiyu, and Volkan Kursun. "PMOS-only sleep switch dual-threshold voltage domino logic in sub-65-nm CMOS technologies." *IEEE Transactions on Very Large Scale Integration (VLSI) Systems* 15, no. 12 (2007): 1311–1319.
22. Akl Charbel J., and Magdy A. Bayoumi. "Single-phase SP-domino: a limited-switching dynamic circuit technique for low-power wide fan-in logic gates." *IEEE Transactions on Circuits and Systems II: Express Briefs* 55, no. 2 (2008): 141–145.
23. D.J.R. Gnana, David, J. and N. Bhat. "A low power, process invariant keeper for high speed dynamic logic circuits." *2008 IEEE International Symposium on Circuits and Systems* (2008): 1668–1671.
24. Volkan Kursun, and Eby G. Friedman. "Domino logic with variable threshold voltage keeper." U.S. Patent 7,388,399, issued (June 17, 2008).
25. R. Gnana, D. Jeyasingh, N. Bhat, and B. Amrutur. "Adaptive Keeper Design for Dynamic Logic Circuits Using Rate Sensing Technique." *IEEE Transactions on Very Large Scale Integration (VLSI) Systems* 19, no. 2 (2009): 295–304.
26. S.H. Rasouli, H.F. Dadgour, K. Endo, H. Koike, and K. Banerjee. "Design optimization of FinFET domino logic considering the width quantization property." *IEEE transactions on electron devices* 57, no. 11 (2010): 2934–2943.
27. Meher Preetisudha, and K.K. Mahapatra. "Ultra low-power and noise tolerant CMOS dynamic circuit technique." *IEEE TENCON 2011-2011 IEEE Region 10 Conference* (2011): 1175–1179.
28. A. Peiravi and M. Asyaei. "Robust low leakage controlled keeper by current-comparison domino for wide fan-in gates." *Integration The VLSI Journal* 45, no. 1 (2012): 22–32.
29. A. Peiravi and M. Asyaei. "Current-comparison-based domino: New low-leakage high-speed domino circuit for wide fan-in gates." *IEEE transactions on Very Large Scale Integration (VLSI) Systems* 21, no. 5 (2012): 934–943.
30. R. Singh, G.M. Hong, M. Kim, J. Park, W.Y. Shin, and S. Kim. "Static-switching pulse domino: A switching-aware design technique for wide fan-in dynamic multiplexers." *INTEGRATION, the VLSI Journal* 45, no. 3 (2012): 253–262.

31. D. Sarma, V.S. Srinivasa, and K.K. Mahapatra. “Improved techniques for high performance noise-tolerant domino CMOS logic circuits.” *IEEE 2012 Students Conference on Engineering and Systems* (2012): 1–6.
32. S. Mehrotra, S. Patnaik, and M. Pattanaik. “Design technique for simultaneous reduction of leakage power and contention current for wide fan-in domino logic based 32:1 multiplexer circuit.” In 2013 IEEE Conference on Information & Communication Technologies, IEEE, (2013): 905–910.
33. F. Moradi, T. Vu Cao, E.I. Vatajelu, A. Peiravi, H. Mahmoodi, and D.T. Wisland. “Domino logic designs for high-performance and leakage-tolerant applications.” *INTEGRATION, the VLSI journal* 46, no. 3 (2013): 247–254.
34. I.C. Wey, T.C. He, H.C. Chow, P.H. Sun, and C.C. Peng. “A high-speed, high fan-in dynamic comparator with low transistor count.” *International Journal of Electronics* 101, no. 5 (2014): 681–690.
35. M. Asyaei and A. Peiravi. “Low power wide gates for modern power efficient processors.” *INTEGRATION, the VLSI Journal* 47, no. 2 (2014): 272–283.
36. M. Asyaei. “A new leakage-tolerant domino circuit using voltage-comparison for wide fan-in gates in deep sub-micron technology.” *INTEGRATION, the VLSI Journal* 51 (2015): 61–71.
37. A. Dadoria, K. Khare, T.K. Gupta, and R.P. Singh. “A novel high-performance lekage-tolerant, wide fan-in domino logic circuit in deep-submicron technology.” *Circuits and Systems* 6, no. 04 (2015): 103.
38. P. Meher and K. Mahapatra. “Modifications in CMOS dynamic logic style: A review paper.” *Journal of the Institution of Engineers (India): Series B* 96, no. 4 (2015): 391–399.
39. Vikas Mahor, and Manisha Pattanaik. “Low leakage and highly noise immune FinFET-based wide fan-in dynamic logic design.” *Journal of Circuits, Systems and Computers* 24, no. 05 (2015): 1550073.
40. Patnaik Satwik , Uthara Hari, Mitali Ahuja, and Soumya Narang. “A modified variation-tolerant keeper architecture for evaluation contention & leakage current minimization for wide fan-in domino structures.” *IEEE 2015 International Conference on Circuits, Power and Computing Technologies [ICCPCT-2015]* (2015): 1–7.
41. P.K. Pal, A.K. Dubey, S.R. Kassa, and R.K. Nagaria. “Voltage comparison based high speed & low power domino circuit for wide fan-in gates.” *IEEE 2016 International Conference on Electron Devices and Solid-State Circuits (EDSSC)* (2016): 96–99.
42. R. Niaraki and M. Nobakht. “A sub-threshold 9T static random-access memory cell with high write and read ability with bit interleaving capability.” *International Journal of Engineering-Transactions B: Applications* 29, no. 5 (2016): 630–636.
43. A.P. Shah, N. Vaibhav, and S. Daulatabad. “DOIND: A technique for leakage reduction in nanoscale domino logic circuits.” *Journal of Semiconductors* 37, no. 5 (2016): 055001.
44. T.K. Gupta, A.K. Pandey, and O.P. Meena. “Analysis and design of lector-based dual-Vt domino logic with reduced leakage current.” *Circuit World* 43, no. 3 (2017): 97–104.
45. Padhi Shyamali , A. Anita Angeline, and V.S. Kanchana Bhaaskaran. “Design of process variation tolerant domino logic keeper architecture.” *IEEE 2017 International Conference on Nextgen Electronic Technologies: Silicon to Software (ICNETS2)* (2017): 301–308.
46. Mohammad Asyaei and Farshad Moradi. “A domino circuit technique for noise-immune high fan-in gates.” *Journal of Circuits, Systems and Computers* 27, no. 10 (2018): 1850151.

47. S. Garg and T.K. Gupta. "Low power domino logic circuits in deep-submicron technology using CMOS." *Engineering Science and Technology, an International Journal* 21, no. 4 (2018): 625–638.
48. Pal Indrajit, and Aminul Islam. "Circuit-level technique to design variation-and noise-aware reliable dynamic logic gates." *IEEE Transactions on Device and Materials Reliability* 18, no. 2 (2018): 224–239.
49. M. Asyaei and E. Ebrahimi. "Low power dynamic circuit for power efficient bit lines." *AEU-International Journal of Electronics and Communications* 83, pp. 204–212.
50. M. Asyaei. "A new low-power dynamic circuit for wide fan-in gates." *Integration the VLSI Journal* 60 (2018): 263–271.
51. V. Mahor and M. Pattanaik "A state-of-the-art current mirror-based reliable wide fan-in FinFET domino OR gate design." *Circuits, Systems, and Signal Processing* 37, no. 2 (2018): 475–499.
52. A.K. Pandey, P.K. Verma, R. Verma, and T.K. Gupta. "Analysis of noise immunity for wide or footless domino circuit using keeper controlling network." *Circuits, Systems, and Signal Processing* 37, no. 10 (2018): 4599–4616.
53. M. Asyaei. "A new circuit scheme for wide dynamic circuits." *International Journal of Engineering-Transactions B: Applications* 31, no. 5 (2018): 699–704.
54. A. Kumar Dadoria, B. Kavita Khare, B. Uday Panwar, and B. Anita Jain. "Performance evaluation of domino logic circuits for wide fan-in gates with FinFET." *Microsystem Technologies* 24, no. 8 (2018): 3341–3348.
55. S.R. Ghimiray, P. Meher, and P.K. Dutta. "Ultralow power, noise immune stacked-double stage clocked-inverter domino technique for ultradeep submicron technology." *International Journal of Circuit Theory and Applications* 46, no. 11 (2018): 1953–1967.
56. Asyaei Mohammad , and Farshad Moradi. "A domino circuit technique for noise-immune high fan-in gates." *Journal of Circuits, Systems and Computers* 27.10 (2018): 1850151.
57. Kumar Ankur, and R.K. Nagaria. "A new leakage-tolerant high speed comparator based domino gate for wide fan-in OR logic for low power VLSI circuits." *Integration* 63 (2018): 174–184.
58. A.P. Shah, V. Neema, S. Daulatabad, and P. Singh. "Dual threshold voltage and sleep switch dual threshold voltage DOIND approach for leakage reduction in domino logic circuits." *Microsystem Technologies* 25, no. 5 (2019): 1639–1652.
59. S. Garg and T.K. Gupta. "FDSTDL: Low-power technique for FinFET domino circuits." *International Journal of Circuit Theory and Applications* (2019). doi:10.1002/cta.2627
60. Angeline A. Anita, and V.S. Kanchana Bhaaskaran. "High speed wide fan-in designs using clock controlled dual keeper domino logic circuits." *ETRI Journal* 41.3 (2019): 383–395.
61. Pandey Amit Kumar et al. "Low power, high speed and noise immune wide-OR footless domino circuit using keeper controlled method." *Analog Integrated Circuits and Signal Processing* 100.1 (2019): 79–91.
62. Garg Sandeep , and Tarun K. Gupta. "SCDNDTDL: a technique for designing low-power domino circuits in FinFET technology." *Journal of Computational Electronics* 19.3 (2020): 1249–1267.
63. Bansal Deepika et al. "Improved domino logic circuits and its application in wide fan-in OR gates." *Micro and Nanosystems* 12.1 (2020): 58–67.
64. Garg, Sandeep, Tarun K. Gupta, and Amit K. Pandey. "A 1-bit full adder using CNFET based dual chirality high speed domino logic." *International Journal of Circuit Theory and Applications* 48.1 (2020): 115–133.

65. Bansal Deepika, et al. "Low power wide fan-in domino OR gate using CN-MOSFETs." *International Journal of Sensors Wireless Communications and Control* 10.1 (2020): 55–62.
66. Kumar Ankur, et al. "Leakage-tolerant low-power wide fan-in OR logic domino circuit." *Advances in VLSI, Communication, and Signal Processing*, pp. 631–642. Springer, Singapore, 2021.
67. Singhal Smita, and Anu Mehra. "Gated clock and revised keeper (GCRK) domino logic design in 16 nm CMOS technology." *IETE Journal of Research* (2021): 1–8. doi: 10.1080/03772063.2021.1875269
68. J.C. Park, V.J. Mooney, and P. Pfeiffenberger. "Sleepy stack reduction of leakage power." *International Workshop on Power and Timing Modeling, Optimization and Simulation*, pp. 148–158. Springer: Berlin, Heidelberg, 2004.

13 Potassium Trimolybdates as Potential Material for Fabrication of Gas Sensors

Aditee Joshi, and S.A. Gangal
Department of Electronic Science, Savitribai Phule Pune University, 411 007 Pune

13.1 INTRODUCTION

Sensors have a pivotal role in any control system. They essentially transmit a response with respect to presence of a specific physical parameter. Sensors can be classified according to physical parameters like optical sensors, magnetic sensors, thermal sensors, mechanical sensors and so on. In addition to these parameters, they can also be categorized based on the type of output given by the sensors. Output can be resistive, showing change in resistance values. A capacitive sensor detects change in capacitance, or it can produce change in output voltagesetc. The resistive type of vapor/gas sensor forms a foundation of mainstream gas sensor devices from the chemical domain. The features offered by this device are simple operating procedures, precision and rudimentary design. The sensor involves detection of a target gas by two or three terminal devices. Such chemiresistor sensors have attracted a huge demand being low cost, handy, power savers and extraordinarily sensitive applications in monitoring most domains [1]. This demand brings out the need for resistive devices in the chemical sensors domain.

In industry-based domains, ammonia forms the basis for many procedures. Like in agriculture, it is produced and useful as a nitrogen source for plant growth promoters and also for cattle feedstuff. For the microelectronics sector, crystal growth of GaN/Si_3N_4 involves the use of ammonia gas as a nitrogen-rich source; however, this gas is marked as harmful because it is a lethal gas if humans are exposed above a certain concentration. The acceptable exposure for a short period of time for ammonia is 35 ppm, and the threshold value is around 25 ppm. These limits are set up according to industrial hygiene and occupational health standards. If the limits exceed a large amount of time, then significant hazards can result. Side effects typically include poor functioning of the respiratory system and vision. Prolonged continued exposure causes irreversible damage to all these organs. Moreover, NH_3 sensors have been useful in medical diagnostics as well.

DOI: 10.1201/9781003220350-13

The concentration of NH_3 in human breath is a marking condition for renal and ulcer diseases. For patients with such disorders, the concentration of NH_3 in exhaled breath is in the range of 0.82 to 14.7 ppm (mean value 4.88 ppm) [2–4]. This necessitates the development of a highly sensitive ammonia sensor to detect ultra low concentrations of NH_3. An important characteristic of the ammonia sensor is the ability to exhibit high selectivity, display fast saturation with higher sensitivity and have a reversible recovery under room temperature ambient conditions.

Although many sensors have been reported in earlier literature, it is still a challenge to develop a low cost, portable and highly sensitive sensor at room temperature. In this context, many research outcomes have reported NH_3 sensors. The categorization can be considered as the ones based on inorganic-type of materials like semiconducting oxides and their organic counterparts like polymers or carbon-based materials, along with blends of both groups, as composites. Metal oxides as a sensor material have been mainly synthesized in thin films forms and nanodimensional structures. These nanostructured materials can be obtained in various shapes depending on synthesis conditions, typically like nanorods, nanowires, etc. The gas-sensing ability for these materials happens to be unique owing to key features such as higher surface area available for gas interactions and fast response recovery profile. In addition to this, with tiny dimensions such sensors can be integrated together in the form of an array-like structure that indicates device fabrication possibilities for commercial use. In literature, a NH_3 gas sensor held at 115°C based on nanoclusters of SnO_2, providing sensitivity of 200 ppm, with 500 ppm when an incident of NH_3 is presented [5]. From the group of composites, polyaniline blended with TiO_2 demonstrates trillion parts detecting ability [6]. A flexible nanowire sensor for real-time NH_3 monitoring [7] and reduced graphene oxide composite with conducting polymer as a sensor for diagnostic applications [8] are also reported.

For a sensor structure, the most important component is an active layer or a sensing layer. It is the heart of the sensor device. There are two major groups of materials used in the fabrication of an active layer of chemiresistive gas sensors, inorganic and organic materials. Inorganic materials are comprised of mostly semiconducting and semiconductor oxide materials. The organic part is comprised of a class of polymer materials. Over the last few years, the observation reports application of metal oxides in gas sensors due to characteristics like higher selectivity over other gases at an operating temperature and complete reversibility. The group of materials mostly used for fabrication of gas sensors are oxides of Sn, Zn, α-Fe_2O_3, WO_3, In_2O_3 and TiO_2. The materials possess non-stoichometric structural profiles that facilitate availability of free electrons from oxygen vacancies on the surfaces. This leads to presence of adsorbed oxygen ions in significant abundance, which further contributes to conductivity of the materials. The presence of abundant microstructural structural defects results in good sensing characteristics in oxide materials [9a, 9b].

In addition to the typically reported sensing materials, an element from polyoxymetalate (POM) frameworks group has gained interest. The construction of the material includes a combination of group V and group VI elements of the periodic table. The typical bonding forms large anionic clusters, including sharing of oxide ions. As a result of the structural uniqueness, the materials exhibit some unusual properties [10]. This versatility in all domains makes them prominent in applications like photocatalysis, H_2 evolution and generation of electricity through the photoelectrochemical approach [12,13]. When the cation in these compounds is replaced with K^+, the antitumor effect is similar; therefore, they have a promising potential in antitumoral and antiviral treatments [14]. Some polymolybdates, e.g. $K_2Mo_3O_{10}$, have been used as catalysts for the growth of high-quality single crystals [15, 16], such as $Yb_xY_{1-x}Al_3(BO_3)_4$, a crystal for solid-state lasers. One of the subgroup of POMs is potassium clusters, typically molybdates. Potassium combined with molybdates to give potassium trimolybdate can serve as a template in its nanostructured form for application in soft microfluidics [17].

Throughout the chapter, we discuss the potential of potassium trimolybdates nanowires in fabricating room-temperature operable, fast-detecting and highly selective NH_3 sensors. The chapter highlights synthesis techniques for nanowires, construction of gas sensor configurations, gas response studies and plausible gas-sensing mechanisms. Potassium trimolybdate in nanowire form is synthesized using chemical reaction of ammonium molybdate precursor with potassium chloride under normal ambient conditions. Two sensor structures, namely isolated nanowires and nanowire mat-type sensor structures, will be discussed. The isolated sensor structure is constructed by drop-casting a dilute nanowire dispersion on micro-spaced electrodes, followed by electrophoresis process for alignment. Thin mat-type structures are prepared by drop-casting dispersion of nanowires.

13.2 METHODS

13.2.1 Synthesis of Nanowires of $K_2Mo_3O_{10}.4H_2O$

$K_2Mo_3O_{10}.4H_2O$ nanowires were synthesized using a simple chemical method. This includes mixture of two precursors $(NH_4)_6Mo_7O_{24}.4H_2O$ and KCl in aqueous medium in reported proportion by Gong et al. [18]. In the typical process, deionized water (DI) dissolved solutions of ammonium molybdate $(NH_4)_6Mo_7O_{24}.4H_2O$ and potassium chloride (KCl) (high purity) were blended as per reported ratio in order to get the best yield and uniform nanowires in terms of phase and purity [18]. This reaction can be conducted at different elevated temperatures, but here in our case, the reaction was conducted at room temperature ambient conditions. After a period of fifteen minutes of continuous stirring, the transparent solution turned turbid, showing white deposits with cotton-like appearance. These products were then separated through filtering and subsequent centrifuge processes and cleaning cycles for removing unreacted precursors. After drying at room temperature, the products were investigated for physical and structural characteristics.

13.2.2 Patterning of Electrodes

For alignment of nanowires, substrates with electrodes patterned in the micro scale are very important. The electrode patterning demands several steps through the use of sophisticated instrumentation. In this technique, the main aim was to avoid any laborious microfabrication techniques and subsequent patterning process to get microelectrode geometry. To obtain good precision in microelectrodes, a simple process was developed. In this, a simple silk thread was used as a masking material, and it was tightly glued to glass substrates. All these samples were further loaded in the system for deposition of aluminium metal using standard physical vapor deposition method. Wires for measurement of sensor resistance were glued to aluminium electrodes using silver paste. These sensor structures were further used for aligning nanowires by dielectrophoresis (DEP) process.

13.2.3 Alignment of Nanowires Across Microelectrodes

A dilute aqueous dispersion of nanowires (5 mg/5 ml) was prepared. Agilent signal generator with 1 Hz to 20 MHz frequency range and 10 V peak-to-peak amplitude was used for the DEP technique. A 2 µl droplet from dispersion was further diluted in 1 ml DI water to make a superdilute dispersion. From this, a 2 µl drop was placed on the microgap of microelectrodes. The sample was further subjected to 20 MHz and 10 V peal-to-peak signal for DEP process.

13.2.4 Gas Sensor Response

Gas sensors based on isolated and mat-type fabricated nanowires were used further for studying gas-sensing characteristics. The sensor resistance in air and in the presence of gas was used to measure gas response. In the measurement set-up, a one liter air tight container was used for mounting the sensors, and a specific concentration of gas was injected. For resistance measurement, a potential divider arrangement was used. In this, a high value resistance (order of MΩ) was connected with the sensor. Voltage applied to this system gives a voltage drop across the fixed-value resistor. When gas was adsorbed on the sensor, the change in sensor resistance introduced a change in potential difference across the fixed resistor. This change with time was measured on a 3½ digit multimeter. Using the voltage drop values, change in gas sensor base resistance with time was calculated. Normalized resistance is sensor resistance in gas to resistance in air. Sensitivity for sensors is ratio of resistance in air to resistance in gas.

13.3 RESULTS AND DISCUSSION

13.3.1 Nanowire Characterization and Sensor Fabrication

Nanowires of $K_2Mo_3O_{10}.4H_2O$ were characterized using electron microscopy and powder X-ray diffraction (XRD). For initial imaging, optical microscope (Nikon: MM 40) (Biotech labs: ~2 µm resolution) was used. The floccule-like deposits indicated that nanowires bunch within a dense population. For measurement of dimensions of nanowires, electron microscopy was performed on drop-casted nanowire dispersions. Figure 13.1 shows dimensions of nanowires with the diameter ~300 nm (Figure 13.1a,b) and length in the range of 10–30 µm.

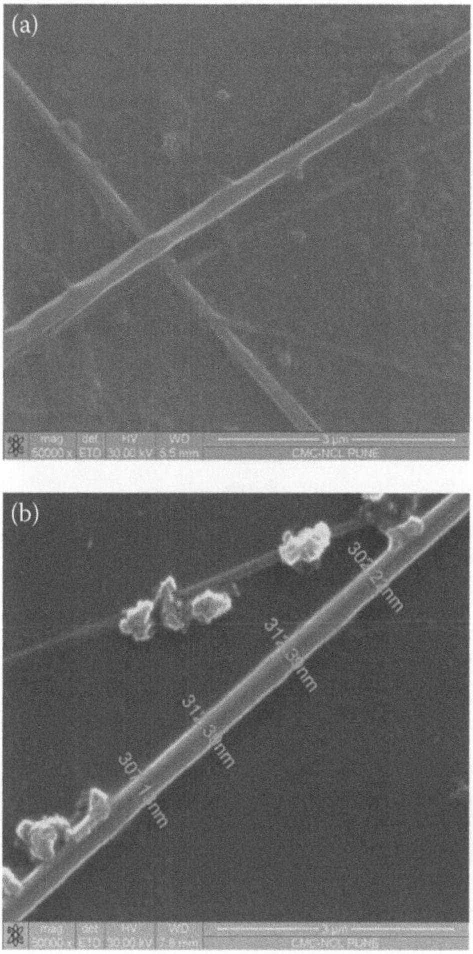

FIGURE 13.1 SEM micrograph of at room temperature synthesized nanowires—$K_2Mo_3O_{10}.4H_2O$. a) clusters of nanowires b) isolated nanowire.

Nanowire structure and orientation of planes were analyzed through powder XRD. Figure 13.2 indicates the diffractogram for $K_2Mo_3O_{10}.4H_2O$, and it was compared to the standard available database for XRD data.

Crystal structure and planes assigned to the standard structure of the nanowire database file (JCPDS 051-1628) were matched with present nanowire results, indicating the pure orthorhombic phase of $K_2Mo_3O_{10}.4H_2O$ nanowires. These results were consistent per reported XRD data for $K_2Mo_3O_{10}.4H_2O$ nanowires [18, 19].

The microelectrode-patterned substrates were imaged under optical microscope to observe the formation of a microgap. As stated in an earlier section, aluminium microelectrodes were deposited by thermal evaporation technique by keeping premasked substrates. Figure 13.3a shows the image of microelectrodes under optical microscope (Nikon: MM 40). It shows uniform formation of microgap

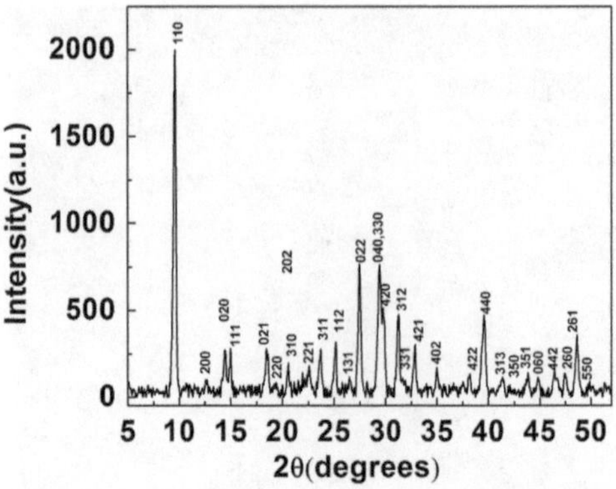

FIGURE 13.2 Powder X-ray diffractogram of potassium trimolybdates nanowires.

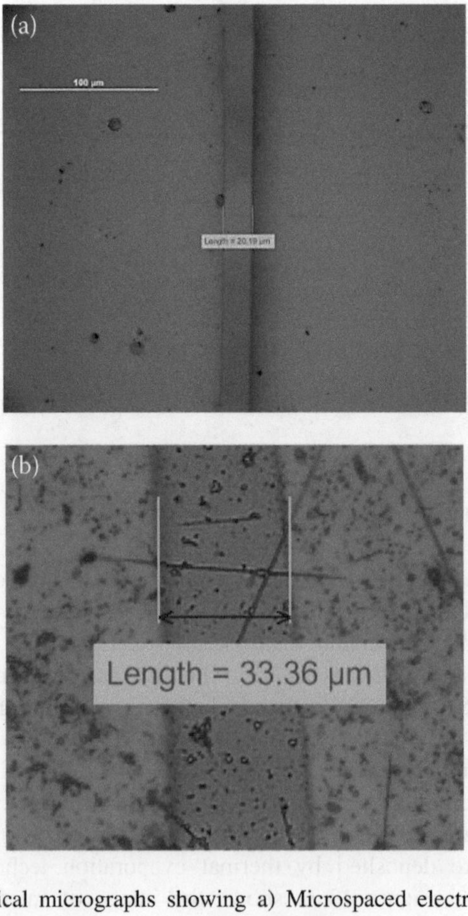

FIGURE 13.3 Optical micrographs showing a) Microspaced electrodes b) Isolated and aligned nanowires.

between two aluminium contacts with a value of 20 μm. The isolation and alignment of nanowires between micro-spaced electrodes was carried out using the DEP technique, as mentioned in a previous section, after the DEP process substrates were checked under an optical microscope for possible alignment of nanowire(s). Figure 13.3b shows an image of a substrate with aligned nanowires across two electrodes. These samples were prepared and used for gas-sensing results. For comparison, a plane mat film of nanowire was also drop-cast on coated substrates for gas-sensing studies.

13.3.2 Gas Response Studies and Mechanism

Gas response of nanowire sensors was first measured by exposing the sensors to lower concentrations of different vapors. Figure 13.4 shows normalized response curves of isolated nanowire samples for 12.5 ppm concentration of NH_3, acetone, methanol and ethanol. The logarithmic scale plot shows higher change for NH_3 than any other vapors. This is almost three orders change for NH_3, compared to any other response. From this, it was concluded that nanowires are selective for NH_3 over any other vapor.

For evaluation of sensor performance, static response was measured. As shown in Figure 13.5, the sensor responds to 12.5 ppm of NH_3 in a few seconds and recovers back to base value in a few minutes. Response and recovery time are 40 s and 10 min, respectively. Rapid detection of NH_3 is indicative of potential of potassium trimolybdate nanowires for fabricating NH_3 gas sensors. Sensor sensitivity was calculated and plotted for 12.5 ppm concentration of NH_3, acetone, methanol and ethanol, as given in Figure 13.6. Note that the sensitivity for NH_3 is 20 times higher than other vapors.

Further sensor response was recorded for low concentrations of NH_3 for judgment about the lower limit of detection. The attempt was made in order to use the sensor for lower NH_3 concentrations, below its toxic limit. The sensor was exposed to a range of concentrations between 0.5 ppm and 10 ppm, and the response was monitored, as seen in Figure 13.7.

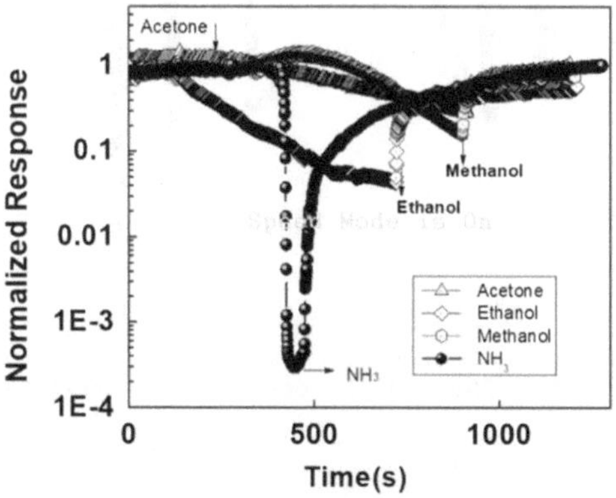

FIGURE 13.4 Sensing characteristics for 12.5 ppm NH_3, ethanol, methanol and acetone.

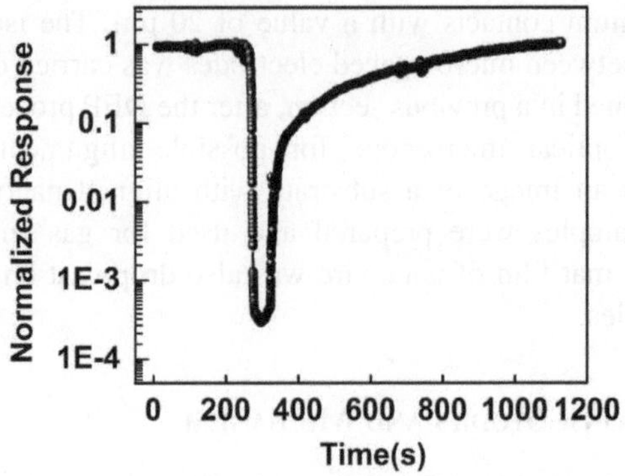

FIGURE 13.5 Response recovery curve for 12.5 ppm NH_3.

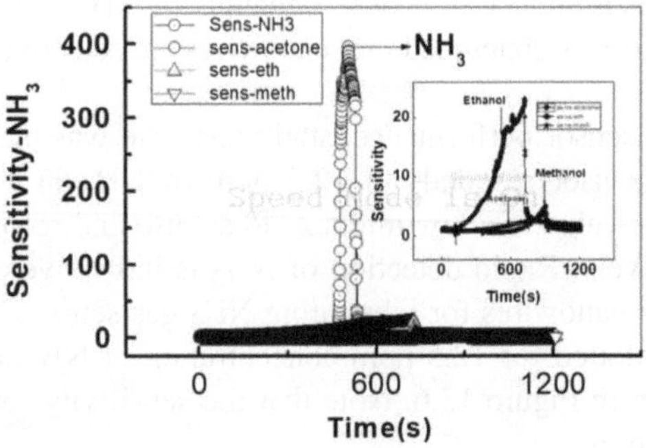

FIGURE 13.6 Sensitivity curves for 12.5 ppm NH_3.

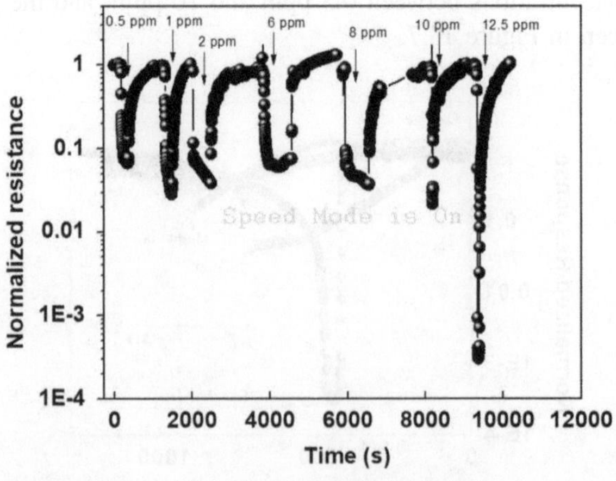

FIGURE 13.7 Gas response curves for lower NH_3 concentrations.

For all the concentrations, the response and recovery were observed. The response showed good linearity and detection of 0.5 ppm concentration with substantial sensitivity. Figure 13.8 indicates the sensitivity plot for all concentrations of NH_3. Note that linear region is observed for 0.5–10 ppm concentration.

Various samples were prepared and tested for NH_3 response. All the sensors were highly reproducible. Figure 13.9 shows a histogram plot of sensitivity of different sensors exposed to 12.5 ppm of NH_3 concentration. The yielded sensitivity of the sensors is almost similar to each other. The sensors were significantly reproducible with ±10% error.

Sensors based on dilute dispersion-coated nanowire mat film on microspaced electrodes were investigated for sensing characteristics. The purpose was to compare performance of the same material configured in two different types of structures. Exposure to the same concentration (12.5 ppm NH_3) took a few minutes to reach to saturated value with small response (sensitivity–180, data not shown). For a thicker film of nanowires, it took one hour to reach to the steady state.

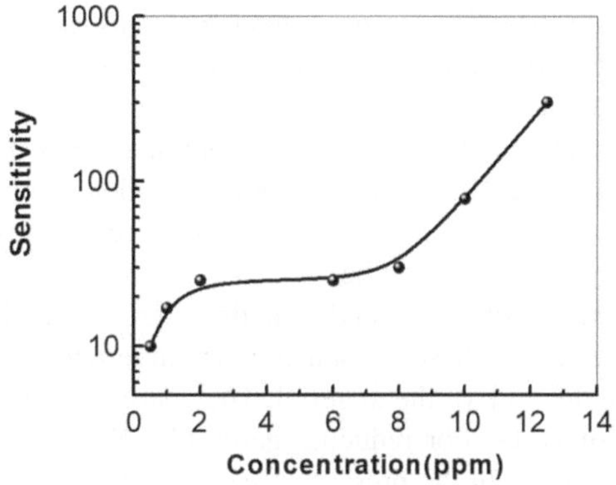

FIGURE 13.8 Sensitivity variation with NH3 concentrations.

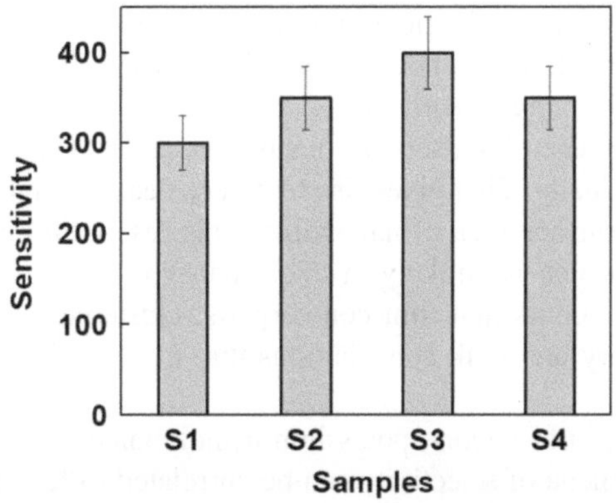

FIGURE 13.9 Gas response curves for 12.5 ppm NH_3 for multiple samples.

These observations outline the importance of isolated nanowire configuration and capacity to detect concentrations well below the toxic limit.

Now, we would like to put forward the governing phenomena in the gas-sensing process of potassium trimolybdate nanowires.

The nanowires belong to a POM group consisting of clusters of anions along with group V and VI metals. Structures of this group are made up of oxides of molybdenum, tungsten or vanadium with the basic formula of {MOx}n. The nature of compounds is anionic and metal cations like K^+ act as a linker [20a]. Such compounds lead to formation of ionic crystals with different alkali metals. Our present nanowire materials are formed by linking K^+ with an MO_x group. Specifically, K^+ is combined with anionic cluster chains of $(Mo_3O_{10})^{2-}$. This has been reported to be catalytic and useful in the growth of a crystal [15,18,21]. For such crystals with an ionic nature having alkali metal ions, the cavities formed by water desorption lead to porous surface structure. The resultant porosity can play a vital role in surface-based adsorption-desorption reactions and also for separation of different alcohol, nitriles, esters and water mixtures [22]. Effectually, sensing applications based on such materials can also be investigated.

The sensor structure provides the main lead for analyzing plausible reactions. The electrode spacing is of the order of a few tens of microns with aligned nanowires. When gas molecules are adsorbed on sensors, interactions happen with individual nanowire surfaces, and responses are integrated together. As stated earlier, the anionic clusters give rise to an abundant number of delocalized electrons on the surface and formation of redox states through electrocatalytic mechanisms. Exceptional sensitivity of sensors can be associated with physisorption reactions on the surface. The presence of oxide and a hydrate group on the surface are responsible for higher values of base resistance at room temperature as oxide anions accept electrons from the bulk of the material. The charge transfer reaction depends on the redox nature of gases. For reducing gases like NH_3, electrons are donated during gas film interactions. In the present case, NH_3 adsorption results in reacting with surface functional groups and physisorbed oxygen on nanowire surface. The nanowires accept electrons; the reaction with oxide groups relieves the electrons back to the bulk of the material. This electron donation process makes the nanowires electron rich and results in increasing the conductivity. When more such NH_3 molecules are adsorbed, the concentration of electrons donated becomes more substantial, leading to more sensitivity.

Through this structure of sensors, all nanowires are isolated and are being exposed to NH_3 individually. That gives rise to a very fast and high response through the entire available surface area of nanowires. This results in a very fast response and recovery time, which is unlikely in thick mat-type films and a dispersed network of nanowires. For its mat film counterparts, clustered nanowires happen to have a grain boundary and bulk contributions that may result in sluggishness and lower sensitivity.

Among all the types of vapors, potassium trimolybdate nanowires are selective for NH_3. The phenomena of selectivity can be correlated to fundamental properties of vapors and their mechanism of interaction with nanowires. Higher selectivity is a result of the difference in electron affinity characteristics of vapors. Typically, for

TABLE 13.1
Comparison of Nanowire Sensor Results with Literature Reports

Material	Sensitivity	Lowest Detectable Concentration	Operating Temperature	Reference
Polyaniline-DBSA composite	3.3	100 ppm	RT	Kumar et al., [26, 2015]
V_2O_5	2 %	80 ppb	350°C	Huotari et al., [27, 2015]
NiOnanocones-Zinc Oxide nanothorns	20 %	15 ppm	RT	Wang et al., [28, 2015]
Cu-PANi	30 %	5 ppm	RT	Patil et al., [29, 2015]
rGO: Ag nanowires	3 %	15 ppm	RT	Tran et al., [30, 2014]
SnO_2:SnS_2 hybrids	1.2	10 ppm	RT	Xu et al., [31, 2015]
ZnO -nanorods	2	20 ppm	RT	Anantachaisilp et al., [32, 2014]
Single Walled Carbon NanoTubes	4 %	25 ppm	RT	Han et al., [33, 2014]
Graphene- Doped	1 %	200 ppb	RT	Seyedeh et al., [34, 2016]
Chemically reduced GO	5 %	200 ppm	RT	Ghosh et al., [35, 2013]
Potassium Trimolybdate Nanowires	**15**	**500 ppb**	**Room Temperature**	**Joshi et al., [36, 2016]**

NH_3 this value is 211 Kcal/mol over the other vapors (~185 Kcal/mol) [23, 24]. Selectivity mechanism can also be attributed to NH_3 vapors' adsorption kinetics. In this case, NH_3 has a superior value of 6.4 mMol/g [25], suggestive of higher adsorption capacity. Another characteristic feature is higher work function and electronegativity. In this way, the selectivity mechanism of sensors for NH_3 is complete.

Now, a comparison of gas sensors for NH_3 reported in literature is considered. Here, we present a comparison of previously reported NH_3 sensors along with reference to detection limit, sensitivity and materials used, as indicated in Table 13.1. Observations demonstrate the highest sensitivity offered by our work related to nanowires. Sensors operate at room temperature ambient conditions and also possess the capability for detection of sub ppm levels of NH_3.

13.4 CONCLUSION

Chemiresistor sensors fabricated using isolated and aligned nanowires of potassium trimolybdate possess superior gas-sensing abilities. They demonstrate the potential for use as room temperature operable NH_3 gas sensors. The performance parameters of sensors are found to be excellent in terms of exceptional selectivity, fast response and reversibility. For low concentrations of NH_3, sensors can exhibit significant responses, even below the toxic limit, and can respond to ppb levels of concentration. From this, they can be envisioned as a strong candidate for employment in NH_3 gas sensors.

ACKNOWLEDGMENTS

The authors would like to acknowledge The Head, Department of Electronic Science for providing infrastructural support. Additionally, central material characterization facility, NCL, Pune is also acknowledged for characterization support.

REFERENCES

1. Li, L., Gao, P., Baumgarten, M., Mullen, K., Lu, N., Fuchs, H. and Chi, L. "High performance field-effect ammonia sensors based on a structured ultrathin organic semiconductor film". *Adv. Mater.* 25 (2013): 3419–3425.
2. Bevc, S., Mohorko, E., Kolar, M. et al. "Measurement of breath ammonia for detection of patients with chronic kidney disease". *Clin. Nephrol.* 88 (2017): 14–17.
3. Turner, C., Španˇel, P., and Smith, D. "A longitudinal study of ammonia, acetone and propanol in the exhaled breath of 30 subjects using selected ion flow tube mass spectrometry, SIFT-MS". *Physiol. Meas.* 27 (2006): 321–337.
4. Davies, S., Spanel, P. and Smith, D. "Quantitative analysis of ammonia on the breath of patients in end-stage renal failure". *Kidney Int.* 52 (1997): 223–228.
5. Liu, X., Chen, N., Han, B. et al. "Nanoparticle cluster gas sensor: Pt activated SnO_2 nanoparticles for NH_3 detection with ultrahigh sensitivity". *Nanoscale.* 7 (2015): 14872–14880.
6. Gong, J., Li, Y., Hu, Z. and Deng, Y. "Ultrasensitive NH_3 gas sensor from polyaniline nanograin enchased TiO_2 Fibers". *J. Phys. Chem. C.* 114 (2010): 9970–9974.
7. Ning, T., Cheng, Z., Lihuai, X. et al. "A fully integrated wireless flexible ammonia sensor fabricated by soft nano-lithography". *ACS. Sens.* 4 (2019): 726–732.
8. Tan, N. Ly and Sangkwon, P. "Highly sensitive ammonia sensor for diagnostic purpose using reduced graphene oxide and conductive polymer". *Sci. Rep.* 8 (2018): 18030(1-12).

9a. Chung, Y.K., Kim, M.H., Um, W.S. et al. "Gas sensing properties of WO_3 thick film for NO gas dependent on process condition." *Sens. Actuators B* 60 (1999): 49–56.

9b. Qi, Q., Wang, P.P., Zhao, J. et al. "SnO_2 nanoparticle-coated In_2O_3 nanofibers with improved NH_3 sensing properties". *Sens. Actuators B* 194 (2014) : 440–446.

10. Pope, M.T. and M¨uller, A. (ed). (1994). *Polyoxometalates: From Platonic Solid to Anti-retroviral Activity*. Dordrecht: Kluwer–Academic.
11. Papaconstantinou, E. "Photocatalytic oxidation of organic compounds using heteropoly electrolytes of molybdenum and tungsten". *J. Chem. Soc. Chem. Commun.* 1 (1982): 12–13.
12. Hiskia, A. and Papaconstantinou, E. "Photocatalytic oxidation of organic compounds by polyoxometalates of molybdenum and tungsten. Catalyst regeneration by dioxygen". *Inorg. Chem.* 31 (1992): 163–167.
13. Papaconstantinou, E., Ioannidis, A., and Hiskia, A. "Photocatalytic processes by polyoxometalates. Splitting of water. The role of dioxygen". *Mol. Eng.* 3 (1993): 231–239.

14. Yamase, T. "Polyoxometalates for molecular devices: Antitumor activity and luminescence". *Mol. Eng.* 3 (1993): 241–262
15. Tian, J.Y., Wang, Q.C., Guan, J.Q. et al. "Study on growth and optical properties of YbYAB crystal". *J. Chin. Rare Earth Soc.*, 2 (1999): 172–175.
16. Teshima, K., Kikuchi, Y., Suzuki, T. and Oishi, S. "Growth of $ErAl_3(BO_3)_4$. Single crystals from a $K_2Mo_3O_{10}$ flux". *Cryst Growth Des.* 6 (2006): 1766–1768.
17. Joshi, A., Morarka, A., Gangal, S. and Bodas, D. "A simple and low cost method for fabrication of nanochannels using water soluble nanowires". *NSTI Nanotech Proceedings* 2 (2011): 382–385.
18. Gong, W., Xue, J., Zhang, K., Wu, Z. et al. "Room temperature synthesis of $K_2Mo_3O_{10}{\cdot}3H_2O$ nanowires in minutes". Nanotechnol 20 (2009): 215603(1–6).
19. Lasocha, W., Jansen, J. and Schenk, H. "Crystal structure of fibrillar trimolybdate $K_2Mo_2O_{10}{\cdot}3H_2O$". *Solid State Chem.* 115 (1995): 225–228.
20a. Pope, M.T. and Müller, A. "Chemie der polyoxometallate: Aktuelle Variationen über ein altes Thema mit interdisziplinären Bezügen". *Angew. Chem.* 103 (1991): 56–70.
20b. Pope, M.T. and Müller, A. "Polyoxometalate chemistry: An old field with new dimensions in several disciplines." *Angew. Chem. Int. Ed. Engl.* 30: 34–48.
20c. Müller, A. and Roy, S. *The Chemistry of Nanomaterials: Synthesis, Properties and Applications*. Wiley-VCH, Weinheim, 2004.
21. Papaconstantinou, E. "Photochemistry of polyoxometalates of molybdenum and tungsten and/or vanadium". *Chem. Soc. Rev.* 18 (1989): 1–31.
22. Uchida, S., Kawamoto, R. and Mizuno, N. "Recognition of small polar molecules with an ionic crystal of α-Keggin-type polyoxometalate with a macrocation". *Inorg. Chem.* 45 (2006): 5136–5144.
23. Peterson, K.A., Xantheas, S.S., Dixon, D.A. and Dunning, T.H. Jr. "Predicting the proton affinities of H_2O and NH_3". *J. Phys. Chem. A* 102 (1998): 2449–2454.
24. Wr′oblewski, T., Ziemczonek, L., Alhasan, A.M. and Karwasz, G.P. "Ab initio and density functional theory calculations of proton affinities for volatile organic compounds". *Eur. Phys. J. Special Topics.* 144 (2007): 191–195.
25. Saha, D. and Deng, S. "Adsorption equilibrium and kinetics of CO_2, CH_4, N_2O, and NH_3 on ordered mesoporous carbon". *J. Colloid Interface Sci.* 345 (2010): 402–409.
26. Kumar, J., Shahabuddin, Md, Singh, A., Singh, S.P., Saini, P., Dhawan, S.K. and Gupta, Vinay. "Highly sensitive chemo-resistive ammonia sensor based on DBSA doped polyaniline thin film". *Sci. Adv. Mater.* 7 (2015): 518–525.
27. Huotari, J., Bjorklund, R., Lappalainen, J. and Lloyd Spetz, A. "Pulsed laser deposited nanostructured vanadium oxide thin films characterized as ammonia sensors". *Sens. Actuators B* 217 (2015): 22–29.
28. Wang, J., Yang, P. and Wei, X. "High-performance, room-temperature, and no-humidity-impact ammonia sensor based on heterogeneous nickel oxide and zinc oxide nanocrystals". *ACS Appl. Mater. Interfaces* 7 (2015): 3816–3824.
29. Patil, U.V., Ramgir, N.S., Karmakar, N., Bhogale, A., Debnath, A.K., Aswal, D.K., Gupta, S.K. and Kothari, D.C. "Room temperature ammonia sensor based on copper nanoparticle intercalated polyaniline nanocomposite thin films". *Appl. Surf. Sci.* 339 (2015): 69–74.
30. Tran, Q.T., Hoa, H.T.M., Yoo, D.H. et al. "Reduced graphene oxide as an over-coating layer on silver nanostructures for detecting NH3 gas at room temperature". *Sens. Actuators B* 194 (2014): 45–50.
31. Xu, K., Li, N., Zeng, D. et al. "Interface bonds determined gas-sensing of SnO_2–SnS_2 hybrids to ammonia at room temperature". *ACS Appl. Mater. Interfaces.* 7 (2015): 11359–11368.

32. Anantachaisilp, S., Smith, S.M., That, C.T., Osotchan, T., Moon, A.R. and Phillips, M.R. "Tailoring deep level surface defects in ZnO nanorods for high sensitivity ammonia gas sensing". *J. Phys. Chem. C* 118 (2014): 27150–27156.
33. Han, J.W., Kim, B., Li, J. and Meyyappan, M. "A carbon nanotube based ammonia sensor on cellulose paper" *RSC Adv.*, 4 (2014): 549–553.
34. Seyedeh, M.M.Z., Mir, M.S., Milo, H., Chowdhury, Sk. Fahad, Tao, L. and Akinwande, D. "Enhanced sensitivity of graphene ammonia gas sensors using molecular doping". *Appl. Phys. Lett.* 108 (2016): 033106(1–5).
35. Ghosh, R., Midya, A., Santra, S., Ray, S.K. and Guha, P.K. "Chemically reduced graphene oxide for ammonia detection at room temperature". *ACS Appl. Mater. Interfaces*, 5 (2013): 7599–7603.
36. Joshi, A. and Gangal, S. "Implication of potassium trimolybdate nanowires as highly sensitive and selective ammonia sensor at room temperature", *Materials Research Express*, 3(2016): 095008(1–6).

14 Carbon Allotropes-Based Nanodevices
Graphene in Biomedical Applications

Sugandha Singhal, and Meenal Gupta
University School of Basic & Applied Sciences, Guru Gobind Singh Indraprastha University, Dwarka Sector 16C, New Delhi-110078, India

Md. Sabir Alam
NIMS Institute of Pharmacy, NIMS University, NH-11C, Delhi-Jaipur Expy, Sobha Nagar, Jaipur, Rajasthan, India

Md. Noushad Javed
Department of Pharmacy, School of Medical and Allied Sciences, KR Mangalam University, Gurugram, India

Jamilur R. Ansari
Faculty of Physical Sciences, PDM University, Sector 3A, Bahadurgarh-124507, Haryana, India

14.1 INTRODUCTION

Carbon is a nonmetallic, tetravalent element that is also a building block of life on earth. The unique property of self-binding and binding other elements results in the formation of structurally diverse compounds possessing varied physical and chemical properties. The applicative importance is well observed in versatile technical drug applications (Figure 14.1).

Material scientists have succeeded in tailoring the structural and functional properties of existing materials to create new and modified versions by altering synthetic routes [1]. The two naturally occurring carbon allotropes are graphite with sp^2 hybridized and diamond with sp^3 hybridized carbon networks. Graphite is soft, ductile and conductive, whereas diamond is hard and insulating; thus, they show unique opposing properties. In 1985, the first synthetic carbon allotrope, zero dimension (0D) fullerene was discovered by Kroto et al. [2], followed by one

DOI: 10.1201/9781003220350-14

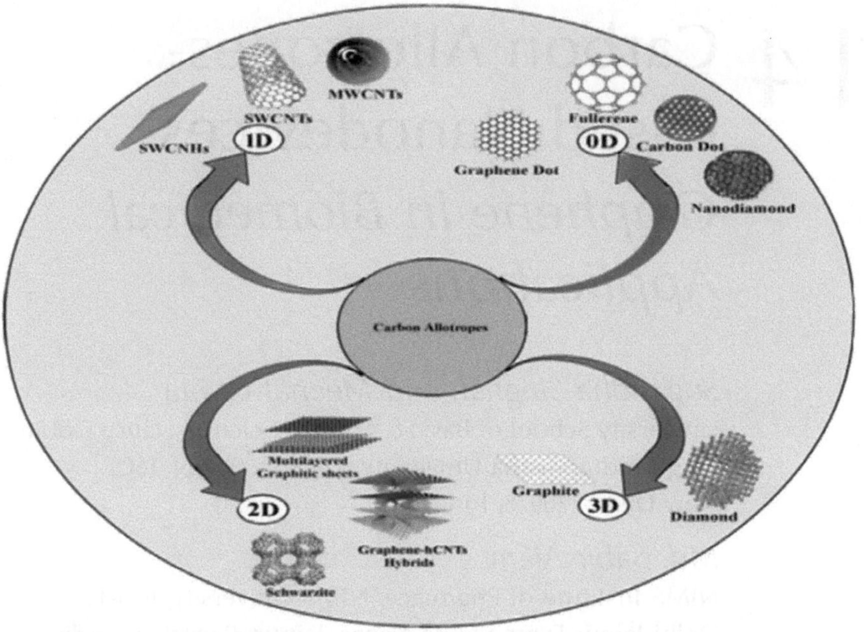

FIGURE 14.1 Different dimensional allotropes of carbon. Fullcrenes, carbon dot and graphene dot are examples of (0D). SWCNT, MWCNTs and SWCNHs are examples of (1D). Graphene-hCNTs hybrid, Schwarzite and multi-layered graphitic sheets are examples of (2D). Graphite and diamond are examples of (3D).

dimension (1D) carbon nanotubes (sp hybridized) by Iijima [3] and zero dimension (2D) grapheme (sp^2 hybridized) by Novoselov [4]. This extensive property of carbon helps in the fabrication of newer allotropes that could consist of extended rings or polygons [1].

There is ongoing research in modifying the physical properties of current allotropes using different structural and binding combinations. The field of medicine and clinical biology requires tools to enhance its scope; however, such nanomaterials are being suitably groomed [5]. Here comes to play theoretical tools, such as *ab-initio* DFT calculations to predict more stable allotropes by altering binding and determining various properties [6]. The enhanced physical properties obtained for the same also provides insight to find routes for their synthesis in macroscopic quantities. The synthesis of 1D and 2D carbon nanostructures or their hybrids is carried out by the substrate chemical vapor deposition (CVD) method or by using electron irradiation on the surface of graphene. Some of the CVD techniques, like plasma-enhanced CVD, spray pyrolysis, as well as magnetron sputtering deposition, have been used to synthesize carbon nanotubes (CNTs) doped with nitrogen, graphite nanosheets and their assemblies on conducting and nonconducting substrates [7]. The doping provides additional electrical and mechanical stability. Also, 2D carbon nanostructures from nanosheets and their hybrids have been reported with CNT bases. Carbon nanowalls (CBNs) with high surface area that are synthesized using the magnetron sputtering method, irrespective of any catalyst, find applications in electrodes for hydrogen fuel. These carbon nanomaterials, due to their differing electron-conducting properties, possess a great potential for the fabrication of electronic devices [8]. However, because these fullerenes are hollow, cage-like, soclathrin-coated vesicle cells, they also may be rationalized for their potential applications for peptide delivery, either as a therapeutical or as a diagnostic agent [9]. Other biological actions could be observed, mainly because of strain in structure due to sp^2 carbon bending. Fullerenes are electron donors and acceptors, which produce reactive oxygen species on irradiation and could cause photodynamic damage to biological systems, including DNA cleavage. Among them, C_{60} is the most effective radical scavenger due to the presence of conjugated double bonds in fullerenes. Thus, they quench oxygen radicals instead and show anti-inflammatory and anti-allergic properties [1].

Among carbon-based nanomaterials, graphene-based nanomaterials (GBNs) possess unique physical and chemical properties that widely attract researchers owing to their various biomedical uses, which include optical imaging, magnetic resonance imaging and delivery of genes and drugs, and for their theranostic applications in cancer. Research findings emphasized that these types of nanomaterials must exhibit certain inherent properties, such as good solubility, low toxicity and higher biological compatibility; otherwise, their inability would make them inefficient against tumors [10]. Therefore, the composite system of GBNs has been designed for potential future biomedical applications [11].

14.2 TAILORED GRAPHENE ALLOTROPES

Graphene is a 2D carbon allotrope with sp^2 hybridization with remarkable properties (Landau levels and high carrier mobility) obtained due to the Dirac cone structure. It is also the most stable form, well known for its greater carrier mobility, quantum Hall effect and electronic and spintronic properties. The dispersion cones of linear nature are obtained due to the honeycomb lattice of hexagonal ring symmetry.

Similarly, the other sp^2 and sp^3 carbon allotropes also have these Dirac cones. This concludes that Dirac cones as well as linear carrier dispersion are essential 2D carbon allotrope features [12].

Few materials explored by researchers, keeping in view the Dirac cone feature, are allotropes of graphene, silicene, so-MoS_2, *Pmmn* boron, Cd_3As_2 and FeB_6 monolayers [6]. The common accepted necessary conditions for Dirac allotropes include 2D, chemical equivalence, honeycomb-like structure of bi-lattice, hexagon symmetry and sp^2 hybridization. However, recently discovered alternative Dirac allotropes violate a few of these conditions, thereby paving the way for carbon lattices and a possible ending of Dirac allotropes. So, a suitable technique for Dirac carbon allotropes in 2D lattices needs to be designed [13]. The structural features of allotropes that alter electronic properties for graphene are the needs of the current scenario for its applications in electronic devices.

14.2.1 Types of 2D Dirac Allotropes

Liu et al. [14] theoretically predicted T graphene, a 2D metallic graphene allotrope with buckled carbon sheets of tetragonal symmetry. The dynamically and energetically more stable buckled T form of graphene can exhibit other properties because of its band structure and liberal 2D features compared to graphene. It also has a high vF and Dirac fermions, the same as graphene. The bonds are non-equivalent, without honeycomb structure. So, various multi-member rings containing Dirac graphene allotropes can be predicted. The two nanoribbons with dissimilar band structures have better properties than graphene [14]. Similarly, Xu et al. [13] designed three 2D low-energy rectangular graphene allotropes, namely S, D and E graphene. Among these, metastable allotropes S and D are 2D, and the third form E is neither 2D nor puckered. The synthetic route for the preparation of these allotropes includes chemical vapor deposition (CVD) and electron irradiation on the surface of grapheme [15, 16]. The D and E forms show anisotropic properties due to Dirac points. Hence, by increasing the unit cell of graphene, Dirac allotropes can be obtained (Figure 14.2).

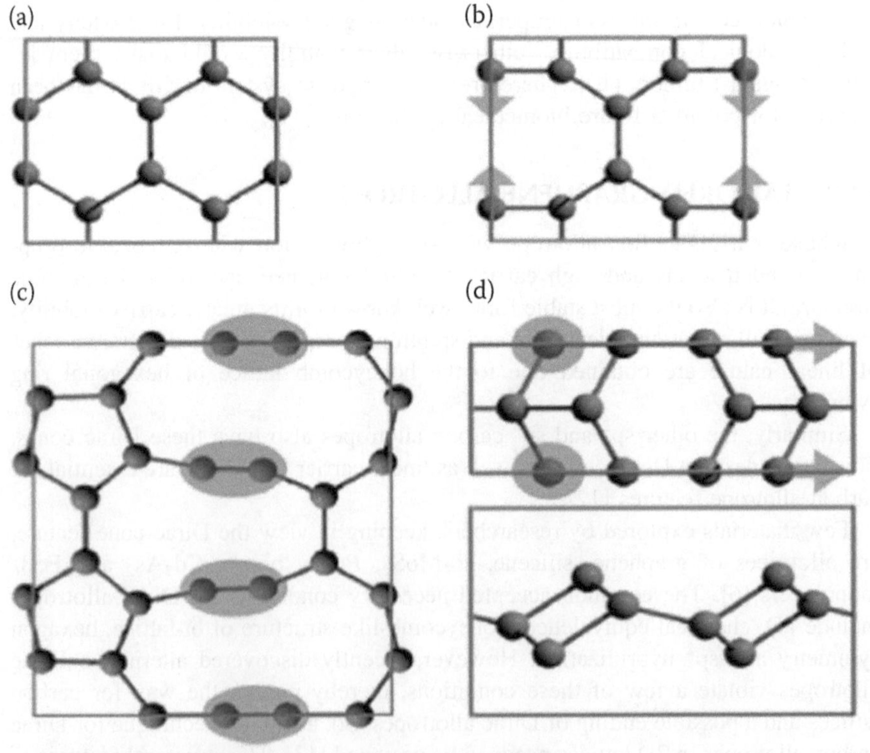

FIGURE 14.2 Structures of (a) (2X1) graphene cell and selected low-energy Dirac carbon allotropes using CALYPSO search, (b) S-graphene form, (c) D-graphene form and (d) E-graphene form, respectively.

This concludes that the common features that provide fundamental insights toward graphene structural features are Dirac cones and linear carrier dispersion [12]. Regarding carbon pentagons, the well-known isolated pentagon rule (IPR) states that pentagons should be isolated rather than fused, which provides kinetic stability to fullerenes. Synthesis of fullerenes paved the way for further research to stabilize non-IPR fused pentagon-based carbon nanomaterials across various dimensions [17,18]. Singh et al. (2014) reported a stable 2D allotrope named pentahexocite, a continuous sheet of 5, 6, 8 carbon atom rings. The modes of phonon spectra are unstable, giving rise to planar sheet structure. Moreover, it has mechanical stability as that of graphene. The sheet is of metallic character, which has flat dispersive Fermi-level bands owing to the anisotropic properties. The above sheet becomes a precursor for designing and development of CNTs with chiral properties. However, due to the acoustical and optical mode mixing as well as reduced symmetry over graphene, the conductivity is lower than graphene. The nanowires (1D) with 5,8 carbon chains [15] have been put to use in valley-tronic devices [19]. The photonic bandgap has 40 THz value for the pentahexoctite sheet, making it a suitable candidate for nano-electronic devices [20]. On similar lines, Jena et al. (2015) predicted with theoretical calculations a dynamically and mechanically stable 2D allotrope pentagraphene with pentagon rings and a resemblance to Cairo pentagonal tiling. Also, its ability to withstand high temperatures of 1000K proves thermal stability. The mechanical strength is greater than graphene; it has a negative Poisson's ratio and a large bandgap with value 3.25 eV. The semiconducting pentagon-based nanotubes are formed by rolling pentagraphene sheets. The rolled pentagraphene sheets form semiconducting carbon nanotubes, and the stable 3D T12-carbon structure is built by sheet stacking with a large bandgaps are both thermally and dynamically stable. These extensive features of graphene and its derivatives can help in fields of device fabrication in nanomechanics and nano-electronics [16]. Oganov et al.(2015) predicted a low-energy rectangular 2D allotrope, namely phagraphene, consisting of 5-, 6-, and 7-membered carbon rings. The low energy value for phagraphene of –9.03 compared to graphene; -9.23 eV/atom, is due to sp^2 bonding. The calculated value for the planar atomic density of phagraphene is 0.37 units, which is similar to graphene, and the plane group is Pmg. The electronic structure suggests that Dirac cones are distorted to reduce external strain without the effect of Fermi velocity. With the help of these structural observations, further research studies on mass-free Dirac fermions for 2D systems and construction of new phases in the artificial lattices of photons [21] were possible. The above examples indicate that pentagon-containing carbon materials possess varied mechanical and electronic properties [18]. These examples indicate that neither the structural features of graphene (among it are hexagon symmetry and bi-lattice backbone) nor sp^2 hybridization is a prerequisite for Dirac carrier dispersion.

14.2.2 D Dirac Graphene Allotropes

The 3D metallic allotropes exhibit remarkable physical characteristics, like highly porous structure and a large ratio for surface area to volume. With the help of experimental and theoretical DFT calculation study, various porous carbon 3D structures have been predicted, such as K6 carbon, T6 carbon, H18 carbon, M carbon, 3D sp^2 carbon network and 3D sp^2-sp^3 networks with catalytic applications [6]. However, when moving from 2D to 3D, the question arises whether they can exhibit Dirac property or not.

Carbon honeycombs (CHCs), also known as 3D graphene, are the growing interests of researchers, some of which have been theoretically predicted and experimentally synthesized. Pang et al. (2016) theoretically predicted the structure of 3D CHCs. The walls of CHCs have sp^2 binding, and its triple junction has sp^3 bonding. The two new CHCs exhibited remarkable mechanical as well as thermal properties, along with metallic character [6]. Similarly, Chen et al. (2017) predicted two stable CHCs, Cmcm and Cmmm-CHCn with thermal, dynamic and mechanical stability. The two allotropes had Dirac cones of direction-dependent nature under strain. The energies for the two were lower than previously reported results. Also, physical absorption efficiency is more than CNT and fullerenes [6]. The next category of 3D materials includes superhard materials that possess superior physical properties with exceptional mechanical stability. Thus, they have industrial applications, such as cutting, polishing and drilling tools, as well as coatings in surface protection [22, 23].

14.2.3 Superhard 3D Metallic Carbon Allotropes

Wang et al. (2014) predicted various structural properties of K6 carbon, suggesting its applications in the field of electronics using DFT formula. The parameters, like the elastic constant value of greater than 0, indicate mechanical stability. Bulk modulus value of 228 GPa and shear modulus value of 52 GPa are both lower than diamond. High Poisson's ratio value of 0.39 compared to a diamond (0.08) provides ductility. It has a higher value for the density of states, 0.10 states/eV per atom, than CNTs, and its low density is responsible for observed aerospace and hydrogen storage applications [24].

Jia et al. (2016) proposed a new superhard allotrope, C20-T with cubic T symmetry, a large cavity leading to a porous structure. Vickers' hardness factor determines superhardness for materials. The value was found to be 72.76 GPa, whereas for a diamond it is 93.27, and for already predicted BC12 carbon, it is 65.65. The value for C20-T was higher than the renowned superhard material c-BN. The bandgap's value of 5.44 eV indicates it to be an insulator. Other calculated parameters include brittle nature, as indicated by Poisson's ratio of 0.11, and isotropic nature. The material also exhibited mechanical strength similar to c-BN [25]. Further adding to the category of superhard allotropes, Zhong et al. (2012) proposed four new carbon lattice allotropes, two with 5-6-7 chain (Z-ACA, Z-CACB) and others with 4-6-8 carbon chains (Z4-A3B1, A4-A2B2). In comparison to graphite, all four materials were found to be metastable, along with great mechanical strength comparable to diamond. As far as electronic properties are concerned, the 5-6-7 type show direct bandgaps, whereas the 4-6-8 types show indirect bandgaps close to Z-carbon, which are responsible for their semiconducting nature. These were synthesized by cold compression of graphite. Among these, the 4-6-8 are the most stable [26].

14.2.4 Carbon Allotropes as Electrochemical Detectors

The major multidimensional carbon allotropic forms have been used as electrochemical detectors for microchip electrophoresis (ME) and capillary electrophoresis (CE). The 0D fullerenes find applications as electrochemical and biosensors, whereas the 1D CNTs and 2D graphene have been extensively studied for their action as detectors in CE and ME due to high surface area.

The major characteristics of carbon nanomaterials responsible for the electrocatalytic effect are as follows:

1. A low detection potential: Low current densities prevail due to the large surface area.
2. A high faradaic ampero-metric current: The enhanced analytical sensitivity arises due to multiple redox reactions occurring at the surface area of carbon nanomaterials.
3. The electrodes have high resistance and stability and couple with analytes. This results in stability and reproducibility via detection.
4. A high rate of electron transfer of heterogeneous nature gives sharp and intense peaks in the separation of analytes.

Therefore, based on selectivity, sensitivity and reproducibility, these carbon nanomaterials provide enhanced performance [27].

14.3 BIOMEDICAL APPLICATIONS OF GRAPHENE-BASED NANOMATERIALS

Recently, mechanisms involved in the progression of several inflammatory and infectious disorders, leading to life-threatening conditions, have been found to be highly complicated and often cross-talked [28]. Although the current approach of clinical management of such disorders is based on monotherapy, such approaches have a compellingly high chance of drug resistance [29]. Apart from this, there are also relatively higher incidents of high-dose mediated adverse impacts [30]. In infectious disorders, adaptive defense mechanisms in microbes against such anti-infectious agents are the main reason for such resistance. However, in neuronal disorders, highly cross-linked mechanisms of different neuro-modulators are the main reason that directly impacted therapeutical performance [31]. Hence, shifts are toward the integration of more than one drug to avoid failure of monotherapy as well as to enable multiple sites of targeting [32]. Such approaches favor reducing the requirement of high dose of any single drugs with polytherapeuticals because different sites of drug actions and relatively lower side effects are being reported from these combinations [33]. Integration of nanotechnology-based approaches with such therapeuticals provides tangiable industrially applicable and commercial solutions for various healthcare issues [34]. Subsequently, there is a lot of interest toward the exploitation of nanomaterial attributes and nanoparticle-based systems for simultaneous site-targeted delivery of loaded drugs with optimal rate, from the fabricated dosage forms [35]. Other important facts about these nanoparticles is their role as a biosensor for diagnostic applications and in bio-electronics for monitoring and controls. For tissue engineering purposes, the role of carbon-based nanoparticles is growing at a high rate because of their low cost, high reproducibility and ease of process controls during process optimization [36]. Similar to the role of metallic nanoparticles, surfaces of these carbon nanomaterials, such as GBNs graphene oxide (GO) and reduced grapheneoxide (RGO), are highly active and being exploited for their imaging role and therapeutic potentials [37]. GO and RGO have an extensively large surface area for interaction of molecular substrates, such as doxorubicin (DOX) and polyethylenimine (PEI), which is a path followed in gene delivery and gene transfection purposes [38–43].

The biomedical applications for GBNs have been widely studied by researchers [18,44]. For cancer therapy functionalized, GO or RGO using nanomaterials such as iron oxide, gold and quantum dots, show greater affinity in vitro as well as in vivo [45,46]. Similar to other nanoparticles, targeting a tumor through ligand-based conjugation GO/RGO shows potential against tumor accumulation [47–49]. Surface enhancement using PEG not only improves the properties of GO but also reduces the potential toxicity effects [50,51]. The graphene-based nanofillers help to alter the mechanical properties as well as provide additional binding sites for the functionalization of biomolecules (Figure 14.3). Other important properties (Table 14.1) of these nanoparticles include the alteration of cell proliferation and differentiation for the regeneration of tissue [52].

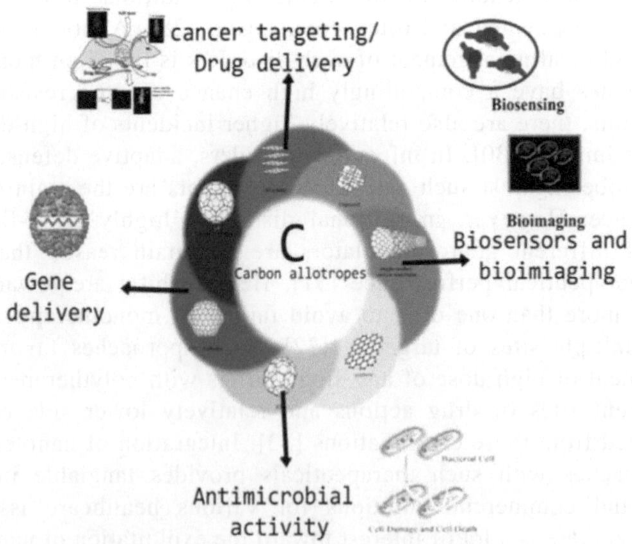

FIGURE 14.3 Biomedical applications of carbon allotropes.

TABLE 14.1
Role of Carbon Allotropes in Drug Delivery Applications

S. No.	Carbon Allotropes Family	Particle Size (nm)	Drug	Activity/Applications	Reference
Graphene					
1	$CoFe_2O_4$/GO, CoFe2O4/GO-DOX	20–100	Doxorubicin hydrochloride (DOX)	In vitro cytotoxicity of HeLa cell lines and in vitro MRI, and fluorescence activity	[25]
2	GO-DEX-Apt-CUR	200	Curcumin (CUR)	Cytotoxicity activity on 4T1 and MCF-7 cell line	[77]
3	GQD-BTN-DOX	5–100	Doxorubicin (DOX)	Cytotoxicity activity on A549 cells	[78]
4	DOX/DNA-GA, DOX/DNA-GC	30	Doxorubicin (DOX)	In vitro cytotoxicity activity on HeLa cells	[79]
5	PPa-NGO-mAb	200	PPa-NGO-mAb	Cytotoxicity activity on U87-MG and MCF-7 cells and photodynamic therapy	[80]
6	GO-CMC-FI-HA/DOX	500	Doxorubicin (DOX)	In vitro cytotoxicity of HeLa cell line	[81]
7	GO-CMC-FI-LA-Ac-DOX	500	Doxorubicin (DOX), fluorescein isothiocyanate (FITC)	Cytotoxicity activity on liver cancer cell line SMMC-7721 andL929	[82]
8	GO-PVP-FA-CPT, AND GO-CS-FA-CPT	100–200	Camptothecin (CPT)	Hemolytic activity, anti-inflammatory, MCF-7 cells	[83]
9	RC–DOX–GO	30	Doxorubicin (DOX)	Bel-7402, SMMC-7721, HepG2	[84]
10	DOX -GO-based magnetic fluorescent hybrid	100–200	Doxorubicin (DOX)	In vitro cytotoxicity of HepG2 cells	[85]
Fullerenes					
11	MSN-C60-TEG-COOH- DOX	50–500	Doxorubicin (DOX)	pH-responsive drug delivery and bio-imaging system	[86]
12	C60-PEI-FA/DTX	50–200	Docetaxel (DTX)	PC3 human prostate cancer cell line	[52]
13	Glycination of C60-TAM	500	Tamoxifen	In vitro cytotoxicity of MCF7 cell line	[87]

(Continued)

TABLE 14.1 (Continued)
Role of Carbon Allotropes in Drug Delivery Applications

S. No.	Carbon Allotropes Family	Particle Size (nm)	Drug	Activity/Applications	Reference
14	FLU-MTX	< 100	Methotrexate (MTX)	In vitro cytotoxicity activity on MDA-MB-231 cells	[88]
15	DOXO-C82-Crgd	0.9–1.1	Doxorubicin (DOX)	Cytotoxicity of non-small lung cancer cells NCl-H2135	[89]
16	C60 + Dox	40–1000	Doxorubicin (Dox)	HCT-116 colon carcinoma, human T-leukemia cells, HL-60 line, MCF-7 line	[90]
17	5FU-$B_{24}N_{24}$-C60 nanocluster	–	5-fluorouraci (5FU)	Anti-cancer drug delivery by density functional theory (DFT)	[91]
18	5FU-$C_{60}R_5Cl$	10–100	5-fluorouracil, cyclophosphamide, and cisplatin	Anti-cancer drugs activity used for delay release	[92]
19	DTX-C60-fullerenes	530.7–1472	Docetaxel	In-vitro cytotoxicity activity MCF-7 and MDA-MB-231 cells and ex-vivo hemolysis activity	[93]
20	C60-OH-APA-DTX	25–325	Docetaxel (DTX)	In vitro cytotoxicity activity on MDA-MB-231 cells,	[94]
Carbon nanotubes (CNT)					
21	PPD-DOX-CNT		Doxorubicin	BTB, MR, and NR	[95]
22	HA -SWNT-GO- C60	50–100	NIR laser 808 nm	In vitro laser Irradiation and In vitro cytotoxicity of MCF-7 cells	[96]
23	DTX- MWCNTs	287.9 ± 10.1	Docetaxel (DTX)	In-vitro anti-cancer activity on MCF-7 and MDA-MB-231 and ex-vivo haemolysis	[97]
24	BRB-C-P-MWCNT and BRB-C-Ph-MWCNT	500	berberine (BRB)	In vitro and in vivo study in animals, haemolytic and DPPH assay activity	[98]

TABLE 14.1 (Continued)
Role of Carbon Allotropes in Drug Delivery Applications

S. No.	Carbon Allotropes Family	Particle Size (nm)	Drug	Activity/Applications	Reference
25	SWCNT-PEG-GEM	10–50	Gemcitabine (GEM	In vivo and in vitro cytotoxicity assay on A549 and MIA PaCa-2 cell line	[99]
26	DOX/PEG-FA-MWCNTs	300–500	Doxorubicin	In vitro cytotoxicity activity on MCF7 cell line	[100]
27	SWCNT-Cur	170	Curcumin,NIR 808 nm	in vitro cytotoxicity and photothermal activity on PC-3 cells,	[101]
28	DOX/TPGS-MWCNTs	~250 nm	doxorubicin (DOX)	In vivo and in vitro release studies	[102]
29	DOX/ES-PEG-MWCNTs	–	Doxorubicin (DOX)	in vitro, ex vivo and in-vivo studies on MCF-7 cell line	[103]
30	DOX/FA/CHI-SWNT		DOX	In vivo and in vitro cytotoxicity activity on HeLa cells	[104]

14.3.1 Structural Properties of Graphene Nanomaterials

14.3.1.1 Graphene Oxide (GO)

The scaling of graphite oxide (GO) results in the formation of GO. Usually, the acid-base treatment followed by sonication is used for its production. GO consists of several functional groups on its surface, such as oxygen, carbonyl, epoxide, hydroxyl and phenol, which makes it different from graphene. Also, the GO surface is activated due to the presence of aliphatic (sp^3) and aromatic (sp^2) domains [53,54] and [55]. Hummer's method has been used for its synthesis. Also, no specific structure for GO is known until now. However, morphology and characterization techniques provide a basic idea of its structure.

14.3.1.2 Reduced Graphene Oxide (RGO)

RGO, on the other hand, is derived from GO using reduction methods (chemical or thermal nature). Structurally, it is intermediate between both the graphene as well as GO [11].

Apart from the above-mentioned GBNs' structural properties, some other physico-chemical properties are listed in Table 14.2.

TABLE 14.2
Physicochemical Properties of GBNs

Property	Graphene	Graphene Oxide (GO)	Reduced GO (rGO)
Charge carrier concentration	1.4×10^{13} cm^{-2}	Very low	N/A
Electrical conductivity	10^4 S cm^{-1}	10^{-1} S cm^{-1}	200–35,000 S cm^{-1}
Fracture strength	130 GPa	120 MPa	N/A
Optical transmittance	97.7%	Low value	60-90%
Mobility at room temperature	Near 200,000 cm^{2}V^{-1}s^{-1}	Lower than 15,000	Intermediate b/w two
Thermal conductivity	Near 5000 W mK^{-1}	2000 W mK^{-1}	0.14-0.87 WmK^{-1}
Young's modulus	1000 GPa	220 GPa	N/A

14.3.2 Synthetic Routes for Graphene-Based Nanomaterials

The basic two methods adopted for graphene synthesis are top-down and bottom-up approaches. The former involves the conversion of a multi-layer graphite sheet into a single layer using different physical, chemical or mechanical strains, while the latter involves the deposition of carbon layer onto substrates via CVD.

The most common method for the development of graphene derivatives specifically GO in high yields is Hummer's method, which involves GO scale-off using a mixture of concentrated sulphuric acid and potassium permanganate. However, the structural properties of GO are altered by surface oxidation from residual radical and carboxylic functionality. The other graphene-based nanomaterial RGO is thus obtained by reducing GO using hydrazine, etc. reducing agents to refurbish the structural properties of graphene. The optical and conducting properties for RGO are enhanced over GO, making it a more favorable candidate for photothermal therapy [56].

Graphene has hydrophobic character, whereas GO has a hydrophilic nature due to two features, pi stacking and electrostatic interactions providing a route for drugs to bind via the physical or chemical route to their surface, giving rise to drug delivery applications.

PEG-activated GO was used by Liu et al. to deliver analog of camptothecin (CPT) [57], after which the use of GO as a drug carrier in antibiotics, anti-cancer, antibodies, genes, etc. saw a stupendous increase.

14.3.3 Action of Graphene Nanomaterials in Cancer Imaging

Fullerenes are often used in imaging for diagnostic purposes but were found to be toxic if administered in large dosage. Graphene, due to its luminescence in the visible region and comparably less toxicity, has been chosen as an alternative.

Therefore, the current demands are for building the new nanocomposite/hybrid forms of GO with other drugs.

14.3.3.1 GO as a Drug Carrier

A nontoxic drug and carrier system was designed with GO and the drug chlorogenic acid (CA) and elicited a highly cytotoxic response from cancer cells. This pH-sensitive system enabled control release of drug from the carrier [58]. Amino group functionalized GO combined with carboxymethyl cellulose was able to release the anti-cancer drug Doxorubicin (Dox) in a controlled manner at the targeted site [38]. Copper complexes of combined carriers made from polyethyleneimine (PEI) and folic acid (FA) were found to be effective in the treatment of nasopharyngeal carcinoma cell line (HNE-1) [59]. The system is water-soluble and biocompatible; thus, it shows well-defined inhibition towards these cell lines and leads to cell death for HNE-1 [60].

GO-based nanofillers act as easy binders and release anti-cancer drugs of hydrophobic nature and are used to make injectable hydrogels [61]. Composites of G/GO hydrogels with Camptothecin (CPT) and Doxorubicin (XDR) incorporated systems were able to achieve controlled release of anti-cancer drugs gradually compared to F-127-drug combination [62,63].

14.3.3.2 rGO as Carrier

Subsequently, nontoxic rGO-Ag nanocomposites showed anti-cancer effects by increasing production of ROS (greater than its precursors) at a dose of 30g/ml and resulted in apoptosis of A549 lung cancer cells. Also, their action towards ovarian cancer cells (OvSCs) was studied. A colony reduction in OvSCs is observed in the application of these composites. They also enhance the gene expression for pro-apoptotic cells and suppress anti-ones [63].

14.3.3.3 GO Chips as a Carrier

The spread of cancer to secondary sites is known as metastasis due to CTCs, eventually leading death. Therefore, the isolation of these cells is crucial for the prevention of tumor metastasis. GO nanosheets functionalized with phospholipid-polyethylene-glycolamine (PL-PEG-NH_2) on a gold-patterned chip was used to eventually isolate highly sensitive and low cell concentration CTCs from cancer-afflicted cells of breast, pancreas and lungs. This is a significant development regarding graphene-based chips as biosensors. The important advantage is that they are cost-effective compared to previously reported ones [63].

14.3.3.4 Quantum Dots as A Carrier

Graphene quantum dots (GQDs) add another category of carriers for drug delivery in cancer. The arginyl-glycylaspartic acid (RGD) functionalized GQDs bind to cancer cell lines. The GQDs-RGD-Dox composite system enhances the uptake of Dox by human prostate cell lines, PC-3 and DU-145 as compared to pure Dox. Herceptin-labeled GQD nanocarrier composites loaded with Dox were used on breast cancer cells in specific conditions of temperature and pH [64].

14.3.4 Action in Tumor Targeting

GBNs show tumor-targeting applications, specifically angiogenesis, which is preferred over passive and tumor cell-based targeting. One of the potential markers is CD105, which is expressed only in contact with endothelial cells. TRC105 is a new promising marker for PET imaging of GO/RGO nano-conjugate. For the 4T1 breast tumor, the expression was twice for the conjugate of TRC105 and GO than the nonconjugated GO form. These GO conjugates can further be used for cancer treatment or PTT [65].

14.3.5 Action in Drug and Gene Delivery

The two factors giving rise to gene delivery applications for graphene-based nanomaterials include strong π-π and hydrophobic interactions, which are responsible for loading DOX and Camptothecin (CPT) and binding hydrophilic drugs like paclitaxel, respectively. Also, functionalized graphene materials obtained via coating with PEI or chitosan could be employed for the same. The nGO-PEG-DOX nanocomposite is formed as a result of π-π stacking. Similarly, one more water-soluble complex, nGO-PEGSN38, has been synthesized with remarkable chemotherapeutic action compared to the free SN38 molecules. Multiple drugs can also be loaded onto nGO for chemotherapy to provide better inhibition of cancer cells. As it is observed that a combination of DOX- and CPT-loaded nGO shows better action than their counterparts [66]. DA functionalized GO changed the negatively charged moiety to positive moiety with a change in pH from 7.4 to 6.8, making way for pH-responsive drug release. Thus, GBNs demonstrate a significant role in drug delivery action of aromatic as well as hydrophobic drugs (Table 14.3).

14.3.6 Action of Graphene Nanomaterials in Bioimaging

The two main aspects of bio-imaging are:

1. The tools used for detection should be fast and sensitive.
2. The nanomaterials used as contrasting agents should be biodegradable, cross the blood-brain barrier (BBB), biocompatible, sensitive and specific. In addition to the above, they must be able to target the affected tissue within a stipulated time. These contrasting agents can be modified using G, GO and their composites. For example, G/GO-based quantum dots (QD) result in surface and size dependable fluorescence.

Moreover, surface engineering using various biopolymers, such as polyethylene glycol (PEG), polystyrene and polypeptides, etc., help to increase biocompatibility [67]. An (rGO-IONP-PEG) composite formed by covering rGO/iron oxide with PEG finds application in photothermal therapy (PTT) and other imaging sources. MRI uses fluorine-based GO as a drug carrier with a magnetic response.

TABLE 14.3
Carbon Allotrope-Based Biosensors used in Various Detection Techniques of Diseases

S. No.	Carbon allotropes family	Characterization techniques	Detection techniques	Applications	Reference
Graphene					
1	GO/PDDA/AgNPs	AFM, TEM, Zeta-Potential,	Surface-enhanced Raman scattering (SERS) detection of folic acid	Applications of medicine and biotechnology	Ren et al. (2011)
2	Metallic NPs attached to graphene oxide	TEM, EM	SERS probe of 670 nm laser	Sensing of HIV DNA and bacteria	Fan et al. (2013)
3	(G/Ag) grapheme/silver nanocomposite	AFM, UV–vis measurement, confocal microscopy	Surface-enhanced Raman scattering (SERS) sensors	G/Ag SERS model exhibited potential application for detecting prohibited colorants	Xie et al. (2012)
4	GO@SiO_2@Ag NPs	UV-Vis spectrum, TEM, DXR™ Raman Microscope	4-mercaptobenzoic acid (4-MBA) as a SERS chemical	e SERS substrate for detection of trace amounts of chemicals in various fields	Pham et al. (2016)
5	rGO-AuNPs/AgNPs	TEM, SEM, EDX, Raman spectrometer	SERS	Prepared surface of planar microelectrode chips using an electric field-guided Ag nanoparticle assembly process	Mohammadi et al. (2018)
6	AgNP–AgNP–GO	AFM, UV/vis-near-infrared (NIR), TEM, XRD	Surface-enhanced Raman scattering (SERS) sensing	4-mercaptobenzoic acid (4-MBA) act as highly efficient Raman probes for DNA capability for multiplex DNA detection	Tsung-Wu Lin et al. (2015)

(Continued)

TABLE 14.3 (Continued)
Carbon Allotrope-Based Biosensors used in Various Detection Techniques of Diseases

S. No.	Carbon allotropes family	Characterization techniques	Detection techniques	Applications	Reference
7	RGO-AgNP-FA	TEM, EDS, UV-Vis spectra, XRD, AFM, Raman spectra	SERS detection	Raman probe for cancer diagnosis in vitro liver cancer cells	Hu et al. (2013)
8	rGO-NS-DOX	UV–vis Spectra, TEM, SEM, EDX, Zeta-potential, XPS, Raman-IR microscope	Surface-enhanced Raman scattering (SERS)	SERS-based applications, such as drug delivery and chemotherapy	Wang et al. (2014)
9	m-GQDs-IFO	UV-Vis, FT-IR, Raman spectra, XPS, SEM, TEM, cyclicvoltammograms	Electrochemical sensors	Electrochemical detection of an anti-cancer ifosfamide drug	Prasad et al. (2017)
10	Grapheme modified disposable pencil graphite electrodes (GME)	EIS, SEM, DPV	Electrochemical sensors	Analyzing mir21 in the cell lysates of mir-21 positive breast cancer cellline (MCF-7)	Kilic et al. (2014)
Fullerenes					
11	AuNPs@C_{60}/GCE	UV–vis and FTIR spectra, FESEM, EDX, EDS	Electrochemical sensor	Non-enzymatic glucose sensing	Sutradhar et al. (2017)
12	Fe3O4@Au NPs-hairpin-SiO_2 NPs -C_{60}-Conjugates	TEM, CV and PEC,	Photo-electrochemical biosensor	DNA detection	Wang et al. [6]
13	QD-C60-MNPs complex	TEM, PL spectra, fluorescence microscopy	Photochemical-based molecular beacon nanosensors	Detection of pathogen DNA or RNA	Liu et al. (2018)

14	Au/4-ATP/ C_{60} /PAMAM	SEM and EIS, ELISA	Electrochemical biosensors	Biosensor to detect Fetuin-A in real blood samples	Zihni, Onur, Uygun et al. (2018)
15	GO_x/C_{60}-Fc-CS-IL	Cyclic voltammograms, amperometry	Electrochemical biosensors	Biosensor was applied to the determination of glucose in blood serum samples	Zhilei et al. (2010)
16	C_{60}-$TOAB^+$	FT-IR, SEM, cyclic voltammetry (CV)	Non-enzymatic electrochemical biosensor	Glucose detection method by monitoring salivary glucose level	Ye et al. (2016)
17	anti-HSP70-fullerene C_{60}	EIS, CV, chrono-amperometry, SEM	Electrochemical biosensor	Real human blood serum samples were analyzed by the biosensor and validated by using Elisa	Demirbakan et al. (2016)
18	Hb/C60-NCNT/GCE	FTIR, TEM, EDS, cyclic voltammetry	Electro-chemical sensor	Electro-chemistry of hemoglobin and its application in bio-sensing to determination of H_2O_2 in cell extracts or urine	Shen et al. (2013)
19	GCE/AuNPs/L-Cys–C_{60}–APBA/HT/DA–Cu_2O	SEM, ECL	Electro-chemi-luminescence biosensor	Prepare electro-chemi-luminescence biosensor for dopamine based exhibited high sensitivity and stability, good reproducibility and repeatability	Haijun Wang et al. [12]
20	c-C_{60}/CeO_2/PtNPs, PtNPs, c-C_{60}/ PtNPs and c-C_{60}/CeO_2/AuNPs	FE-SEM, EDS and zeta potential, XPS, amperometry,	Ultrasensitive electro-chemical biosensor	Biosensor achieved accurate quantitative detection of CYP2C19*2 gene in human serum samples and correlations with standard DNA sequencing	Zhang et al. (2018)

(*Continued*)

TABLE 14.3 (Continued)
Carbon Allotrope-Based Biosensors used in Various Detection Techniques of Diseases

S. No.	Carbon allotropes family	Characterization techniques	Detection techniques	Applications	Reference
Carbon nanotubes (CNT)					
21	PP/MWCNTs/PGE	EIS, XPS, AFM, SEM, FTIR, DPV	Electrochemical DNA biosensor	PP/MWCNTs/PGE biosensor was used to investigate the interaction of 6-MP with ds-DNA	Maleh et al. (2015)
22	MWCNTs/FcMe/CS/HRP /BSA /LOx/SPBGE	SEM, Cyclic voltammogram, Chrono-amperometric, HPLC	Electro-chemical lactate biosensor	Electro-chemical lactate biosensor for the determination of lactate in a commercial embryonic cell culture medium	Ibáñez et al. (2016)
23	Ag@MH/MWCNT	Cyclic Voltammetry, FT-IR, XRD, FESEM,	Electro-chemical glucose biosensor	Ag@MH/MWCNT nano-composite-based glucose biosensor excellent activity for electrocatalytic oxidation of glucose	Baghayeri et al. (2016)
24	MWCNT–DNA conjugates	TEM	Lateral flow biosensor (LFB)	MWCNT-based LFB can be extended to visually detect protein biomarkers using MWCNT–antibodyconjugates and detection of DNA sequence	Qiu et al. (2015)
25	GOQDs/CMWCNTs/PGE	SEM, XRD, HPLC, confocal, EIS, Raman spectrometer, cyclic voltammograms	Cell-based electro-chemical biosensor	Cell-based electro-chemical toxicity biosensor to assess the toxicity of priority pollutants in the aquatic environment	Zhu et al. (2016)

26	BC-PAni/SWCNTs nanocomposites	FE-SEM, FT-IR, XRD, TGA, linear sweep voltammetry	BC-PAni/SWCNTs electro-chemical biosensors	Potential applications in the biomedical field such as biosensors, biofuel cells, and bioelectronic devices	Jasim et al. (2016)
27	AChE&ChOx/Pt-Au/ MWCNT/GCE	CV, ampero-metric apparatus,EIS, FESEM, EDX	Electro-chemi-luminescence (ECL) biosensor	ECL bio-sensing system was designed and developed for individual detection of different organo-phosphorous pesticides (OPs) in food samples	Miao et al. (2016)
28	AChE/pristineCNTs/ GCE,AChE/ CNT–OH/ GCE,AChE/ CNT–COOH/GCE	DPV, EIS, XPS, GC, UV-Vis spectro-photometer, IR spectra, amperometric apparatus,	Electrochemical based organophosphorus pesticide (OPs) biosensors	Electro-chemical based pesticide biosensor is successfully employed for the direct analysis of vegetable samples	Yu et al. (2015)
29	CH-GOx/PPy-Nf-fMWCNTs/GCE	CV, EIS, FESEM, ampero-metry, EDS, FTIR, Raman spectroscopy, UV-visible	Electro-chemical based glucose biosensor	Biosensor-exhibited excellent anti-interference ability, reproducibility, long-term storage stability, repeatability, and acceptable glucose detection in real serum samples	Shrestha et al. (2016)
30	GR-MWNTs /AuNPs film	SEM, XRD, ATR–FTIR, cyclic voltammogram, amperomety,	Electro-chemical based glucose biosensor	Accurate determination of glucose present in human serum sample, potential applications in immobilization of enzymes and fabrication of biosensors	Devasenathipathy et al. (2015)

The majorly used imaging techniques include magnetic resonance imaging (MRI), electron paramagnetic resonance imaging (EPRI), photo-acoustic, ultrasound, fluorescence imaging, surface-enhanced Raman scattering (SERS) and coherent anti-stokes Raman scattering imaging (CARS). Among these, the most prominent technique used is fluorescence and particularly Forster Resonance Energy Transfer (FRET) imaging [68]. The resolution of the spatial image is high, and the signals generated show sensitivity towards the association, conformational change and separation in range within 10 nm. They employ the role of G/GO-based sensors. This technique involves a donor dye whose role is to absorb the laser energy and transfer it to acceptor for final fluorescence detection. While using the EPRI technique for oxidative sensing, triphenylmethyl radical derivatives as well as other radicals and carbon-based ink are used.

As far as CARS is concerned, the use is in cell cultures, tissues and in vivo studies on animal models with a largely sensitive signal. On the contrary, the use of imaging with Raman spectroscopy due to low signal intensity is limited to cell cultures and ex vivo tissues [69].

14.3.7 Action as Biosensors

The biosensors are used to detect chemical analytes or biomolecules, which help in diagnosis and therapeutic measures. The ability of GBNs to absorb aromatic molecules due to pi interactions helps in making new biosensors. They can also be used for the detection of macromolecules. The GO-based biosensors are known for their quick response time, larger sensitiveness, and they can even lower detection limits [40]. The properties of GBN nanocomposites are enhanced at electronic levels by layering them with secondary biomolecules, bio-enzymes or metal nanoparticles. Some of the electrochemical biosensors with an enzymatic backbone have been fabricated like alcohol dehydrogenase, tyrosinase, catalase, organophosphorus hydrolase, etc. [70,71].

14.3.8 Action of Graphene Nanomaterials in Tissue Engineering

The tissues may get damaged by tumor or other deformations, and many materials have been employed to subside the effects. A few bioactive materials are hybrid and are being used for artificial tissue generation. Hydroxyapatite is utilized in different forms and shapes. Subsequently, hydrogels and nano-TiO_2 have been used. GBNs have multifarious physical and chemical properties, such as fine flexibility, elasticity and adaptability, so that they can be used to tailor materials for this aspect of biomedical application. GBNs could be modified as they have functional groups present on their surface that interact with biomolecules, including DNA, proteins, peptides and enzymes. In tissue engineering applications, GBNs have been used in neural, cartilage, skin, skeletal muscle, cardiac or bone tissue engineering. Over the years, GBN-based stem cells have also been used for treating bone diseases [72, 73, 74].

14.4 OTHER CARBON ALLOTROPES AND THEIR APPLICATIONS

A newer category of 2D carbon allotropes, i.e graphyne and graphyne family members (GFMs), has been under study. Baughman et al. (1987) theoretically predicted the structural features, mechanical strength, bandgaps, thermal and electronic properties of graphyne and GFMs [106]. Further, their relative applications, including water desalination and purification, ion battery anode material, storage of hydrogen gas, separation of gas and many others, have been well studied. Graphyne has a direct bandgaps compared to silicon, thus increasing its light-absorbing efficiency and radiation-induced recombination [75].

14.5 CONCLUSION

The major multidimensional carbon allotropic forms have been used as electrochemical detectors for ME and CE. The 0D fullerenes find applications as electrochemical and biosensors, whereas the 1D CNTs and 2D graphene have been extensively studied for their action as detectors in CE and ME due to high surface area. The major characteristics of carbon nanomaterials responsible for the electro-catalytic effect are low current densities prevailing due to large surface area; a high faradaic amperometric current, with enhanced analytical sensitivity, which arises due to multiple redox reactions occurring at the surface area of carbon nanomaterials; and electrodes with high resistance and stability that couple with analytes. This results in stability and reproducibility via detection. A high rate of electron transfer of heterogeneous nature gives sharp and intense peaks in the separation of analytes.

Drugs, gene delivery, biosensing, bio-electronics, tissue engineering, anti-cancer and antimicrobial are a few of the varied applications of carbon nanomaterials. Among GBNs, GO and RGO are majorly studied for their medicinal use. GO and RGO have an extensively large surface area for interaction of molecular substrates, such as doxorubicin (DOX) and polyethylenimine (PEI), which is a path followed in gene delivery and gene transfection. A nontoxic drug and carrier system was designed with GO and the drug chlorogenic acid (CA) and elicited a highly cytotoxic response from cancer cells. This pH-sensitive system enabled controlled release of drugs from the carrier. GO-based nanofillers act as easy binders and release anti-cancer drugs of hydrophobic nature and are used to make injectable hydrogels. Composites of G/GO hydrogels with camptothecin (CPT) and doxorubicin (DXR) incorporated systems were able to achieve controlled release of anti-cancer drugs gradually compared to F-127-drug combination. GBNs show tumor-targeting applications, specifically angiogenesis, which is preferred over passive and tumor cell-based targeting. One of the potential markers is CD105, which is expressed only in contact with endothelial cells. TRC105 is a new promising marker for PET imaging of GO/RGO nanoconjugate. For the 4T1 breast tumor, the expression was twice for the conjugate of TRC105 and GO than the nonconjugated GO form. These GO conjugates can further be used for cancer treatment or PTT. The two factors giving rise to gene delivery applications for graphene-based nano-materials include strong π-π and hydrophobic interactions are responsible for loading DOX and camptothecin (CPT) and binding hydrophilic drugs like paclitaxel respectively. Also, functionalized graphene materials obtained via coating with PEI or chitosan could be employed for the same. The nGO-PEG-DOX nanocomposite was formed as a result of π-π stacking.

ACKNOWLEDGMENTS

J.R. Ansari gratefully acknowledges the Vice Chancellor, Pro Vice Chancellor and Dean, Faculty of Physical Sciences, PDM University, Bahadurgarh, India, for their co-operation during the process of writing this chapter. ***M.S. Alam*** acknowledges the Vice Chancellor and Dean, NIMS Institute of Pharmacy, NIMS University, Jaipur, Rajasthan, India, for their kind co-operation. Dr. M.N. Javed expresses thanks to the KR Mangalam University organization for all kinds of supports.

ABBREVIATION

DFT	Density Functional Theory
CVD	Chemical Vapour Deposition
CNTs	Carbon Nanotubes
CNBs	Carbon Nanowalls
GBNs	Graphene-Based Nanomaterials
IPR	Isolated Pentagon Rule
CHCs	Carbon Honeycombs
ME	Microchip Electrophoresis
CE	Capillary Electrophoresis
GO	Graphene Oxide
RGO	Reduced Graphene Oxide
DOX	Doxorubicin
PEI	Polyethylenimine
PEG	Polyethylene Glycol
CPT	Camptothecin
CA	Chlorogenic Acid
FA	Folic Acid
HNE-1	Nasopharyngeal Carcinoma Cell Line
CTCs	Circulating Tumor Cells
GQDs	Graphene Quantum Dots
BBB	Blood-Brain Barrier
QD	Quantum Dots
PTT	Photothermal Therapy
MRI	Magnetic Resonance Imaging
EPRI	Electron Paramagnetic Resonance Imaging
SERS	Surface-Enhanced Raman Scattering
CARS	Coherent Anti-stokes Raman Scattering
FRET	Forster Resonance Energy Transfer
GFMs	Graphyne Family Members

REFERENCES

1. Hirsch, A. "The era of carbon allotropes". *Nature Materials* 9, no. 11 (2010): 868–871.
2. Kroto, H.W., Heath, J.R., O'Brien, S.C., Curl, R.F., and Smalley, R.E. "C60: Buckminsterfullerene". Nature 318, no. 6042 (1985): 162–163.
3. Iijima, S. "Helical microtubules of graphitic carbon". *Nature* 354, no. 6348 (1991): 56–58.
4. Novoselov, K.S. "Electric field effect in atomically thin carbon films". *Science* 306, no. 5696 (2004): 666–669.
5. Pottoo, F.H., Barkat, M.A., Ansari, M.A., Javed, M.N., Jamal, Q.M., and Kamal, M.A. (2019) Nanotechnologoical based miRNA intervention in the therapeutic management of neuroblastoma, *Seminars in Cancer Biology*, 69, 100–109, doi: 10.1016/j.semcancer.2019.09.017. Epub 2019 Sep 25. PMID: 31562954.
6. Wang, S., Wu, D., Yang, B., Ruckenstein, E., and Chen, H. "Semimetallic carbon honeycombs: New three-dimensional graphene allotropes with Dirac cones". *Nanoscale* 10, no. 6 (2018): 2748–2754.
7. Koziol, K., Boskovic, B.O., and Yahya, N. "Synthesis of carbon nanostructures by CVD method". In: *Carbon and Oxide Nanostructures*, pp. 23–49. Springer Berlin Heidelberg. 2010.
8. Dyakonov, P., Mironovich, K., Svyakhovskiy, S., Voloshina, O., Dagesyan, S., Panchishin, A., Suetin, N., Bagratashvili, V., Timashev, P., Shirshin, E., and Evlashin, S. "Publisher correction: Carbon nanowalls as a platform for biological SERS studies". *Scientific Reports* 8, no. 1 (2018).
9. Pottoo, F.H., Javed, N., Rahman, J., Abu-Izneid, T., and Khan, F.A. "Targeted delivery of miRNA based therapeuticals in the clinical management of Glioblastoma Multiforme". *Seminars in Cancer Biology* (2020a). doi: 10.1016/j.semcancer.2020.04.001.
10. Pottoo, F.H., Sharma, S., Javed, M.N., Barkat, M.A., Harshita, Alam, M.S., Naim, M.J., Alam, M.O., Ansari, M.A., Barreto, G.E., and Ashraf, G.M. "Lipid-based nanoformulations in the treatment of neurological disorders". *Drug Metabolism Reviews* 52, no. 1 (2020b): 185–204.
11. Tadyszak, K., Wychowaniec, J., and Litowczenko, J. "Biomedical applications of graphene-based structures". *Nano-materials* 8, no. 11 (2018): 944.
12. Wang, Z., Zhou, X.-F., Zhang, X., Zhu, Q., Dong, H., Zhao, M., and Oganov, A.R. "Phagraphene: A low-energy graphene allotrope composed of 5–6–7 carbon rings with distorted dirac cones". *Nano Letters* 15, no. 9 (2015): 6182–6186.
13. Xu, L.-C., Wang, R.-Z., Miao, M.-S., Wei, X.-L., Chen, Y.-P., Yan, H., Lau, W.-M., Liu, L.-M., and Ma, Y.-M. "Two dimensional dirac carbon allotropes from graphene". *Nanoscale* 6, no. 2 (2014): 1113–1118.
14. Liu, Y., Wang, G., Huang, Q., Guo, L., and Chen, X. "Structural and electronic properties of T graphene: A two-dimensional carbon allotrope with tetrarings". *Physical Review Letters* 108, no. 22 (2012).
15. Lahiri, J., Lin, Y., Bozkurt, P., Oleynik, I.I., Batzill, M. "An extended defect in graphene as a metallic wire". *Nature Nanotechnology* 5, no. 5 (2010): 326–329.
16. Kotakoski, J., Krasheninnikov, A.V., Kaiser, U., and Meyer, J.C. "From point defects in graphene to two-dimensional amorphous carbon". *Physical Review Letters* 106, no. 10 (2011).
17. Tan, Y.-Z., Xie, S.-Y., Huang, R.-B., and Zheng, L.-S. "The stabilization of fused-pentagon fullerene molecules". *Nature Chemistry* 1, no. 6 (2009): 450–460.

18. Zhang, S., Zhou, J., Wang, Q., Chen, X., Kawazoe, Y., and Jena, P. "Penta-graphene: A new carbon allotrope". *Proceedings of the National Academy of Sciences*. 112, no. 8 (2015): 2372–2377.
19. Chen, J.-H., Autès, G., Alem, N., Gargiulo, F., Gautam, A., Linck, M., Kisielowski, C., Yazyev, O.V., Louie, S.G., and Zettl, A. "Controlled growth of a line defect in graphene and implications for gate-tunable valley filtering". *Physical Review B* 89, no. 12 (2014).
20. Sharma, B.R., Manjanath, A., and Singh, A.K. "Pentahexoctite: A new two-dimensional allotrope of carbon". *Scientific Reports* 4, no. 1 (2015).
21. Chen, Y., Xie, Y., Yang, S.A., Pan, H., Zhang, F., Cohen, M.L., and Zhang, S. "Nanostructured Carbon Allotropes with Weyl-like Loops and Points". *Nano Letters* 15, no. 10 (2015): 6974–6978.
22. Krainyukova, N.V., and Zubarev, E.N. "Carbon honeycomb high capacity storage for gaseous and liquid species". *Physical Review Letters* 116, no. 5 (2016).
23. Zhong, C., Chen, Y., Xie, Y., Yang, S.A., Cohen, M.L., and Zhang, S.B. "Towards three-dimensional Weyl-surface semimetals in graphene networks". *Nanoscale* 8, no. 13 (2016): 7232–7239.
24. Niu, C.-Y., Wang, X.-Q., and Wang, J.-T. "K 6 carbon: A metallic carbon allotrope in sp 3 bonding networks". *The Journal of Chemical Physics* 140, no. 5 (2014): 054514.
25. Wang, J.-Q., Zhao, C.-X., Niu, C.-Y., Sun, Q., and Jia, Y. "C 20 – T carbon: A novel superhardsp 3 carbon allotrope with large cavities. "*Journal of Physics: Condensed Matter* 28, no. 47 (2016): 475402.
26. He, C., Sun, L., Zhang, C., Peng, X., Zhang, K., and Zhong, J. "Four superhard carbon allotropes: A first-principles study". *Physical Chemistry Chemical Physics* 14, no. 23 (2012): 8410.
27. Martín, A., López, M.Á., González, M.C., and Escarpa, A. "Multidimensional carbon allotropes as electrochemical detectors in capillary and microchip electrophoresis: General electrophoresis" 36, no. 1 (2015): 179–194.
28. Pottoo, F.H., Javed, M., Barkat, M., Alam, M., Nowshehri, J.A., Alshayban, D.M., and Ansari, M.A. "Estrogen and Serotonin: Complexity of interactions and implications for epileptic seizures and epileptogenesis". *Current Neuropharmacology* 17, no. 3 (2019b): 214–231.
29. Pottoo, F.H., Tabassum, N., Javed, M.N., Nigar, S., Rasheed, R., Khan, A., Barkat, M.A., Alam, M.S., Ansari, M.A., Maqbool, A., Barreto, G.E., and Ashraf, G.M. "The synergistic effect of raloxifene, fluoxetine, and bromocriptine protects against pilocarpine-induced status epilepticus and temporal lobe epilepsy". *Molecular Neurobiology* 56, no. 2 (2019c): 1233–1247.
30. Nigar, S., Pottoo, F.H., Tabassum, N., Verma, S., and Javed, M.N. "Molecular insights into the role of inflammation and oxidative stress in epilepsy". *J Adv Med Pharm Sci.* 10 (2016): 1–9.
31. Pandey, M., Saleem, S., Nautiyal, H., Pottoo, F.H., and Javed, M.N. "PINK1/Parkin in neurodegenerative disorders: Crosstalk between mitochondrial stress and neurodegeneration". In *Quality Control of Cellular Protein in Neurodegenerative Disorders*, 282–301. IGI Global, 2020.
32. Pottoo, F.H., Tabassum, N., Javed, M.N., Nigar, S., Sharma, S., Barkat, M.A., Alam, M.S., Ansari, M.A., Barreto, G.E., and Ashraf, G.M. "Raloxifene potentiates the effect of fluoxetine against maximal electroshock induced seizures in mice". *European Journal of Pharmaceutical Sciences* 146 (2020c): 105261.
33. Raman, S., Mahmood, S., Hilles, A.R., Javed, M.N., Azmana, M., and Al-Japairai, K.A.S.(2020). "Polymeric nanoparticles for brain drug delivery: A review". *Current drug metabolism*, 21(9), https://doi.org/10.2174/1389200221666200508074348

34. Hasnain M.S., Javed M.N., Alam M.S., Rishishwar P., Rishishwar S., Ali S., Nayak A.K., and Beg S. "Purple heart plant leaves extract-mediated silver nanoparticle synthesis: Optimization by Box-Behnken design". *Materials Science and Engineering: C* 99 (2019): 1105–1114.
35. Javed, M.N., Kohli, K., and Amin, S. "Risk assessment integrated QbD approach for development of optimized bicontinuous mucoadhesive limicubes for oral delivery of rosuvastatin". *AAPS PharmSciTech* 19, no. 3 (2018): 1377–1391.
36. Javed, M.N., Alam M.S., Waziri, A., Pottoo, F.H., Yadav, A.K., Hasnain, M.S., and Almalki, F.A. "QbD applications for the development of nanopharmaceutical products". In *Pharmaceutical Quality by Design*, pp. 229–253. Academic Press. 2019.
37. Alam, M.S., Javed, M, Pottoo F.H., Wajiri F., Almalki F., Hasnain M.S., Garg A., and Saifullah K. "Qbd approached comparison of reaction mechanism in microwave synthesized gold nanoparticles and their superior catalytic role against hazardous nirto-dye". *Applied Organometallic chemistry* 33, no. 9 (2019): e5071.
38. Feng, L., Zhang, S., and Liu, Z. "Graphene based gene transfection". *Nanoscale* 3, no. 3 (2011): 1252.
39. Kim, H., Lee, D., Kim, J., Kim, T., and Kim, W.J. "Photothermally triggered cytosolic drug delivery via endosome disruption using a functionalized reduced graphene oxide". *ACS Nano* 7, no. 8 (2013): 6735–6746.
40. Liu, B., Tang, D., Tang, J., Su, B., Li, Q., and Chen, G. "A graphene-based Au (111) platform for electrochemical biosensing based catalytic recycling of products on gold nanoflowers". *The Analyst* 136, no. 11 (2011): 2218.
41. Miao, W., Shim, G., Kang, C.M., Lee, S., Choe, Y.S., Choi, H.-G., and Oh, Y.-K. "Cholesteryl hyaluronic acid-coated, reduced graphene oxide nanosheets for anticancer drug delivery". *Biomaterials* 34, no. 37 (2013): 9638–9647.
42. Tang, L.A.L., Wang, J., and Loh, K.P. "Graphene-based SELDI probe with ultrahigh extraction and sensitivity for DNA oligomer". *Journal of the American Chemical Society* 132, no. 32 (2010): 10976–10977.
43. Zhang, L., Lu, Z., Zhao, Q., Huang, J., Shen, H., and Zhang, Z. "Enhanced chemotherapy efficacy by sequential delivery of siRNA and anticancer drugs using PEI-grafted graphene oxide". *Small* 7, no. 4 (2011): 460–464.
44. Yang, K., Feng, L., Shi, X., and Liu, Z. "Nano-graphene in biomedicine: Theranostic applications". *Chem. Soc. Rev.* 42, no. 2 (2013): 530–547.
45. Williams, K.J., Nelson, C.A., Yan, X., Li, L.-S., and Zhu, X. "Hot electron injection from graphene quantum dots to TiO2". *ACS Nano* 7, no. 2 (2013): 1388–1394.
46. Yang, K., Wan, J., Zhang, S., Zhang, Y., Lee, S.-T., and Liu, Z. "In vivo pharmacokinetics, long-term biodistribution, and toxicology of PEGylated graphene in mice". *ACS Nano* 5, no. 1 (2011): 516–522.
47. Hong, H., Yang, K., Zhang, Y., Engle, J.W., Feng, L., Yang, Y., Nayak, T.R., Goel, S., Bean, J., Theuer, C.P., Barnhart, T.E., Liu, Z., and Cai, W. "In vivo targeting and imaging of tumor vasculature with radiolabeled, antibody-conjugated nanographene". *ACS Nano* 6, no. 3 (2012): 2361–2370.
48. Hong, H., Zhang, Y., Engle, J.W., Nayak, T.R., Theuer, C.P., Nickles, R.J., Barnhart, T.E., and Cai, W. "In vivo targeting and positron emission tomography imaging of tumor vasculature with 66Ga-labeled nano-graphene". *Biomaterials* 33, no. 16 (2012): 4147–4156.
49. Shi, S., Yang, K., Hong, H., Valdovinos, H.F., Nayak, T.R., Zhang, Y., Theuer, C.P., Barnhart, T.E., Liu, Z., and Cai, W. "Tumor vasculature targeting and imaging in living mice with reduced graphene oxide". *Biomaterials* 34, no. 12 (2013): 3002–3009.

50. Mishra, S., Sharma, S., Javed, M.N., Pottoo, F.H., Barkat, M.A., Alam, M.S., Amir, M., and Sarafroz, M. "Bioinspirednanocomposites: Applications in disease diagnosis and treatment". *Pharmaceutical nanotechnology* 7, no. 3 (2019): 206–219.
51. Sharma, S., Javed, M.N., Pottoo, F.H., Rabbani, S.A., Barkat, M., Sarafroz, M., and Amir, M. "Bioresponse inspired nano-materials for targeted drug and gene delivery". *Pharmaceutical nanotechnology* 7, no. 3 (2019): 220–233.
52. Shi, S., Chen, F., Ehlerding, E.B., and Cai, W. "Surface engineering of graphene-based nano-materials for biomedical applications". *Bioconjugate Chemistry* 25, no. 9 (2014): 1609–1619.
53. Kuila, T., Bose, S., Mishra, A.K., Khanra, P., Kim, N.H., and Lee, J.H. "Chemical functionalization of graphene and its applications". *Progress in Materials Science* 57, no. 7 (2012): 1061–1105.
54. Schniepp, H.C., Li, J.-L., McAllister, M.J., Sai, H., Herrera-Alonso, M., Adamson, D.H., Prud'homme, R.K., Car, R., Saville, D.A., and Aksay, I.A. "Functionalized single graphene sheets derived from splitting graphite oxide". *The Journal of Physical Chemistry B* 110, no. 17 (2006): 8535–8539.
55. Georgakilas, V., Tiwari, J.N., Kemp, K.C., Perman, J.A., Bourlinos, A.B., Kim, K.S., and Zboril, R. "Noncovalent functionalization of graphene and graphene oxide for energy materials, biosensing, catalytic, and biomedical applications". *Chemical Reviews* 116, no. 9 (2016): 5464–5519.
56. Dasari Shareena, T.P., McShan, D., Dasmahapatra, A.K. and Tchounwou, P.B. "A review on graphene-based nano-materials in biomedical applications and risks in environment and health". *Nano-Micro Letters* 10, no. 3 (2018a).
57. Dembereldorj, U., Choi, S.Y., Ganbold, E.-O., Song, N.W., Kim, D., Choo, J., Lee, S.Y., Kim, S., and Joo, S.-W. "Gold nanorod-assembled PEgylated graphene-oxide nanocomposites for photothermal cancer therapy". *Photochemistry and Photobiology* 90, no. 3 (2014): 659–666.
58. Jin, Y., Wang, J., Ke, H., Wang, S., and Dai, Z. "Graphene oxide modified PLA microcapsules containing gold nanoparticles for ultrasonic/CT bimodal imaging guided photothermal tumor therapy". *Biomaterials* 34, no. 20 (2013): 4794–4802.
59. Viraka Nellore, B.P., Pramanik, A., Chavva, S.R., Sinha, S.S., Robinson, C., Fan, Z., Kanchanapally, R., Grennell, J., Weaver, I., Hamme, A.T., and Ray, P.C. "Aptamer-conjugated theranostic hybrid graphene oxide with highly selective biosensing and combined therapy capability". *Faraday Discuss* 175 (2014): 257–271.
60. Li, J., Tan, S., Kooger, R., Zhang, C., and Zhang, Y. "MicroRNAs as novel biological targets for detection and regulation". *Chem. Soc. Rev.* 43, no. 2 (2014): 506–517.
61. Paul, A., Hasan, A., Kindi, H.A., Gaharwar, A.K., Rao, V.T.S., Nikkhah, M., Shin, S.R., Krafft, D., Dokmeci, M.R., Shum-Tim, D., and Khademhosseini, A. "Injectable graphene oxide/hydrogel-based angiogenic gene delivery system for vasculogenesis and cardiac repair". *ACS Nano* 8, no. 8 (2014): 8050–8062.
62. Banerjee, A.N. "Graphene and its derivatives as biomedical materials: Future prospects and challenges". *Interface Focus* 8, no. 3 (2018): 20170056.
63. Dasari Shareena, T.P., McShan, D., Dasmahapatra, A.K., and Tchounwou, P.B. "A review on graphene-based nano-materials in biomedical applications and risks in environment and health". *Nano-Micro Letters* 10, no. 3 (2018b).
64. Debele, T., Peng, S., and Tsai, H.-C. "Drug carrier for photodynamic cancer therapy". *International Journal of Molecular Sciences* 16, no. 9 (2015): 22094–22136.
65. Jares-Erijman, E.A. and Jovin, T.M. "FRET imaging". *Nature Biotechnology* 21, no. 11 (2003): 1387–1395.

66. Gao, L., Li, F., Thrall, M.J., Yang, Y., Xing, J., Hammoudi, A.A., Zhao, H., Massoud, Y., Cagle, P.T., Fan, Y., Wong, K.K., Wang, Z., and Wong, S.T.C. "On-the-spot lung cancer differential diagnosis by label-free, molecular vibrational imaging and knowledge-based classification". *Journal of Biomedical Optics* 16, no. 9 (2011): 096004.
67. Alam, M.S., Garg, A., Pottoo, F.H., Saifullah, M.K., Tareq, A.I., Manzoor, O., Mohsin, M., and Javed, M.N. "Gum ghatti mediated, one pot green synthesis of optimized gold nanoparticles: Investigation of process-variables impact using Box-Behnken based statistical design". *International Journal of Biological Macromolecules* 104 (2017): 758–767.
68. Chen, D., Feng, H. and Li, J. "Graphene oxide: Preparation, functionalization, and electrochemical applications". *Chemical Reviews* 112, no. 11 (2012a): 6027–6053.
69. Lin, D., Wu, J., Ju, H., and Yan, F. "Nanogold/mesoporous carbon foam-mediated silver enhancement for graphene-enhanced electrochemical immunosensing of carcinoembryonic antigen". *Biosensors and Bioelectronics* 52 (2014) 153–158.
70. Nguyen, K.T. and Zhao, Y. "Integrated graphene/nanoparticle hybrids for biological and electronic applications". *Nanoscale* 6, no. 12 (2014): 6245–6266.
71. Tian, J., Huang, T., Wang, P., Lu, J. "GOD/HRP bienzyme synergistic catalysis in a 2-D graphene framework for glucose biosensing". *Journal of The Electrochemical Society* 2015, 162, no. 12, B319–B325.
72. Lee, W.C., Lim, C.H.Y.X., Shi, H., Tang, L.A.L., Wang, Y., Lim, C.T., and Loh, K.P. "Origin of enhanced stem cell growth and differentiation on graphene and graphene oxide". *ACS Nano* 5, no. 9 (2011): 7334–7341.
73. Chen, G.-Y., Pang, D.W.-P., Hwang, S.-M., Tuan, H.-Y. and Hu, Y.-C. "A graphene-based platform for induced pluripotent stem cells culture and differentiation". *Biomaterials*.33, no. 2 (2012b) 418–427.
74. Wang, Y., Lee, W.C., Manga, K.K., Ang, P.K., Lu, J., Liu, Y.P., Lim, C.T., and Loh, K.P. "Fluorinated graphene for promoting neuro-induction of stem cells". *Advanced Materials*. 24, no. 31 (2012): 4285–4290.
75. Kang, J., Wei, Z., and Li, J. "Graphyne and its family: Recent theoretical advances". *ACS Applied Materials & Interfaces* 11, no. 3 (2019): 2692–2706.
76. Wang, G., Ma, Y., Wei, Z., & Qi, M. (2016). Development of multifunctional cobaltferrite/graphene oxide nanocomposites for magnetic resonance imaging andcontrolled drug delivery. *Chemical Engineering Journal*, 289, 150–160.
77. Alibolandi, M., Mohammadi, M., Taghdisi, S.M., & Ramezani, M. (2017). Fabrication of aptamer decorated dextran coated nano-graphene oxide for targeted drug delivery. *Carbohydrate polymers*, 155, 218–229.
78. Iannazzo, D., Pistone, A., Salamò, M., Galvagno, S., Romeo, R., Giofré, S.V., Branca, C., Visalli, G., & Di Pietro, A. (2017). Graphene quantum dots for cancer targeted drug delivery. *International journal of pharmaceutics*, 518(1–2), 185–192.
79. Mo, R., Jiang, T., Sun, W., & Gu, Z. (2015). ATP-responsive DNA-graphene hybrid nanoaggregates for anticancer drug delivery. *Biomaterials*, 50, 67–74.
80. Wei, Y., Zhou, F. , Zhang, D., Chen, Q., & Xing, D. (2016). A graphene oxide based smart drug delivery system for tumor mitochondria-targeting photodynamic therapy. *Nanoscale*, 8(6), 3530–8.
81. Yang, H., Bremner, D.H., Tao, L., Li, H., Hu, J., & Zhu, L. (2016). Carboxymethyl chitosan-mediated synthesis of hyaluronicac id-targeted graphene oxide for cancer drug delivery. *Carbohydrate polymers*, 135, 72–78.
82. Pan, Q., Lv, Y., Williams, G.R., Tao, L., Yang, H., Li, H., & Zhu, L. (2016). Lactobionicacid and carboxymethyl chitosan functionalized graphene oxidenano composites as targeted anticancer drug delivery systems. *Carbohydrate polymers*, 151, 812–820.

83. Deb, A., & Vimala, R. (2018). Natural and synthetic polymer for graphene oxide mediated anticancer drug delivery-A comparative study. *International journal of biological macromolecules*, 107, 2320–2333.
84. Wang, C., Chen, B., Zou, M., & Cheng, G. (2014). Cyclic RGD-modifiedchitosan/graphene oxide polymers for drug delivery and cellular imaging. *Colloids and Surfaces B: Biointerfaces*, 122, 332–340.
85. Gao, Y., Zou, X., Zhao, J.X., Li, Y., & Su, X. (2013). Graphene oxide-based magnetic fluorescent hybrids for drug delivery and cellular imaging. Colloids and surfaces B: biointerfaces, 112, 128–133.
86. Tan, L., Wu, T., Tang, Z.W., Xiao, J.Y., Zhuo, R.X., Shi, B., & Liu, C.J. (2016). Water-soluble photoluminescent fullerene cappe dmesoporoussilica for pH-responsive drug deliveryand bioimaging. Nanotechnology, 27, 315104.
87. Misra, C., Kumar, M., Sharma, G., Kumar, R., Singh, B., & Katare, O.P. (2017). and Raza, K. Glycinated fullerenesfor tamoxifen i ntracellular delivery with improvedanticanceractivityand pharmacokinetics. Nanomedicine, 12, 1011–1023.
88. Bahuguna, S., Kumar, M., Sharma, G., Kumar, R., Singh, B., & Raza, K. (2018). Fullerenol-based intracellular delivery of met hotrexate: a water-solublenanoconjugate for enhanced cytotoxicityand improved pharmacokinetics. *AAPSpharmscitech*, 19, 1084–1092.
89. Zhao, L., Li, H., & Tan, L. (2017). A novel fullerene-based drug delivery system delivering doxorubicin for potential lung cancer therapy. *Journal of Nanoscience and Nanotechnology*, 17, 5147–5154.
90. Panchuk, R.R., Prylutska, S.V., Chumak, V.V., Skorokhyd, N.R., Lehka, L.V., Evstigneev, M.P., Prylutskyy, Y.I., & Berger, W. (2015). Application of C60 Fullerene-Doxorubicin Complex for Tumor Cell Treatment InVitro and InVivo. Journal of biomedical nanotechnology, 11, 1139–1152.
91. Hazrati, M.K., Javanshir, Z., & Bagheri, Z. (2017). B24N24 fullerene as a carrier for 5-fluorouracil anti-cancer drug delivery: DFT studies. Journal of Molecular Graphics and Modelling, 77, 17–24.
92. Lin, M.S., Chen, R.T., Yu, N.Y., Sun, L.C., Liu, Y., Cui, C.H., Xie, S.Y., Huang, R.B., & Zheng, L.S. (2017). Fullerene-based amin o acid ester chloridesself-assembled asspherical nano-vesiclesfor drug delayed release. Colloids and Surfaces B: Biointerfaces, 159, 613–619.
93. Raza, K., Thotakura, N., Kumar, P., Joshi, M., Bhushan, S., Bhatia, A., Kumar, V., Malik, R., Sharma, G., & Guru, S.K. (2015). C 60 -fullerenes for delivery of docetaxel to breast cancer cells: A promising approach for enhanced efficacy and better pharmacokinetic profile. International journal of pharmaceutics, 495, 551–559.
94. Thotakura, N., Sharma, G., Singh, B., Kumar, V., & Raza, K. (2018). Aspartic acid derivatized hydroxylated fullerenes as drug delivery vehicles for docetaxel: An explorative study. Artificialcells, nanomedicine, and biotechnology, 46(8), 1763–1772.
95. Panczyk, T., Wolski, P., & Lajtar, L. (2016). Coadsorption of doxorubicin and selected dyes on carbon nanotubes. theoretical investigation of potential application as a pH-controlled drug delivery system. Langmuir, 32(19), 4719–4728.
96. Hou, L., Yuan, Y., Ren, J., Zhang, Y., Wang, Y., Shan, X., & Liu, Q. (2017). In vitro and in vivo comparative study of the phototherapy anticancer activity of hyaluronic acid-modified single-walled carbon nanotubes, graphene oxide, and fullerene. Journal of Nanoparticle Research, 19(8), 1–18.
97. Raza, K., Kumar, D., Kiran, C., Kumar, M., Guru, S.K., Kumar, P., Arora, S., Sharma, G., Bhushan, S., & Katare, O.P. (2016). Conjugation of Docetaxel with Multiwalled Carbon Nanotubes and Codelivery with Piperine: Implications on Pharmacokinetic Profile and Anticancer Activity. Molecular pharmaceutics, 13(7), 2423–2432.

98. Lohan, S., Raza, K., Mehta, S.K., Bhatti, G.K., Saini, S., & Singh, B. (2017). Anti-Alzheimer's potential of berberine using surface decorated multi-walled carbon nanotubes: a preclinical evidence. International journal of pharmaceutics, 530(1–2), 263–278.
99. Razzazan, A., Atyabi, F., Kazemi, B. and Dinarvand, R. (2016). In vivo drug delivery of gemcitabine with PEGylated single-walled carbon nanotubes. Materials Science and Engineering: C, 62, 614–625.
100. Sharma, S., Mehra, N.K., Jain, K., & Jain, N.K. (2016). Effect of functionalization on drug delivery potential of carbon nanotubes. Artificial cells, nanomedicine, and biotechnology, 44(8), 1851–1860.
101. Li, H., Zhang, N., Hao, Y., Wang, Y., Jia, S., Zhang, H., & Zhang, Y. (2014). Formulation of curcumin delivery with functionalized single-walled carbon nanotubes: characteristics and anticancer effects in vitro. Drug delivery, 21(5), 379–387.
102. Mehra, N.K., Verma, A.K., Mishra, P.R., & Jain, N.K. (2014). The cancer targeting potential of D-α-tocopheryl polyethylene glycol 1000 succinate tethered multi walled carbon nanotubes. Biomaterials, 35(15), 4573–4588.
103. Mehra, N.K., & Jain, N.K. (2015). One platform comparison of estrone and folic acid anchored surface engineered MWCNTs for doxorubicin delivery. Molecular pharmaceutics, 12, 630–643.
104. Meng, L., Zhang, X., Lu, Q., Fei, Z., & Dyson, P.J. (2012). Single walled carbon-nanotubesas drug deliveryvehicles: targeting doxorubicin totumors. Biomaterials, 33(6), 1689–1698.
105. Alibolandi, M., Mohammadi, M., Taghdisi, S.M., & Ramezani, M. (2017). Fabrication of aptamer decorated dextran coated nano-graphene oxide for targeted drug delivery. Carbohydrate polymers, 155, 218–229.
106. Baughman, R.H., Eckhardt, H., & Kertesz, M. (1987). Structure-property predictionsfor new planar forms ofcarbon: Layere d phasescontaining sp 2 and sp atoms. Journal of chemical physics, 87(11), 6687–6699.

15 C-Dot Nanoparticulated Devices for Biomedical Applications

Ritesh Kumar, and Gulshan Dhamija
University School of Basic & Applied Sciences, Guru Gobind Singh Indraprastha University, Dwarka Sector 16C, New Delhi-110078, India

Jamilur R. Ansari
Faculty of Physical Sciences, PDM University, Sector 3A, Bahadurgarh-124507, Haryana, India

Md. Noushad Javed
Department of Pharmacy, School of Medical and Allied Sciences, K.R. Mangalam University, Gurgaon, India

Md. Sabir Alam
NIMS Institute of Pharmacy, NIMS University, NH-11C, Delhi-Jaipur Expy, Sobha Nagar, Jaipur, Rajasthan, India

15.1 INTRODUCTION

Among the nanomaterials, carbon and its allotropes have tremendous potential in biomedical applications and other research areas . The carbon family consists of fullerene, nanotubes, nano diamonds, carbon nanofibers, graphene and carbon dots (C-dots). Xu et al. in 2004 accidentally discovered C-dots with superior properties, like prominent biocompatibility, water solubility, energy conservation ability, photo stability, color photoluminescence, and low cost [1–4]. C-dots is a new category of quantum dots, quasi-spherical particles having a size in the nano range (below 10 nm), and it makes them effective in many research areas, such as sensing, photo catalysis, bio-imaging, gene delivery and drug delivery, etc [5–7]. These useful properties are helpful to provide a substitute for substantial metal-based semiconductor quantum dots [8]. C-dots can be easily prepared from carbon sources, and their properties can be tuned by doping with heteroatoms and functionalizing them as desired. They can be used in many applications, such as sensors, imaging, nanovehicles, etc. [9, 10]. Researchers are mainly focused on excitation wavelength-dependent and excitation wavelength-independent photo-luminescent behavior of C-dots . There are two approaches mainly used to prepare

DOI: 10.1201/9781003220350-15

C-dots, i.e. top-down approach (chemical oxidation, laser ablation, electrochemical synthesis) and bottom-up approach (hydrothermal approach, microwave assisted synthesis) [9, 11]. C-dots mainly consist of carbon, oxygen and hydrogen elements. Nitrogen, boron and sulfur have been used as dopants to enhance the properties of C-dots. Recently, nitrogen-doped C-dots have been an interesting area of research as nitrogen enhances the PL property of C-dots and makes them useful for cellular imaging and biomedical applications [3,12–14]. In previous years, many metallic nanoparticles and semiconductor quantum dots (CdS, CdSe, PbSe) have been used for imaging of brain tumor cells, but this has many disadvantages, such as toxicity, functionalization and aqueous dispersity, which is not good for bio-imaging. There are several surface modification and fabrication approaches available with C-dots, such as providing better fluorescence behavior and good biocompatibility ([15]). Hence, these C-dots are extensively replacing the traditional semiconductor in several bio-imaging applications [16–18]. The occurrence of carboxyl, hydroxyl and aldehyde groups on the shallow of C-dots is helpful in surface functionalization or passivation and change of the physical and fluorescent property of C-dots (Indian Patent 2091/DEL/2015: 2016; PCT patent application WO/2017/060916: 2017). So, application areas of C-dots are not limited to bio-imaging, but similar to other nanoparticles, are also being explored in other areas of research, such as drug delivery, bio-sensing, gene delivery, supercapacitor, catalysis, etc. [19,20]. Also, C-dots with sp^2 conjugate core enhance their interaction with hydrophobic materials. In this way, C-dots have an amphiphilic surfactant-like property due to the presence of hydrophilic functionality as well as hydrophobic functionality [21, 22]. Fluorescent property of C-dots is better than any other fluorescent carbon nanoparticles because of its high quantum yield (QY), facile synthesis and aqueous solubility [23–25].

Fe^{3+} ions play a very important role in our biological system, including oxygen transport, transportation of nerve signals, homeostasis, cellular metabolism and so on. Its excess and deficiency both cause various diseases, like heart attack, hepatitis, anemia and Parkinson's disease. Moreover, a particular amount of Fe^{3+} ions are essential for water quality. Now, C-dot-based chemprobe has been synthesized to measure Fe^{3+} ion. N-doped C-dots show outstanding selectivity and sensitivity for Fe^{3+} ions [11, 26, 27]. The present study aims to summarize the synthesis process, property and present research work in the area of C-dots. Synthesis process, chemical doping, surface functionalization or passivation, proper precursors are useful in size control and tunable emission of C-dots. Next, our focus is on its wider applications, in particular, sensing, bio-imaging, anti-microbial activities and drug delivery, as depicted in Figure 15.1.

15.2 PRECURSOR TO SYNTHESIZE C-DOTS

After the discovery of C-dots in 2004, due to its nontoxic, biocompatible, water-soluble behavior, it got great attention from researchers. First, the source of C-dots was confined to carbonaceous compounds, such as graphite, candle shoot, carbon powder and C_{60}. QY of C-dots obtained from these sources was very low, even less than 10% [28–30]. Around hundreds of natural and chemical products are described

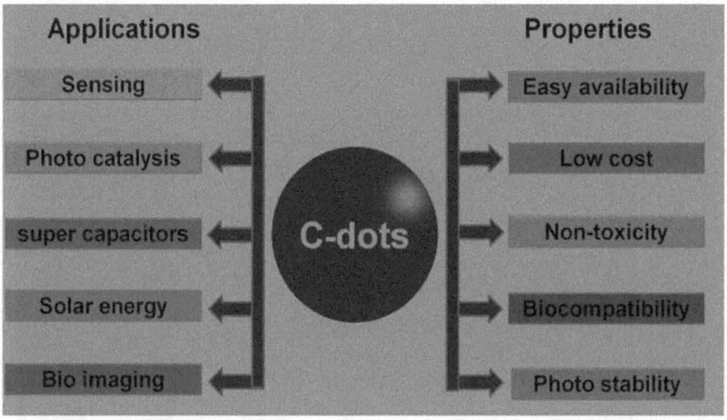

FIGURE 15.1 Application and Properties of C-dots.

for the synthesis of C-dots. Doping of heteroatom such as N, B and S will tune the properties of C-dots and increase its QY and photoluminescent property. Dong et al. in 2013 synthesized the nitrogen- and sulphur-doped C-dots from L-cysteine and citric acid, and QY reached 73%[31, 32]. Moreover, many other metals and non-metals used as doping materials in C-dots to alter its properties for special causes, such as doping of Mg in C-dots, tuned its properties and QY of 83% for cell imaging. Phosphorous doping in C-dots is particularly used for cell labeling. Doping of Mn(II) and Tb(III) in C-dots is used as chemical sensors ([33],). Natural raw materials required to prepare C-dots are garlic [34], orange juice [35], coffee grounds [36], eggs [37], eggshell membrane [38], soy milk [39], bananas [40], flour [41], sweet red papers [42, 43], honey [44], rose flowers [28], waste frying oil [45–47], waste paper [48], humid substances [49, 50], hair [51], petroleum coke [52], etc. It shows that an abundance of raw material is available for the production of C-dots ().

15.3 APPROACH FOR THE SYNTHESIS OF C-DOTS

The source of synthesis of C-dots is abundant in nature, and there are a number of synthetic routes that have been used in order to pursue cost-effective, simple, size-controllable, high-quality, large-scale synthesis of C-dots. These synthetic routes come in the category of top-down and bottom-up approaches that have been used for the preparation of C-dots [5–7]. In the top-down approach, large carbonaceous materials are converted into nanosized C-dots using various physical and chemical methods. Methods for these types of nanoparticles usually involve electrochemical synthesis, solvothermal treatment, chemical oxidation-based processing of materials, etc. (Figure 15.2) . Bottom-up approaches involve the synthesis method, such as hydrothermal and microwave-assisted pyrolysis for the preparation of C-dots in the desired range from smaller particles [53–55].

15.3.1 Top-Down Approach

As discussed, in the top-down approach macroscopic carbonaceous materials like graphene, carbon nanotubes and carbon powder are used to synthesize C-dots.

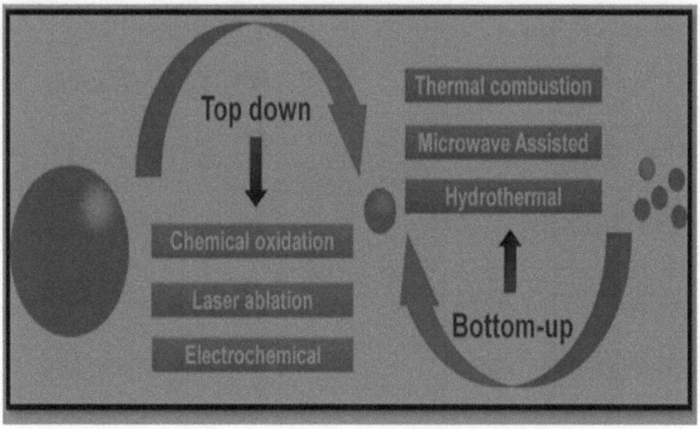

FIGURE 15.2 Top-down and bottom-up approaches.

In this approach, size and morphology of C-dots will not be precise because non-selective cutting tools have been used to prepare C-dots. Variable size and surface state give us a chance to adjust the properties of C-dots for desired application.

15.3.1.1 Chemical Oxidation

H_2SO_4, HNO_3, phosphoric acid, etc. are used as oxidizing agents for the oxidation of carbonaceous material and to provide an effective and convenient method to produce C-dots. Tian et al. and Quo et al. reported the preparation of C-dots via the oxidation of natural gas soot and commercial activated carbon in the presence of nitric acid. Liu et al. used phosphoric acid as oxidant to synthesize C-dots from sucrose. Chemical oxidation provides oxygen containing functional group, i.e. OH, COOH, which is essential to tune the properties of C-dots [56, 57].

15.3.1.2 Electrochemical Synthesis

Electrochemical cells consist of multiwalled carbon nanotubes as working electrodes, $Ag/AgClO_4$ as reference electrodes, and platinum wires as counter electrodes for the fabrication of C-dots. The syntheses of C-dots are confirmed by changing the color of colorless electrolytes from transparent to dark brown as oxidation proceeds [58, 59]. Wang et al. reported the production of C-dots from organic precursor glycine via electrochemical approach with high QY of 55%. Some researchers focused on changing the electrode for tuning the fluorescence feature of C-dots. As far as nonfluorescent C-dots are concerned, they have been synthesized via irradiation of a carbon target using laser source and were finally functionalized to achieve the fluorescent property. The electrochemical synthesis method is one of the best methods used to fabricate C-dots because of the low cost of production and the ability to change the behavior of C-dots by changing the synthetic condition required for the preparation with adjustable emission [56, 60].

15.3.1.3 Laser Ablation

C-dots were synthesized via laser ablation technique using carbon as a target in liquid media by irradiation with a laser beam at certain wavelengths. Preparation of

C-dots using laser ablation is cheap and easy to precede, whereas other methods are difficult and time-consuming. V. Thong Pool et al. synthesized carbon particles via laser ablation of bulk graphite using Nd: YAG laser in ethanol media with the wavelength of 1064 nm. Sun et al. synthesized C-dots using a laser ablation technique using graphite powder in the presence of water vapor in argon atmosphere, which was used as a carrier gas, and thus obtained nonfluorescent CNPs. So, the bright luminescent-emission surface of C-dots will be passivated or functionalized with organic species like amino propyl ethylene poly (ethylene glycol) with luminous efficiency of 4–10%. Yang et al. prepared fluorescent C-dots from ^{13}C powder as the carbon source via laser ablation process with QY of 20%. However, low yield of C-dots obtained by laser ablation is the main drawback of this method. Size of nanoparticles obtained by laser ablation is in the varied range, and most of the large-size particles are easily lost at the time of centrifugation. Thus, low QY limits its many potential applications [14, 61, 62].

15.3.2 Bottom-Up Approach

In the bottom-up approach, smaller particles get converted into C-dots of desirable size range under certain conditions of heat and microwave. From the bottom-up approach thermal/combustion, solvothermal, microwave-assisted pyrolysis and hydrothermal methods have been used, but among all these, microwave-assisted and hydrothermal methods are commonly used to fabricate C-dots. Sucrose, ascorbic acid, amino acids and various natural materials are used as raw materials for C-dot fabrication.

15.3.2.1 Microwave-Assisted Synthesis

Microwave-assisted synthesis is an efficient and time-saving technique for the preparation of C-dots. It improves product yield and quality, as shown in Figure 15.3 [25, 28].

Recently, Liu et al. reported the use of a microwave-assisted pyrolysis approach to develop C-dots from glutaraldehyde and poly (ethylenimine) as a precursor. Varying the ratio of glutaraldehyde and PEI property of C-dots can be tuned [63]. Zhu et al. used the microwave-assisted pyrolysis method to prepare C-dots by mixing PEG_{200} with glucose in distilled water in a 5000 W microwave oven for 2–10 min [64, 65].

15.3.2.2 Hydrothermal Method

Hydrothermal process is an easy and efficient way to fabricate C-dots through carbonization and polymerization reaction with low energy consumption and little harm to the environment. Hydrothermal treatment of glucosamine hydrochloride at 140°C produces nitrogen-doped carbon nanoparticles of around 50–70 nm with hydroxyl, carboxyl and aromatic amine group on the surface because of dehydration, polymerization and nucleation of glucosamine molecule [66, 67]. In 2010, Zhang et al. developed a process for the preparation of 2 nm C-dots by hydrothermal approach using L-ascorbic acid with relatively higher PL efficiency (6.79%). Yang et al. used hydrothermal approach to produce amino functionalized

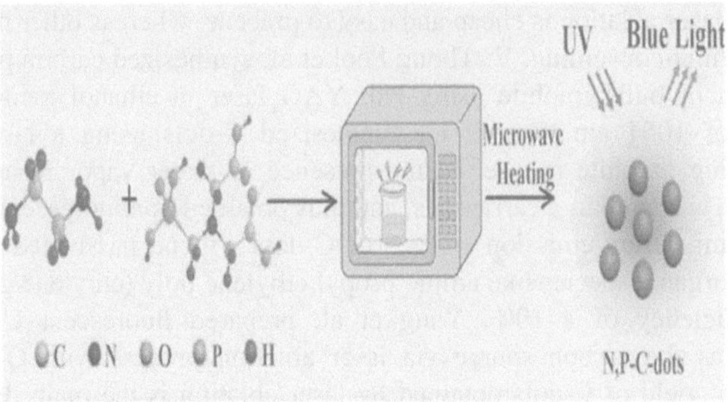

FIGURE 15.3 Microwave-assisted synthesis of carbon dots.

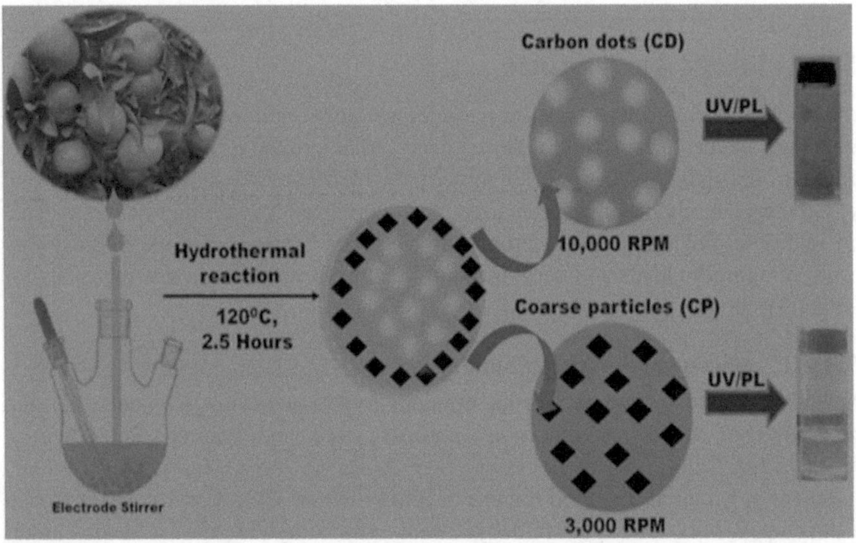

FIGURE 15.4 Preparation of C-dots via hydrothermal method using orange juice.

C-dots from chitosan at 180°C with a QY of 43% [68]. Green organic precursor, such as papaya, peach, limeade and lemon peel, are also used to synthesize C-dots via the hydrothermal route. Solvent, temperature and precursor affect the size of the C-dots as well as their photoluminescent property. Many researchers are working in the area of C-dots using different precursors with longer wavelengths and tunable PL behavior (Figure 15.4).

15.4 FUNCTIONALIZATION OF C-DOTS

To modulate the electronic and optical property of C-dots, it was reported that nitrogen, boron and sulphur were used as dopants, which enhances the behavior of the QY, in particular for C-dots for various applications such as bio-imaging, sensing, photocatalysis, optoelectronic device, etc. Zhang et al. developed nitrogen-doped C-dots from CCl_4 and $NaNH_2$ as a starting material via the solvothermal route. The authors claimed that by adjusting the nitrogen concentration, the photoluminescent property of C-dots as well as the size can be tuned. The variation in the reaction time influences the behavior of the QY for C-dots. Nitrogen-doped C-dots show better photoluminescence and high QY than non-doped C-dots [32]. The microwave-supported approach was used for the preparation of nitrogen-doped C-dots from poly-bromopyrroles [69]. The hydrothermal route was used to synthesize highly luminescent N-doped C-dots with 15.7% QY, as reported by Zhang and Chen [70].

Shan et al. claimed to developed bright blue fluorescent boron-doped C-dots via solvothermal approach with the use of hydroquinone and BBr_3 as the carbon and boron source with a QY of 14.8%, which is superior to the QY of C-dots [29]. Bourlinos et al. claimed the preparation of boron-doped C-dots via microwave-assisted approach using aqueous solution of boric acid urea and citric acid [71].

Recently, sulfur-doped C-dots have been used to amend the electronic properties and structure of C-dots. It was reported that raw material has been used to prepare sulfur-doped C-dots, such as waste frying oil, sodium thiosulfate, ethane-sulfonic acid, do-decane-thiol, sodium hydrosulfide, sulfuric acid, etc. It was observed that C-dots doped with sulphur indicate a relatively high photostability, low toxicity and water dispersibility, and they exhibit a higher QY of 11.8% than undoped C-dots [31]. Yang et al. claimed the synthesis of S-doped C-dots via hydrothermal reduction and in situ doping of sodium hydrosulfide and oxidized C-dots [72]. Xu et al. reported the synthesis process of C-dots doped with sulphur from sodium thiosulphate and sodium citrate as a precursor via a hydrothermal process [73].

Phosphorous acts as an N-type dopant for C-dots and shows remarkable modification in electronic and optical properties of C-dots. Phosphoric acid, mono-sodium phosphate, phosphorous tribromide, etc. were used as a precursor material for synthesis of P-doped C-dots. Zhou et al. developed P-doped C-dots using hydroquinone and phosphorus tribromide as a precursor by the solvothermal method. Prepared P-doped C-dots show QY up to 25%, with strong blue fluorescence and also show low toxicity and excellent biolabeling [74].

Like previous elements, Si, Ge, Gd, Tb, Se and many others have been also used as dopants to modify the electrochemical, biological and optical properties of C-dots for various applications [75–79].

Qian et al. synthesized Si-doped C-dots via the solvothermal method from $SiCl_4$ and hydroquinone and showed QY of 19.2% with maximum fluorescence at 382 nm and excellent selectivity towards H_2O_2, melamine and Fe (iii) [79]. Yuan et al. performed first-time synthesis of Ge-doped C-dots from bis-(2-carboxyethylgermanium (iv) sesquioxide) (Ge-132) from citric acid via carbonation synthesis. Ge-doped C-dots exhibit excellent biocompatibility, high intracellular delivery efficiency, low cell toxicity and excellent selectivity towards Hg ion [25]. Yang et al. synthesized Se-doped C-dots by reducing oxidized C-dots in persence of NaHSe in a hydrothermal based process [80]. Bourlinos et al. used the combination of gadopentetic acid mixed with tris (hydroxymethyl) aminomethane and betain hydrochloride in the atmosphere at around 250°C to synthesize Gd-doped C-dots by pyrolysis route for medical application [75]. Chen et al. developed Tb-doped C-dots using citric acid and terbium (m) nitrate pentahydrate at 190°C. It shows good performance in determination of 2,4,6-trinitrophenol (TNP) in aqueous medium. Sachdeva et al. prepared C-dots using chitosan via hydrothermal technique and observed that the surface was passivated or functionalized using polymeric passivating agent polyethyleneimine and polyethylene glycol to enhance bio-imaging competence and optical performance of c-dots [76, 81].

15.5 CHARACTERIZATION TECHNIQUES

UV-Vis spectroscopy, photoluminescence (PL), X-ray diffraction (XRD) and transmission electron microscopy (TEM/HRTEM) are used to analyze the optical absorbance, excitation and emission properties by varying wavelengths, crystalline nature and size of the nanoparticles and study the surface morphology and size of the C-dots.

15.5.1 Transmission Electron Microscopy

Transmission electron microscopy (TEM) study was used to analyze the surface morphology and size of C-dots. It was also observed that such C-dots have possible application in the fields of medical science, physics, biology, chemistry and other associated research areas. The average diameter using 325 nanoparticles of C-dots was plotted and was observed to be around 2 nm. The high resolution transmission electron microscopy (HRTEM) shows that the crystallinity of C-dots corresponds to two different diffraction planes, namely in-plane lattice spacing and the interlayer spacing, respectively [82, 83]. Shinde et al. synthesized C-dots from multiwall carbon nanotube via the electrochemical method, and the HRTEM study shows that two types of lattice spacing were observed simultaneously (Figure 15.5).

15.5.2 X-ray Diffraction Study

X-ray diffraction study is a useful technique to analyze the crystalline nature and size of C-dots. Liu et al. observed that synthesized C-dots having thickness ~2–3 nm and ~60 nm diameter have a fluorescent QY of 3.8% using hexa-peri-hexabenzo coronene as the starting material via pyrolysis at high temperature, reduction treatment, surface functionalization and oxidative exfoliation, etc. XRD spectrum of C-dots shows a high degree of crystallinity, with the peak corresponding to (002) planes at 22.4° (Figure 15.6) [84].

15.5.3 Optical Properties

UV-Vis spectroscopy was used to analyze the optical absorbance of C-dots, mainly in the UV region (270–390) nm, which was observed due to л–л* transitions of the C=C bond [85]. After surface passivation and doping, absorption wavelength will increase. Sun et al. observed the fabrication of N, S co-doped GQDs having two absorption bands in the range 550–600. In particular, 550/595 nm corresponds

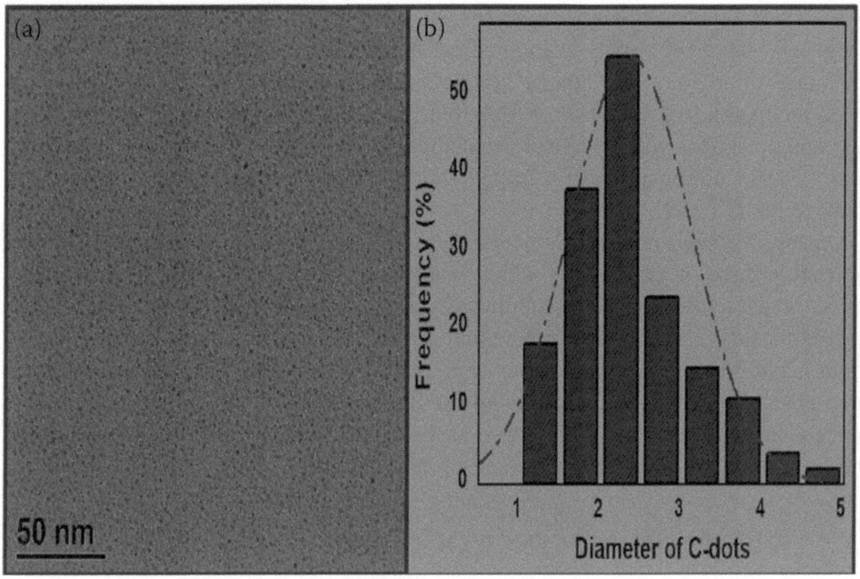

FIGURE 15.5 TEM image and histogram of C-dots.

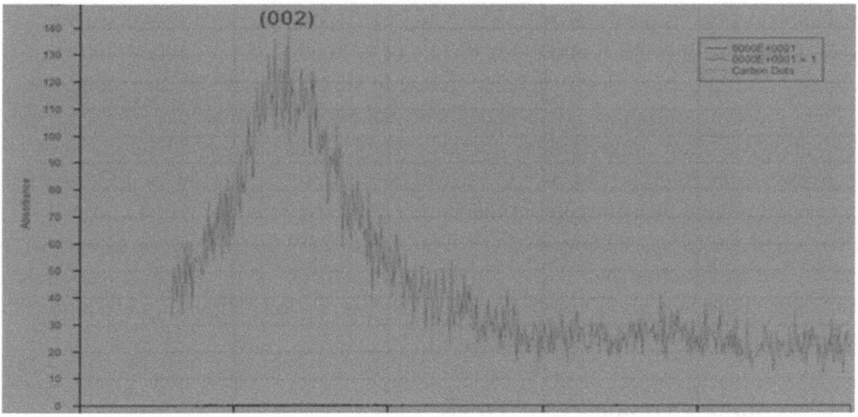

FIGURE 15.6 XRD spectra of C-dots.

to л–л* and n–л* of S=O, and further it was found that C=S corresponds to doping of sulphur [86].

15.6 QUANTUM YIELD

QY is mainly used to express the fluorescence property of C-dots. The QY of C-dots prepared from different resources and different methods has been compared with standard material, commonly quinine sulphate with QY of 54% [87]. B, coumarin and rhodamine have been also used as reference material based on

C-dots fluorescence emission range [87, 88]. Semiconductor quantum dots having a high QY have been used in biomedical applications, but have some major disadvantage of toxicity and non-biocompatibility. In the case of C-dots, they have an easier method of synthesis, lower toxicity and lower QY than semiconductor quantum dots. This will affect the biomedical application of C-dots. Hence, doping with metal or nonmetal and functionalization or surface passivation is required to enhance the QY [89].

During synthesis via different routes, some defects have been formed on the external surface of the C-dots, which results in low QY. So, surface functionalization or passivation provide stability to the C-dots and enhance its QY. C-dots prepared using citric acid and functionalized with hyper-branched poly (amino amine), a fluorescent polymer, show a QY of 17.1%, with low toxicity, easy cellular internalization and potential applications in the field of biomedical application. When C-dots were doped with hetero atoms, such as N, S, and P, it enhanced the QY up to 80% by altering the surface properties [73].

15.7 PHOTOLUMINESCENCE OF C-DOTS

Photoluminescent properties are the most important property of C-dots, including excitation wavelength dependent and excitation wavelength independent, which are due to surface state-related and core-related emissions. The PL behaviors of C-dots are characterized in terms of QY. The QYs of C-dots are very low compared to doped C-dots with S, N, B and other metals or nonmetals. Many other factors also influence the QY of C-dots, such as changing reaction methods and conditions, surface passivation/modification and metal-enhanced fluorescence [53–55, 90]. Kang and co-workers developed C-dots of variable sizes using NaOH/ethanol as an electrolyte and graphite rod as an electrolyte with the emission of blue, green, yellow and orange, respectively, at the same λ_{ex} by varying the size of C-dots [91]. Zheng reported the enhancement of QY from 2% to 24% by treating the C-dots with $NaBH_4$ via a reductive pathway. Conventional organic or inorganic fluorophores will show blinking photoluminescence, whereas C-dots show nonblinking PL property with excellent photostability [92].

15.8 APPLICATIONS OF C-DOTS

15.8.1 Sensing of Ions

C-dots have wider application in the field of sensing due to their fluorescent property and surface functionalization. A number of biological/chemical sensors based on functionalized C-dots have been synthesized for sensing ions in vivo and in vitro: $Pt4^{+}$, Cu^{2+}, Hg^{2+}, Zn^{2+}, Be^{2+}, Cd^{2+}, Fe^{3+}, Cr^{4+}, etc. [93]. Branched (polyethylenimine) (BPEI) have great selectivity towards (Cu^{2+}); therefore, Dong et al. used branched (polyethylenimine) (BPEI) to functionalize C-dots for sensing cu^{2+} ion [39]. Zhang et al. developed C-dots by functionalizing them with quinoline derivative, and functionalized C-dots have great selectivity towards Zn^{2+} ion and can easily detect trace amounts of Zn^{2+} within one minute [94]. Huang's team

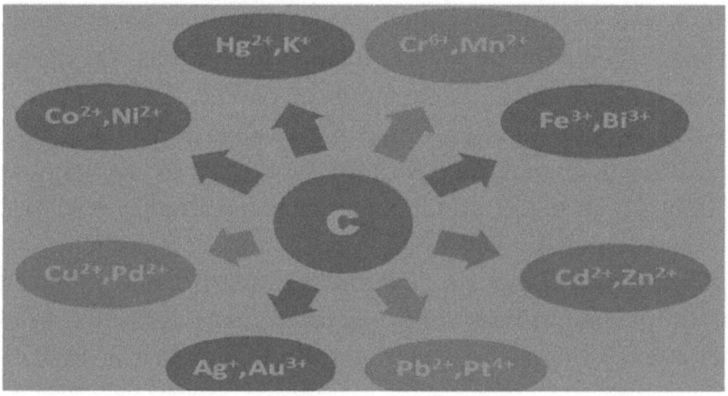

FIGURE 15.7 Sensing of ions by C-dots.

developed a method in which they used a probe of Eu^{3+} ion to detect phosphate with fluorescence "off-on" response [11]. Liu et al. developed a fluorescent probe for fluoride detection by functionalizing C-dots with βcyclodextrin and calixarene-25,26,27,28-tetrol. Shen et al. synthesized a non-enzymatic glucose sensor by functionalizing C-dots with boronic acid via one-step hydrothermal carbonization route, and they were found to be highly sensitive for glucose in the 9–900 μm range [87]. Valcarcel et al. reported the amidation reaction used to functionalize C-dots with amine-terminated group for selective detection of the citrate silver nanoparticle. Valarcel group also developed a C-dots sensor to detect water-soluble C_{60} fullerenes. Tian et al. reported the sensing of superoxide anion by developing a C-dots based ratio metric fluorescent biosensor (Figure 15.7). Wang et al. claimed the sensing of NO_2 by C-dots silica aerogel hybrid material [28].

15.8.2 Catalytic Activity

Photocatalysis is an eco-friendly technique in which a photocatalyst is able to be harnessed as an inexhaustible source of solar energy. C-dots have the property of being inexpensive, low toxic, and easily available, making them a potential candidate to be used as photocatalysts in photocatalysis application. Kang et al. prepared C-dots in the range 1–4 nm via alkali-assisted electrochemical route, and used them as a photocatalyst for oxidation where it was observed that benzaldehyde coverts to benzyl alcohol under near infrared (NIR) light in the presence of H_2O_2. To facilitate the photocatalytic property of C-dots, their optical and electrical property can be enhanced by doping them with another element, such as N, P, S, Cu, etc. [13,95]. Nitrogen-doped C-dots show an excellent photocatalytic property for the degradation of methyl orange (MO) dye in the existence of visible light. Sun et al. claimed the use of gold-doped C-dots as an effective photocatalyst for the photo-reduction of CO_2 into acetic acid. The combination of TiO_2 with C-dots in a composite system enhances the degradation of methylene blue (MB) under visible light, whereas pure TiO_2 and C-dots are not able to degrade MB under similar conditions. This implies that a strong interaction occurs between C-dots and TiO_2 in a catalyst system [19]. Kang et al. observed that, as prepared, C-dots/Fe_2O_3

composite, which was used to degrade methanol and gas phase benzene, fails to show photocatalytic behavior for Fe_2O_3 nanoparticles; thus, they concluded that C-dots play a crucial role in the degradation of photocatalytic activity of C-dots/ Fe_2O_3 composites. Similarly, C-dots/ZnO nanocomposite shows superior photocatalytic property for degradation of harmful gases such as methanol and benzene under visible light [4]. Kang et al. claimed the preparation of C-dots/C_3N_4 nanocomposite for photocatalytic water splitting. Metal nanoparticles, in particular (Au, Ag, Cu)/C-dots nanocomposite, exhibit high performance and selectivity for the photocatalytic oxidation, when cyclo-hexanone was converted to cyclohexane at room temperature (RT) under visible light in the presence of H_2O_2 [11].

15.8.3 C-Dots for Water Purification

With the increase in population and industrialization, water resources get polluted by domestic waste and industrial waste. Industrial waste contains organic pollutants, which pollute the environment and are a major cause of concern. Photocatalytic degradation of organic pollutant is the best way to solve this problem because of its simple operation and relatively low cost. Water is a basic necessity of human beings to fulfill a daily requirement, so to purify the water, many attempts have been made, such as porous clay composite, and it is useful in filtration of synthetic pollutants at the lab scale but not at the commercial scale due to its small permeability [96–98]. Aji et al. synthesized C-dots from discarded used oil via the hydrothermal method at the temperature of 120°C, and C-dots can be used as a co-catalyst/photocatalyst for a water purification system. The density of C-dots is a bit lower compared to water, so they float on the water surface and directly receive solar energy, which enhances their potential to be used as a photocatalyst for the purification of water [96]. Prasannan et al. synthesized the composites of ZnO- and N_2-doped C-dots using the hydrothermal route with orange peel and used them as a photocatalyst for the degradation of organic dyes [99]. Yang et al. reported the synthesis of nitrogen-doped C-dots via hydrothermal protocol using yeast, which was used as a photocatalyst for the evolution of hydrogen from water under UV irradiation. Thus, nitrogen-doped C-dots as a photocatalyst should be used on a large scale in the area of energy and water cleaning [25].

15.8.4 C-Dots for Electronic Application

Many research groups are working in the area of solar cell fabrication, and they found C-dots to be a potential photosensitizer in solar cells. As reported, RhB is used as a photosensitizer in solar cells, and with the incorporation of C-dots, its photoelectric conversion efficiency is increased up to seven times. Nitrogen-doped C-dots also enhance the performance of solar cells. Zhu et al. claimed the insertion of a C-dots layer between TiO_2 and perovskite materials improves the power conversion productivity of the perovskite solar cell. C-dots can also be used in a supercapacitor [100]. Liu et al. used the method of electrodeposition to fabricate C-dots based on a micro supercapacitor with excellent rate capability (1000 VS^{-1}) [29]. Zhu et al. fabricated the C-dots-decorated RuO_2 network with a superior rate capability and specific capacitance of 460 F g^{-1} [78]. Jing et al. claimed the use of C-dots in lithium-ion batteries; hence, they prepared a C-dots-coated Mn_3O_4 composite via electrochemical approach. It can be utilized as electrode material for lithium-ion batteries and shows higher electrochemical performance than pure Mn_3O_4 [101].

15.8.5 Biomedical Application

15.8.5.1 In Vitro Imaging

Surface functionalized C-dots have a remarkable advantage in biomedical application, such as bio-imaging, theranostic, drug delivery, etc., as shown in Figure 15.8.

Various research groups reported the use of functionalized C-dots for in vivo bio-sensing and bio-imaging [58, 59, 102, 102, 103]. Zhu et al. used the solvothermal route for the fabrication of fluorescent C-dots in the range (2–5 nm) for cell imaging with low cytotoxicity . Tang et al. observed that when C-dots were functionalized with folic acid and DOX, it was convenient to identify cancer cells as well as fluorescence imaging [104]. Choi et al. reported improvement in the property of C-dots with zinc phthalocyanine and folic acid to achieve photothermal therapy and tumor targeting [105]. Zhang et al. observed that C_3N_4 nanodots can be well utilized for imaging of the cell nucleus. Li et al. observed that due to passivation of C-dots with varying polymer coating and later on functionalizing with human transferrin (Tf), it was effective to target the in vitro cancer cells. These polymer coatings are of poly (ethylene glycol), poly (ethyleneimine) -b-poly

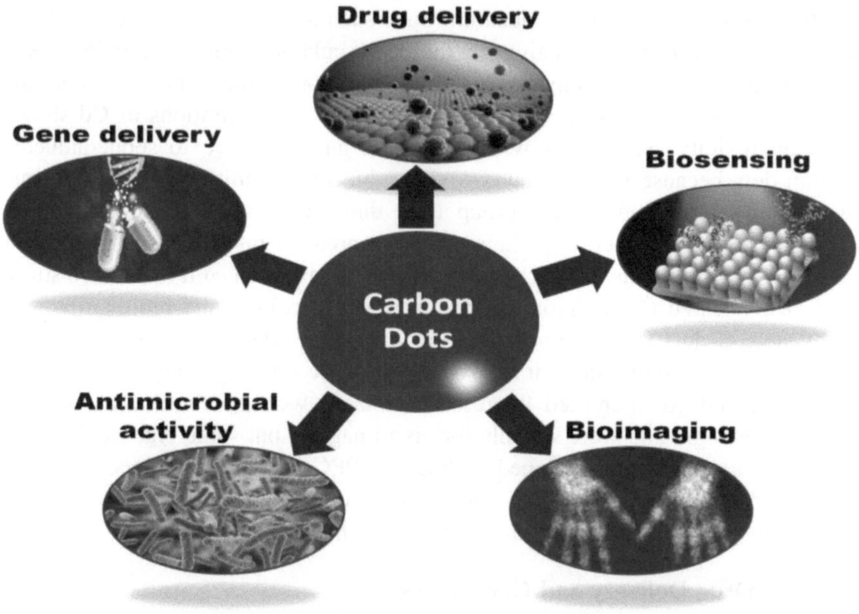

FIGURE 15.8 Biomedical applications of C-dots.

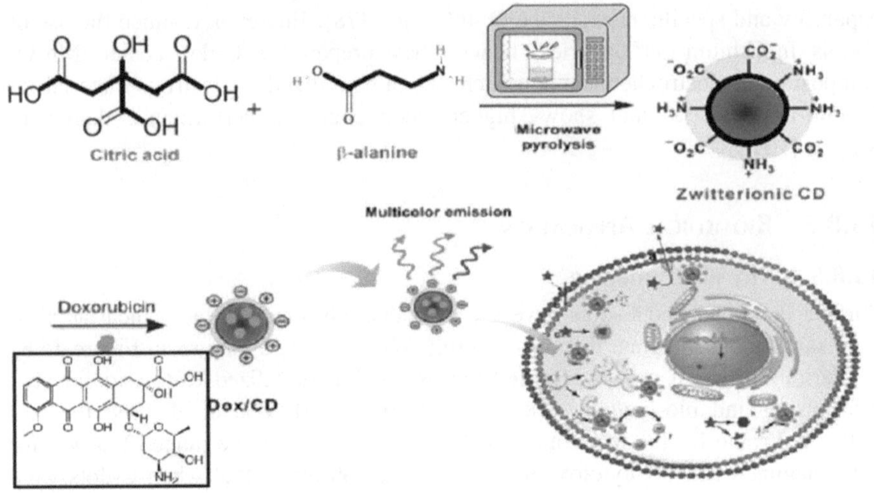

FIGURE 15.9 Cancer-targeting C-dots.

(ethylene glycol) -b-poly (ethyleneimine) and a 4-arm, amine-concluded PEG [106]. The cancer-targeting C-dots are shown in Figure 15.9.

15.8.5.2 In Vivo Imaging

Functionalized C-dots play a very important role as an imaging agent for in vivo optical imaging because of biocompatibility, nontoxicity and stability against photobleaching. Semiconductor quantum dots, such as CdSe/ZnS, have also been used successfully in vivo imaging on tumor vasculature, sentinel lymph nodes, tumor-specific membrane antigens, etc. Cadmium-containing C-dots are not suitable for patient studies because the presence of low concentrations of Cd shows significant toxicity. So, C-dots will be seen as an alternative to semiconductor quantum dots because of their lower toxicity, biocompatibility and physiochemically stability [107–109]. Sun's group used three injection avenues for optical imaging in vivo by C-dots [80]. Kang et al. prepared the un-passivated C-dots via acid oxidation of graphite and carbon nanotubes and investigated it for in vivo study in the near infrared (NIR) fluorescence imaging, toxicology and bio-distribution. Use of these C-dots in mice does not show any toxic effect [110]. Koi et al. observed mice with tumors for their in vivo fluorescence imaging study using C-dots. Red-emitted and green-emitted fluorescence images were obtained at a different exciton wavelength and were suitable for vivo imaging, but some blue- and green-emitted C-dots, such as ZnS-doped C-dots and PEG-passivated C-dots, were not suitable for in vivo imaging [25]. Tao et al. reported that at different excitation wavelengths different color-emitting C-dots have been prepared [10].

15.8.5.3 Drug Delivery and Gene Transform

Solubility, toxicity, resistance of drugs as well as physiological barriers (e.g. BBB) are key concerns for failure of therapy to meet optimal performance [111, 112, 131].

Owing to low toxicity and high biocompatibility, therefore, there is a lot of interest towards integration and exploitation of attributes of C-dots in gene transfer and drug delivery. Liu et al. prepared modified C-dots with polyethyleneimine (PEI) via microwave-assisted pyrolysis [26]. The outer polymeric layer of PEI-modified C-dots were cationic and had the ability to combine with negative DNA for the purpose of gene delivery as well as to properly mediate in the process of gene transfection with HepG2 and COS-7 cells with excellent efficiency. PEI-modified C-dots had the multicolor fluorescent property with the capability to associate dual function of bio-imaging and gene delivery [61, 62]. Karthik et al. developed a quinolone-based photo trigger and a photo responsive, non-drug delivery system (DDS) by C-dots [113]. Lai et al. synthesized polyethylene glycol (PEG)-functionalized C-dots and loaded the doxorubicin (DOX) and its release process in a cell that can be seen by fluorescent image [114, 115]. Wang et al. used the mixture of C-dots, PEG and chitosan to prepare a dual PH/near-infrared light responsive hybrid gel [14]. Yang Shu et al. synthesized organophilic C-dots from imidazolium ionic liquids (ILs) via hydrothermal reaction and used them as a drug delivery carrier [18]. In this way, C-dots have potential application in the field of gene transfer and drug delivery (Table 15.1).

15.8.5.4 Antibacterial Activity and Bacterial Imaging

Functionalized C-dots with certain agents depict powerful antibacterial activity under visible sunlight against a wide spectrum of microorganism. When functionalized C-dots consist of a carboxyl group on the outer surface, they have the ability to associate with an amino group of bacterial membrane and thus facilitate their counting as well as imaging with fluorescence microscopy [126, 127]. Hao Li et al. prepared C-dots via electrochemical route using Vitamin C; they showed wide-spectrum antifungal activity; showed antibacterial activity; and efficiently inhibited the growth of Bacillus subtilis, Staph aureus, Bacillus sp, Escherichia coli, Rhizoctonia Solani, etc. C-dots will entered into the bacteria by the process of diffusion, destroy the wall of the bacteria, combine with the RNA and DNA of the bacteria, alter the genetic process of the bacteria, and kill the bacteria at very minute concentrations [128]. VB Kumar et al., in 2017, for the first time prepared Ga-doped C-dots with the property to inhibit bacterial growth at very low concentrations. Ga-doped C-dots are stable for at least 2 months after synthesis. The activity of C-dots was higher against gram-negative bacteria compared to gram-positive bacteria (Table 15.2) [129].

15.9 TOXICITY

Carbonaceous C-dots are chemically inert, nontoxic, biocompatible and low cytotoxicity, which is the essential property for biomedical application [30]. Wang et al. evaluated the biotoxicity of C-dots in plants (bean sprout) and animals (mice). For this study, they grew bean sprouts in C-dots aqueous solution, and fluorescence study showed that there was no hindrance in plant growth, which means C-dots do not show any toxic effect on plant cells. Similarly, to study the biotoxicity of C-dots in animals, researchers provided

TABLE 15.1
C-Dots Application in Drug Delivery and Cancer Targeting

S. No	C-Dot Derivatives	Particle Size(nm)	Drug	Applications	References
1	CDs-g-HPG	20–200	Quinine sulfate dehydrate	In vitro cell line A549	[116]
2	PPEI-EI-CDs	50	–	In vitro and in vivo of MCF-7 and HT-29	[117]
3	DOX-CDs-$mSiO_2$–PEG	5–100	DOX	HeLa cells	[114, 115]
4	BCD-DOX	20–200	DOX	MCF-7, MDA-MB-231 and BT-549	[118]
5	CDs–Hep/DOX	200	DOX	MCF-7 and A549 cell	[119]
6	CD	8–100	Quinine sulfate	In vitro study of MCF-7 cells	[83]
7	fC-dots–FA–DOX	2–50	DOX	HeLa cells	[50]
8	tris(hydroxymethyl) aminomethane-CD	10–20	–	Optical bio-imaging of HeLa cells	[120]
9	HP-CDs	20	–	Cell imaging and labeling of MGC-803 cells,	[92]
10	PEG-CD	10	Quinine sulfate	MCF-7 cells, Hep G2 cells	[55]
11	OX-CDs-PEPA	–	Oxaliplatin	L929 and HepG2 cell	[121]
12	FRET-CDot-DDS	5–10	DOX	293T cell line and HeLa cell line	[122]
13	CDs-Pt(IV)-PEG-(PAH/DMMA)	10	Cisplatin	A2780 cells and HeLa cells	[123, 124]
14	IL-HCDs	20	Curcumin	HeLa cells	[18]
15	CD-Asp	5–10	–	In vivo and ex vivo imaging and C6 cells, L929 cell	[125]

C-dots solution to 10 mice to drink with normal food intake for 5 days and provided the control group normal drinking water with normal food intake. They found that the mice drinking C-dots solution showed the same behavior as the control group. Wang et al. also studied the biotoxicity of C-dots on different organs of rats, including kidney, lung, brain, stomach, heart, liver, etc. These organs were injected with the highest dose of C-dots and harvested for histopathological analysis. The analysis showed no abnormalities in the organs. The experimental results showed that C-dots have little toxicity and have biocompatibility, making them a potential candidate for bio-imaging and other biomedical applications [28, 145].

TABLE 15.2
C-dot Application in Antimicrobial and Gene Delivery

S.No	C-Dot Derivative	Particles Size (nm)	Application	Drug	References
1	CdS –SNC	4.0 ± 0.6	DNA plasmid	Chloroquin	[115]
2	tRNA-CdS	5–20	E. coli tRNA	–	[130, 131]
3	ct-DNA-CDs	7–12	E. coli and C. albicans, and MCF 7	Tetracyclin	[127]
4	EDA-CDots	5 nm	Light-activated antimicrobial and fluorescence activities	Propidium iodide dye	[6]
5	CDs-S. aureus or CDs-E. coli	50	Bio-imaging of microbial live/dead differentiation	Vancomycin, tetracyclin	[132]
6	CDs-EPS605	5–50	Photostability assay and bio-imaging of microbes	–	[133]
7	CDs-C_{12}	4.0 ± 1.5	Cell imaging of A549 cells and gram-positive/gram-negative bacteria	BS-12 (quaternary ammonium compound lauryl betaine)	[134]
8	Amino guanidine-C-dots (AG/CA-C-dots)	2.0 ± 0.3	HeLa and HEK293Tcell lines and antibacterial activity	Fluoroquinolone, ciprofloxacin, and levofloxacin	[2]
9	C-dots from cigarette smoking with broad spectrum antimicrobial activities	10 ± 0.21	Antimicrobial activities, in vitro cytotoxicity analysis, HeLa cells	Kanamycin,	[135]
10	Degradable C-dots from vitamin C (VC-CDs)	10-20 ± .21	Cellular toxicity, antibacterial and antifugal activity,	–	[91]
11	S-CQDs/N-CQDs	20–200	Antibacterial and cytotoxicity activity	AgMOF-N/S (silver metal-organic frameworks)- heteroatoms	[136]

(Continued)

TABLE 15.2 (Continued)
C-dot Application in Antimicrobial and Gene Delivery

S.No	C-Dot Derivative	Particles Size (nm)	Application	Drug	References
12	Multi-color emission C-dots (M-C dots)	5–100	Antibacterial and fluorescence activity	–	[137, 138]
13	GCDs (gram-shell carbon dots)	2-100 ±.24	Cytotoxicity study, fluorescent imaging	–	[139]
14	CDs-AMP	44 ± 10	Cytotoxicity assay of HeLa cell line, cell imaging, antibacterial activity	Ampicillin	[54]
15	CDs/$Na_2W_4O_{13}$/WO_3	2–10	Photocatalytic, in vitro cytotoxicity of HepG2 and V79 cells	Tungsten oxide	[140]
16	Calf thymus DNA-CDs (ct-DNA-CDs)		DNA binding, antimicrobial activity, cytotoxicity studies of L6 normal and MCF 7	Penicillin–streptomycin	[127]
17	EDA-CDots	4–5	Antibacterial activity	–	[141, 142]
18	CDots/MB or CDots/TB	4–5	Photosensitizing antibacterial activity	Methylene blue (MB)	[143]
19	EDA-CDots or EPA-CDots,	50	Antiviral activity against human norovirus virus	–	[141, 142]
20	Cipro-CDot Conjugate	30	Theranostic agent, bio-imaging, antimicrobial activity	Ciprofloxacin	[144]

15.10 CONCLUSIONS

After a decade of research, researchers are able to develop the most valuable nanomaterial, i.e. C-dots, with the advantage of low cost, low toxicity and outstanding biocompatibility [146]. These properties make them potential candidates to be used for many applications, such as biomedicine, optoelectronics, photocatalysis, photovoltaic and sensing. There are many other reasons that increase the public interest in C-dots. First, precursors used to synthesize C-dots are cost-effective and easily available. Second, there are different routes to synthesize C-dots that affect the size and QY of C-dots. Semiconductor quantum dots are used for bio-imaging probes, but nowadays C-dots have become a good alternative for semiconductor quantum dots due to their biocompatibility and low toxicity. Semiconductor quantum dots contain heavy metal ions; even the presence of a low concentration will be hazardous to humans' health. Due to doping of C-dots with many hetroatoms, functionalization or passivation will improve their QY and photoluminescent property and is useful to enhance their application in drug delivery and bio-imaging. Lots of work has been done so far on C-dots by many research groups, but still some problems remain, like reproducibility of florescence and physical properties of C-dots and the low QY; these problems need to be addressed. Undoubtedly, development of C-dots with their superior properties will be the alternative of organic fluorophores and inorganic nanocrystals and will provide humans a better future.

ACKNOWLEDGMENTS

M.S. Alam gratefully acknowledges the Vice Chancellor and Dean, NIMS Institute of Pharmacy, NIMS University, Jaipur, Rajasthan, India, for their co-operation during the process of writing this chapter. ***J.R. Ansari*** gratefully acknowledges the Vice Chancellor, Pro Vice Chancellor and Dean, Faculty of Physical Sciences, PDM University, Bahadurgarh, India, for their kind co-operation.

REFERENCES

1. Geys, J., Nemmar, A., Verbeken, E., Smolders, E., Ratoi, M., Hoylaerts, M.F., Nemery, B., and Hoet, P.H.M. "Acute toxicity and prothrombotic effects of quantum dots: Impact of surface charge". *Environmental Health Perspectives* 116, no. 12 (2008): 1607–1613.
2. Otis, G., Bhattacharya, S., Malka, O., Kolusheva, S., Bolel, P., Porgador, A., and Jelinek, R. "Selective Labeling and growth inhibition of pseudomonas aeruginosa by aminoguanidine carbon dots". *ACS Infectious Diseases* 5, no. 2 (2018): 292–302. DOI: 10.1021/acsinfecdis.8b00270.
3. Liu, R., Wu, D., Liu, S., Koynov, K., Knoll, W., and Li, Q. "An aqueous route to multicolor photoluminescent carbon dots using silica spheres as carriers". *Angewandte Chemie – International Edition* 48, no. 25 (2009): 4598–4601.
4. Zou, C., Foda, M.F., Tan, X., Shao, K., Wu, L., Lu, Z., Bahlol, H.S., and Han, H. "Carbon-dot and quantum-dot-coated dual-emission core-satellite silica nanoparticles for ratiometric intracellular Cu2+imaging". *Analytical Chemistry* 88, no. 14 (2016): 7395–7403.

5. Mishra, S., Sharma, S., Javed, M.N., Pottoo, F.H., Abul, M.B., Amir, M., and Sarfaroz, M. "Bioinspired nanocomposites: Applications in disease diagnosis and treatment". *Pharmaceutical Nanotechnology*. DOI:10.2174/221173850766190425121509.
6. Al Awak, M.M., Wang, P., Wang, S., Tang, Y., Sun, Y-P., and Yang, L. "Correlation of carbon dots' light-activated antimicrobial activities and fluorescence quantum yield". *RSC Advances* 7 (2017): 30177–30184. DOI: 10.1039/c7ra05397e.
7. Sharma, S., Javed, M.N., Pottoo, F.H., Rabbani, S.A., Abul, M.B., Sarfaroz, M., and Amir, M. "Bioresponse inspired nanomaterials for targeted drug and gene delivery". *Pharmaceutical Nanotechnology*. DOI:10.2174/2211738507666190429103814.
8. Alam, M.S., Garg, A., Pottoo, F.H., Saifullah, M.K., Tareq, A.I., Manzoor, O., Mohsin M., and Javed, M.N. "Gum ghatti mediated, one pot green synthesis of optimized gold nanoparticles: Investigation of process-variables impact using Box-Behnken based statistical design". *International Journal of Biological Macromolecules* 104 (2017): 758–767.
9. Xu, X., Ray, R., Gu, Y., Ploehn, H.J., Gearheart, L., Raker, K., and Scrivens, W.A. "Electrophoretic analysis and purification of fluorescent single-walled carbon nanotube fragments". *Journal of the American Chemical Society* 126, no. 40 (2004): 12736–12737.
10. Liu, H., Ye, T., and Mao, C. "Fluorescent carbon nanoparticles derived from candle soot". *Angewandte Chemie – International Edition* 46, no. 34 (2007): 6473–6475.
11. Zhao, Q.L., Zhang, Z.L., Huang, B.H., Peng, J., Zhang, M., and Pang, D.W. "Facile preparation of low cytotoxicity fluorescent carbon nanocrystals by electrooxidation of graphite". *Chemical Communications*, no. 41 (2008): 5116–5118.
12. Ayala, P., Arenal, R., Loiseau, A., Rubio, A., and Pichler, T. "The physical and chemical properties of heteronanotubes". *Reviews of Modern Physics* 82, no. 2 (2010): 1843–1885.
13. Chandra, S., Laha, D., Pramanik, A., Ray Chowdhuri, A., Karmakar, P., and Sahu, S.K. "Synthesis of highly fluorescent nitrogen and phosphorus doped carbon dots for the detection of Fe3+ ions in cancer cells". *Luminescence* 31, no. 1 (2016): 81–87.
14. Qiao, Z.A., Wang, Y., Gao, Y., Li, H., Dai, T., Liu, Y., and Huo, Q. "Commercially activated carbon as the source for producing multicolor photoluminescent carbon dots by chemical oxidation". *Chemical Communications* 46, no. 46 (2010b): 8812–8814.
15. Alam, M.S., Javed, M., Potoo, F.H., Wajiri, F., Almalki, F., Hasnain, M.S., Garg, A., and Saifullah, K. "Qbd approached comparison of reaction mechanism in microwave synthesized gold nanoparticles and their superior catalytic role against hazardous nirto-dye". *Applied Organometallic Chemistry* 33 (2019): e5071.
16. Hong, G., Diao, S., Antaris, A.L., and Dai, H. "Carbon nanomaterials for biological imaging and nanomedicinal therapy". *Chemical Reviews* 115, no. 19 (2015): 10816–10906.
17. Hota, G., Idage, S.B., and Khilar, K.C. "Characterization of nano-sized CdS–Ag2S core-shell nanoparticles using XPS technique". *Colloids and Surfaces A: Physicochemical and Engineering Aspects* 293, no. 1–3 (2007): 5–12.
18. Shu, Y., Lu, J., Mao, Q.X., Song, R.S., Wang, X.Y., Chen, X.W., and Wang, J.H. "Ionic liquid mediated organophilic carbon dots for drug delivery and bio-imaging". *Carbon* 114 (2017): 324–333. DOI:10.1016/j.carbon.2016.12.038.
19. Hu, X., Cheng, L., Wang, N., Sun, L., Wang, W., and Liu, W. "Surface passivated carbon nanodots prepared by microwave assisted pyrolysis: Effect of carboxyl group in precursors on fluorescence properties". *RSC Advances* 4, no. 36 (2014): 18818–18826.

20. Javed, M.N., Alam, M.S., Waziri, A., Pottoo, F.H., Yadav, A.K., Hasnain, M.S., and Almalki, F.A. "QbD applications for the development of nanopharmaceutical products". In *Pharmaceutical Quality by Design*, pp. 229–253. Academic Press, 2019.
21. Javed, M.N., Kohli, K., and Amin, S. "Risk assessment integrated QbD approach for development of optimized bicontinuous mucoadhesive limicubes for oral delivery of rosuvastatin". *AAPS PharmSciTech* 19, no. 3(2018): 1377–1391.
22. Jiang, D., Chen, L., Xie, J., and Chen, M. "2014. Ag 2 S/g-C 3 N 4 composite photocatalysts for efficient Pt-free hydrogen production. The co-catalyst function of Ag/Ag 2 S formed by simultaneous photodeposition". *Dalton Trans* 43, no. 12(2014): 4878–4885.
23. Dong, X., Su, Y., Geng, H., Li, Z., Yang, C., Li, X., and Zhang, Y. "Fast one-step synthesis of N-doped carbon dots by pyrolyzing ethanolamine". *Journal of Materials Chemistry C* 2, no. 36 (2014): 7477–7481.
24. Kumar, P., Barrett, D.M., Delwiche, M.J., and Stroeve, P. "Methods for pretreatment of lignocellulosic biomass for efficient hydrolysis and biofuel production". *Industrial and Engineering Chemistry Research* 48, no. 8(2009): 3713–3729.
25. Li, H., Shao, F.Q., Zou, S.Y., Yang, Q.J., Huang, H., Feng, J.J., and Wang, A.J. "Microwave-assisted synthesis of N,P-doped carbon dots for fluorescent cell imaging". *Microchimica Acta* 183, no. 2(2016): 821–826.
26. Liu, Y., Xiao, N., Gong, N., Wang, H., Shi, X., Gu, W., and Ye, L. "One-step microwave-assisted polyol synthesis of green luminescent carbon dots as optical nanoprobes". *Carbon* 68 (2014): 258–264.
27. Qu, K., Wang, J., Ren, J., and Qu, X. "Carbon dots prepared by hydrothermal treatment of dopamine as an effective fluorescent sensing platform for the label-free detection of iron(III) ions and dopamine". *Chemistry – A European Journal* 19, no. 22 (2013): 7243–7249.
28. He, H., Wang, X., Feng, Z., Cheng, T., Sun, X., Sun, Y., Xia, Y., Wang, S., Wang, J., and Zhang, X. "Rapid microwave-assisted synthesis of ultra-bright fluorescent carbon dots for live cell staining, cell-specific targeting and in vivo imaging". *Journal of Materials Chemistry B* 3, no. 24 (2015): 4786–4789.
29. Lu, W., Qin, X., Liu, S., Chang, G., Zhang, Y., Luo, Y., Asiri, A.M., Al-Youbi, A.O., and Sun, X. "Economical, green synthesis of fluorescent carbon nanoparticles and their use as probes for sensitive and selective detection of mercury(II) ions". *Analytical Chemistry* 84, no. 12 (2012): 5351–5357.
30. Mohapatra, S., Sahu, S., Nayak, S., and Ghosh, S.K. "Design of Fe3O4@ SiO2@ carbon quantum dot based nanostructure for fluorescence sensing, magnetic separation, and live cell imaging of fluoride ion". *Langmuir* 31, no. 29 (2015): 8111–8120.
31. Chandra, S., Patra, P., Pathan, S.H., Roy, S., Mitra, S., Layek, A., Bhar, R., Pramanik, P., and Goswami, A. "Luminescent S-doped carbon dots: An emergent architecture for multimodal applications". *Journal of Materials Chemistry B* 1, no. 18 (2013): 2375–2382.
32. Dong, Y., Pang, H., Yang, H. Bin, Guo, C., Shao, J., Chi, Y., Li, C.M., and Yu, T. "Carbon-based dots co-doped with nitrogen and sulfur for high quantum yield and excitation-independent emission". *Angewandte Chemie – International Edition* 52, no. 30 (2013): 7800–7804.
33. Wang, Y., Meng, H., Jia, M., Zhang, Y., Li, H., and Feng, L. "Intraparticle FRET of Mn(II)-doped carbon dots and its application in discrimination of volatile organic compounds". *Nanoscale* 8, no. 39 (2016): 17190–17195.
34. Liu, H., Bai, Y., Zhou, Y., Feng, C., Liu, L., Fang, L., Liang, J., and Xiao, S. "Blue and cyan fluorescent carbon dots: One-pot synthesis, selective cell imaging and their antiviral activity". *RSC Advances* 7, no. 45 (2017): 28016–28023.

35. Sahu, S., Behera, B., Maiti, T.K., and Mohapatra, S. "Simple one-step synthesis of highly luminescent carbon dots from orange juice: Application as excellent bio-imaging agents". *Chemical Communications* 48, no. 70 (2012): 8835–8837.
36. Hsu, P.C., Shih, Z.Y., Lee, C.H., and Chang, H.T. "Synthesis and analytical applications of photoluminescent carbon nanodots". *Green Chemistry* 14, no. 4 (2012): 917–920.
37. Wang, J., Wang, C.F., and Chen, S. "Amphiphilic egg-derived carbon dots: Rapid plasma fabrication, pyrolysis process, and multicolor printing patterns". *Angewandte Chemie – International Edition* 51, no. 37 (2012): 9297–9301.
38. Wang, Q., Liu, X., Zhang, L., and Lv, Y. "Microwave-assisted synthesis of carbon nanodots through an eggshell membrane and their fluorescent application". *Analyst* 137, no. 22 (2012): 5392–5397.
39. Zhu, C., Zhai, J., and Dong, S. "Bifunctional fluorescent carbon nanodots: Green synthesis via soy milk and application as metal-free electrocatalysts for oxygen reduction". *Chemical Communications* 48, no. 75 (2012): 9367–9369.
40. De, B. and Karak, N. "A green and facile approach for the synthesis of water soluble fluorescent carbon dots from banana juice". *RSC Advances* 3, no. 22 (2013): 8286–8290.
41. Ma, G.P., Yang, D.Z., Chen, B. ling, Ding, S.M., Song, G.Q., and Nie, J. "Preparation and characterization of composite fibers from organic-soluble chitosan and poly-vinylpyrrolidone by electrospinning". *Frontiers of Materials Science in China* 4, no. 1 (2010): 64–69.
42. Yin, B., Deng, J., Peng, X., Long, Q., Zhao, J., Lu, Q., Chen, Q., Li, H., Tang, H., Zhang, Y., and Yao, S. "Green synthesis of carbon dots with down- and up-conversion fluorescent properties for sensitive detection of hypochlorite with a dual-readout assay". *Analyst* 138, no. 21 (2013): 6551–6557.
43. Yu, H., Li, X., Zeng, X., and Lu, Y. "Preparation of carbon dots by non-focusing pulsed laser irradiation in toluene". *Chemical Communications* 52, no. 4 (2016): 819–822.
44. Yang, X., Zhuo, Y., Zhu, S., Luo, Y., Feng, Y., and Dou, Y. "Novel and green synthesis of high-fluorescent carbon dots originated from honey for sensing and imaging". *Biosensors and Bioelectronics* 60 (2014): 292–298.
45. Hu, Y., Yang, J., Tian, J., Jia, L., and Yu, J.S. "Waste frying oil as a precursor for one-step synthesis of sulfur-doped carbon dots with pH-sensitive photoluminescence". *Carbon* 77 (2014): 775–782.
46. Javed, M.N., Alam, M.S., and Pottoo, F.H. "Inventors and assignee. Metallic nanoparticle alone and/or in combination as novel agent for the treatment of uncontrolled electric conductance related disorders and/or seizure, epilepsy & convulsions". PCT patent application WO/2017/060916. (2017).
47. Javed, M.N., Alam, M.S., and Pottoo, F.H. "Inventors and Assignee. Metallic nanoparticle alone and/or in combination as novel agent for the treatment of uncontrolled electric conductance related disorders and/or seizure, epilepsy & convulsions". Indian Patent 2091/DEL/2015. (2016).
48. Wei, J., Zhang, X., Sheng, Y., Shen, J., Huang, P., Guo, S., Pan, J., Liu, B., and Feng, B. "Simple one-step synthesis of water-soluble fluorescent carbon dots from waste paper". *New Journal of Chemistry* 38, no. 3 (2014): 906–909.
49. Ali, M.A., Srivastava, S., Solanki, P.R., Reddy, V., Agrawal, V.V., Kim, C., John, R., and Malhotra, B.D. "Highly efficient bienzyme functionalized nanocomposite-based microfluidics biosensor platform for biomedical application". *Scientific Reports* 3 (2013): 1–9.
50. Mewada, A., Pandey, S., Thakur, M., Jadhav, D., & Sharon, M. (2014). Swarming carbon dots for folic acid mediated delivery of doxorubicin and biological imaging. J. Mater. Chem. B, 2, 698–705. doi: 10.1039/c3tb21436b.

51. Liu, S.S., Wang, C.F., Li, C.X., Wang, J., Mao, L.H., and Chen, S. "Hair-derived carbon dots toward versatile multidimensional fluorescent materials". *Journal of Materials Chemistry C* 2, no. 32 (2014): 6477–6483.
52. Ding, H., and Xiong, H.M. "Exploring the blue luminescence origin of nitrogen-doped carbon dots by controlling the water amount in synthesis". *RSC Advances* 5, no. 82 (2015): 66528–66533.
53. Reckmeier, C.J., Schneider, J., Susha, A.S., and Rogach, A.L. "Luminescent colloidal carbon dots: optical properties and effects of doping [Invited]". *Optics Express* 24, no. 2 (2016): A312.
54. Roxana Jijiea, Alexandre Barras, Julie Bouckaert, Nicoleta Dumitrascu, Sabine Szunerits, and Rabah Boukherroub. "Enhanced antibacterial activity of carbon dots functionalized with ampicillin combined with visible light triggered photodynamic effects". *Colloids and Surfaces B: Biointerfaces* 170 (2018): 347–354. DOI:10.1016/j.colsurfb.2018.06.040.
55. Rui-Jun Fana, Qiang Suna, Ling Zhang, Yan Zhang, and An-Hui Lu. "Photoluminescent carbon dots directly derived from polyethylene glycol and their application for cellular imaging". *Carbon* (2014): 87–93. DOI:10.1016/j.carbon.2014.01.016.
56. Bao, L., Liu, C., Zhang, Z.L., and Pang, D.W.; "Photoluminescence-tunable carbon nanodots: Surface-state energy-gap tuning". *Advanced Materials* 27, no. 10 (2015): 1663–1667.
57. Zhang, C., Hu, Z., Song, L., Cui, Y., and Liu, X. "Valine-derived carbon dots with colour-tunable fluorescence for the detection of Hg2+ with high sensitivity and selectivity". *New Journal of Chemistry* 39, no. 8 (2015): 6201–6206.
58. Luo, P.G., Yang, F., Yang, S.T., Sonkar, S.K., Yang, L., Broglie, J.J.U., Liu, Y., and Sun, Y.P. "Carbon-based quantum dots for fluorescence imaging of cells and tissues". *RSC Advances* 4, no. 21 (2014): 10791–10807.
59. M. Zhang, X. Zhao, Z. Fang, Y. Niu, J. Lou, Y. Wu, S. Zou, S. Xia, M. Sun, and F. Du "Fabrication of HA/PEI-functionalized carbon dots for tumor targeting, intracellular imaging and gene delivery". *RSC Advances* 7 (2017): 3369–3375. DOI:10.1039/C6RA26048A.
60. Shinde, D.B., and Pillai, V.K. "Electrochemical preparation of luminescent graphene quantum dots from multiwalled carbon nanotubes". *Chemistry: A European Journal* 18, no. 39 (2012): 12522–12528.
61. Hu, L., Sun, Y., Li, S., Wang, X., Hu, K., Wang, L., Liang, X.J., and Wu, Y. "Multifunctional carbon dots with high quantum yield for imaging and gene delivery". *Carbon* 67(2014): 508–513.
62. Hu, Q., Gong, X., Liu, L., and Choi, M.M.F. "Characterization and analytical separation of fluorescent carbon nanodots". *Journal of Nanomaterials* 2017 (2017): 30–37.
63. Liu, H., He, Z., Jiang, L., and Zhu, J. "Microwave-assisted synthesis of wavelength-tunable photoluminescent carbon nanodots and their potential applications". *ACS Applied Materials and Interfaces* 7, no. 8 (2015): 4913–4920.
64. Zhu, H., Wang, X., Li, Y., Wang, Z., and Yang, X. "Microwave synthesis of fluorescent carbon nanoparticles with electrochemiluminescence properties". *Chemical Communications*, no. 34 (2009): 5118–5120.
65. Zhu, L., Yin, Y., Wang, C.F., and Chen, S. "Plant leaf-derived fluorescent carbon dots for sensing, patterning and coding". *Journal of Materials Chemistry C* 1, no. 32 (2013): 4925–4932.
66. Ryu, J., Jin, D., and Ahn, D.J. "Hydrothermal preparation of carbon microspheres from mono-saccharides and phenolic compounds". *Carbon* 48, no. 7 (2010): 1990–1998.

67. Yang, Y., Cui, J., Zheng, M., Hu, C., Tan, S., Xiao, Y., Yang, Q., and Liu, Y. "One-step synthesis of amino-functionalized fluorescent carbon nanoparticles by hydrothermal carbonization of chitosan". *Chemical Communications* 48, no. 3 (2012): 380–382.
68. Zhang, B., Liu, C.Y., and Liu, Y. "A novel one-step approach to synthesize fluorescent carbon nanoparticles". *European Journal of Inorganic Chemistry*, no. 28 (2010): 4411–4414.
69. Li, Y., Zhao, Y., Cheng, H., Hu, Y., Shi, G., Dai, L., and Qu, L. "Nitrogen-doped graphene quantum dots with oxygen-rich functional groups". *Journal of the American Chemical Society* 134 (2012): 18–21.
70. Zhang, R., and Chen, W. "Biosensors and bioelectronics nitrogen-doped carbon quantum dots: Facile synthesis and application as a 'turn-off' fluorescent probe for detection of Hg 2 þ ions". *Biosensors and Bioelectronic* 55 (2014): 83–90.
71. Shan, X., Chai, L., Ma, J., Qian, Z., Chen, J., and Feng, H. "B-doped carbon quantum dots as a sensitive fluorescence probe for hydrogen peroxide and glucose detection". *Analyst* 139, no. 10 (2014): 2322–2325.
72. Yang, S., Sun, J., He, P., Deng, X., Wang, Z., Hu, C., Ding, G., and Xie, X. "Selenium doped graphene quantum dots as an ultrasensitive redox fluorescent switch". *Chemistry of Materials* 27, no. 6 (2015): 2004–2011.
73. Xu, Q., Pu, P., Zhao, J., Dong, C., Gao, C., Chen, Y., Chen, J., Liu, Y., and Zhou, H. "Preparation of highly photoluminescent sulfur-doped carbon dots for Fe (III) detection". *Journal of Materials Chemistry A* 3, no. 2 (2015): 542–546.
74. Zhou, J., Shan, X., Ma, J., Gu, Y., Qian, Z., Chen, J., and Feng, H. "Facile synthesis of P-doped carbon quantum dots with highly efficient photoluminescence". *Rsc Advances* 4, no. 11 (2014): 5465–5468.
75. Bourlinos, A.B., & Bakandritsos, A. (2012). Koul oumpis, A., Gournis, D., Krysmann, M., Giannelis, E.P., Polakova, K., Safarova, K., Hola, K. and Zboril, R. Gd (III)-dope d carbon dotsasa dual fluorescent-MRI probe. Journal of Materials Chemistry, 22(44), 23327–23330.
76. Chen, B. Bin, Liu, Z.X., Zou, H.Y., and Huang, C.Z. "Highly selective detection of 2,4,6-trinitrophenol by using newly developed terbium-doped blue carbon dots". *Analyst* 9 (2016): 2676–2681.
77. Gong, N., Wang, H., Li, S., Deng, Y., Ye, L., and Gu, W. "Microwave-assisted polyol synthesis of gadolinium-doped green luminescent carbon dots as a bimodal nanoprobe". *Langmuir* 30, no. 36 (2014): 10933–10939.
78. Liu, J.H., Fan, J.B., Gu, Z., Cui, J., Xu, X.B., Liang, Z.W., Luo, S.L., and Zhu, M.Q. "Green chemistry for large-scale synthesis of semiconductor quantum dots". *Langmuir* 24, no. 10 (2008): 5241–5244.
79. Qian, Z., Shan, X., Chai, L., Ma, J., Chen, J., and Feng, H. "Si-doped carbon quantum dots: a facile and general preparation strategy, bio-imaging application, and multifunctional sensor". *ACS Applied Materials & Interfaces* 6, no. 9 (2014): 6797–6805.
80. Sun, Y.P., Zhou, B., Lin, Y., Wang, W., Fernando, K.A.S., Pathak, P., Meziani, M.J., Harruff, B.A., Wang, X., Wang, H., Luo, P.G., Yang, H., Kose, M.E., Chen, B., Veca, L.M., and Xie, S.Y. "Quantum-sized carbon dots for bright and colorful photoluminescence". *Journal of the American Chemical Society* 128, no. 24 (2006b): 7756–7757.
81. Chih-Wei Lai, Yi-Hsuan Hsiao, Yung-Kang Peng, and Pi-Tai Chou "Facile synthesis of highly emissive carbon dots from pyrolysis of glycerol; gram scale production of carbon dots/mSiO2 for cell imaging and drug release". *J. Mater. Chem.* 22, no. 29 (2012): 14403–14409. DOI:10.1039/c2jm32206d.

82. Qu, S., Wang, X., Lu, Q., Liu, X., and Wang, L. "A biocompatible fluorescent ink based on water-soluble luminescent carbon nanodots". *Angewandte Chemie international edition* 51, no. 49 (2012): 12215–12218.
83. Vadivel, R., Thiyagarajan, S.K., Raji, K., Suresh, R., Sekar, R., and Ramamurthy, P. "Outright green synthesis of fluorescent carbon dots from eutrophic algal blooms for in vitro imaging". *ACS Sustainable Chemical Engineering* 4, no. 9 (2016): 4724–4731. DOI:10.1021/acssuschemeng.6b00935.
84. Meiling, T.T., Cywiński, P.J., and Bald, I. "White carbon: Fluorescent carbon nanoparticles with tunable quantum yield in a reproducible green synthesis". *Scientific reports* 6 (2016): 28557.
85. Wu, Z.L., Liu, Z.X., and Yuan, Y.H. "Carbon dots: Materials, synthesis, properties and approaches to long-wavelength and multicolor emission". *Journal of Materials Chemistry B* 5, no. 21 (2017b): 3794–3809.
86. Feng, Y., Zhong, D., Miao, H., and Yang, X. "Carbon dots derived from rose flowers for tetracycline sensing". *Talanta*140 (2015): 128–133.
87. Liu, H., Wang, Q., Shen, G., Zhang, C., Li, C., Ji, W., Wang, C., and Cui, D. "A multifunctional ribonuclease A-conjugated carbon dot cluster nanosystem for synchronous cancer imaging and therapy". *Nanoscale Research Letters* 9, no. 1 (2014): 1–11.
88. Lu, S., Wu, D., Li, G., Lv, Z., Chen, Z., Chen, L., Chen, G., Xia, L., You, J., and Wu, Y. "Carbon dots-based ratiometric nanosensor for highly sensitive and selective detection of mercury(II) ions and glutathione". *RSC Advances* 6, no. 105 (2016): 103169–103177.
89. Wang, J., and Qiu, J. "A review of carbon dots in biological applications". *Journal of Materials Science* 51, no. 10 (2016): 4728–4738.
90. Lim, S.Y., Shen, W., and Gao, Z. "Carbon quantum dots and their applications". *Chemical Society Reviews* 44, no. 1 (2015): 362–381.
91. Li, H., Huang, J., Song, Y., Zhang, M., Wang, H., Lu, F., Huang, H., Liu, Y., Dai, X., Gu, Z., Yang, Z., Zhou, R., and Kang, Z. "Degradable carbon dots with broad-spectrum antibacterial activity". *ACS Applied Materials and Interfaces* 10, no. 32 (2018): 26936–26946.
92. Zhang, Z., Zheng, T., Li, X., Xu, J., and Zeng, H. "Progress of carbon quantum dots in photocatalysis applications". *Particle and Particle Systems Characterization* 33, no. 8 (2016): 457–472.
93. Sharma, V., Tiwari, P., and Mobin, S.M.; "Sustainable carbon-dots: Recent advances in green carbon dots for sensing and bio-imaging". *Journal of Materials Chemistry B* 5, no. 45 (2017): 8904–8924.
94. Zhang, Z., Shi, Y., Pan, Y., Cheng, X., Zhang, L., Chen, J., Li, M.J. and Yi, C. "Quinoline derivative-functionalized carbon dots as a fluorescent nanosensor for sensing and intracellular imaging of Zn 2+". *Journal of Materials Chemistry B.* 2, no. 31 (2014): 5020–5027.
95. Wang, L., Yin, Y., Jain, A., and Susan Zhou, H. "Aqueous phase synthesis of highly luminescent, nitrogen-doped carbon dots and their application as bio-imaging agents". *Langmuir* 30, no. 47 (2014): 14270–14275.
96. Aji, M.P., Wiguna, P.A., Susanto, S., Rosita, N., Suciningtyas, S.A., and Sulhadi, S. "Performance of photocatalyst based carbon nanodots from waste frying oil in water purification". *AIP Conference Proceedings* (2016): 1725.
97. Hamdi, N. and Srasra, E. "Filtration properties of two Tunisian clays suspensions: effect of the nature of clay". *Desalination* 220, no. 1–3 (2008): 194–199.
98. Li, H., Huang, J., Song, Y., Zhang, M., Wang,H., Lu, F., Huang, H., Liu, Y., Dai, X., Gu, Z., Yang, Z., Zhou, R., and Kang, Z. "Degradable carbon dots with broad-spectrum antibacterial activity". *ACS Applied Materials and Interfaces* 10, no. 32 (2018): 26936–26946. DOI:10.1021/acsami.8b08832.

99. Prasannan, A. and Imae, T. "One-pot synthesis of fluorescent carbon dots from orange waste peels". *Industrial and Engineering Chemistry Research* 52, no. 44 (2013): 15673–15678.
100. Wang, Y., Zhu, Y., Yu, S., and Jiang, C.. "Fluorescent carbon dots: Rational synthesis, tunable optical properties and analytical applications". *RSC Advances* 7, no. 65 (2017), 40973–40989.
101. Eriksen, S.S. "Nasjonale interesser i utviklingspolitikken". *Internasjonal Politikk* 65, no. 4 (2007): 113–122.
102. Kong, B., Zhu, A., Ding, C., Zhao, X., Li, B., and Tian, Y."Carbon dot-based inorganic-organic nanosystem for two-photon imaging and bio-sensing of pH variation in living cells and tissues". *Advanced Materials* 24, no. 43 (2012): 5844–5848.
103. Zhu, A., Qu, Q., Shao, X., Kong, B., and Tian, Y. "Carbon-dot-based dual-emission nanohybrid produces a ratiometric fluorescent sensor for in vivo imaging of cellular copper ions". *Angewandte Chemie – International Edition* 51, no. 29 (2012): 7185–7189.
104. Elberg, K., Rozkošný, R., and Knutson, L. (2009) "A review of of the holarctic Sepedon fuscipennis and S. spinipes groups with description of a new species (Diptera: Sciomyzidae)". *Zootaxa*, no. 2288 (1), 51–60.
105. Choi, Y., Choi, Y., Kwon, O.H., and Kim, B.S. "Carbon dots: Bottom-up syntheses, properties, and light-harvesting applications". *Chemistry – An Asian Journal* 13, no. 6 (2018): 586–598.
106. Zhang, J. and Yu, S.H. "Carbon dots: Large-scale synthesis, sensing and bio-imaging". *Materials Today* 19, no. 7 (2016): 382–393.
107. De, M., Ghosh, P.S., and Rotello, V.M. "Applications of nanoparticles in biology". *Advanced Materials* 20, no. 22 (2008): 4225–4241.
108. Devadas, B., and Imae, T. "Effect of carbon dots on conducting polymers for energy storage applications". *ACS Sustainable Chemistry and Engineering* 6, no. 1 (2018): 127–134.
109. Gambhir, S.S., and Weiss, S. "Quantum dots for live cells". *Science* 307 (January) (2005): 538–545.
110. Kang, Y.F., Li, Y.H., Fang, Y.W., Xu, Y., Wei, X.M., and Yin, X.B. "Carbon quantum dots for zebrafish fluorescence imaging". *Scientific Reports* 5 (2015): 1–12.
111. Pottoo, F.H., Javed, M.N., Barkat, M., Alam, M., Nowshehri, J.A., Alshayban, D.M., and Ansari, M.A. "Estrogen and serotonin: Complexity of interactions and implications for epileptic seizures and epileptogenesis". *Current Neuropharmacology* 17, no. 3 (2019b): 214–231.
112. Pottoo, F.H., Tabassum, N., Javed, M.N., Nigar, S., Rasheed, R., Khan, A., Barkat, M.A., Alam, M.S., Maqbool A., Ansari M.A., and Barreto G.E. "The synergistic effect of raloxifene, fluoxetine, and bromocriptine protects against pilocarpine-induced status epilepticus and temporal lobe epilepsy". *Molecular Neurobiology* 56, no. 2 (2019a): 1233–1247.
113. Himaja, A.L., Karthik, P.S., Sreedhar, B., and Singh, S.P. "Synthesis of carbon dots from kitchen waste: Conversion of waste to value added product". *Journal of Fluorescence* 24, no. 6 (2014): 1767–1773.
114. Lai, C.W., Hsiao, Y.H., Peng, Y.K., and Chou, P.T. "Facile synthesis of highly emissive carbon dots from pyrolysis of glycerol; Gram scale production of carbon dots/mSiO 2 for cell imaging and drug release". *Journal of Materials Chemistry* 22, no. 29(2012): 14403–14409.
115. Gao, L., and Ma, N. "DNA-templated semiconductor nanocrystal growth for controlled dna packing and gene delivery". *ACS Nano* 6, no. 1(2012): 689–695. Doi:10.1021/nn204162y.

116. Li, S., Guo, Z., Feng, R., Zhang, Y., Xue, W., and Liu, Z. "Hyperbranched polyglycerol conjugated fluorescent carbon dots with improved in vitro toxicity and red blood cell compatibility for bio-imaging". *RSC Advances* 7 (2017): 4975–4982. DOI:10.1039/c6ra27159f.
117. Wang, Y., Anilkumar, P., Cao, L., Liu, J-H., Luo, P.G., Tackett, K.N. II, Sahu, S., Wang, P., Wang, X., and Sun, Y.-P. "Carbon dots of different composition and surface functionalization: Cytotoxicity issues relevant to fluorescence cell imaging". *Experimental Biology and Medicine* 236 (2011): 1231–1238. DOI:10.1258/ebm.2011.011132
118. Wang, Z., Liao, H., Wu, H., Wang, B., Zhao, H., and Tan, M. "Fluorescent carbon dots from Tsingtao® beer for breast cancer cell imaging and drug delivery". *Anal. Methods* 20, no. 7 (2015): 8911–8917. DOI:10.1039/C5AY01978H.
119. Zhang, M., Yuan, P., Zhou, N., Su, Y., Shaoab, M., and Chi, C. "pH-Sensitive N-doped carbon dots–heparin and doxorubicin drug delivery system: preparation and anticancer research". *RSC Advances* 7 (2017): 9347–9356. DOI:10.1039/c6ra28345d.
120. Zhang, Y-Y., Wu, M., Wang, Y-Q., He, X-W., Li, W-Y, and Feng, X-Z. "A new hydrothermal refluxing route to strong fluorescent carbon dots and its application as fluorescent imaging agent". *Talanta* 117 (2013): 196–202. DOI:10.1016/j.talanta.2013.09.003.
121. Zheng, M., Liu, S., Li, J., Qu, D., Zhao, H., Guan, X., Hu, X., Xie, Z., Jing, X., and Sun, Z. "Integrating oxaliplatin with highly luminescent carbon dots: An unprecedented theranostic agent for personalized medicine". *Advanced Materials* 26, no. 21 (2014): 3554–3560. DOI:10.1002/adma.201306192.
122. Tang, J., Kong, B., Wu, H., Xu, M., Wang, Y., Wang, Y., Zhao, D., and Zheng, G. "Carbon nanodots featuring efficient FRET for real-time monitoring of drug delivery and two-photon imaging". *Advanced Materials* 25, no. 45 (2013): 6569–6574, DOI:10.1002/adma.201303124.
123. Feng, T., Ai, X., Yang, G.A.P., and Zhao, Y. "Charge-convertible carbon dots for imaging-guided drug delivery with enhanced in vivo cancer therapeutic efficiency". *ACS Nano* 10, no. 4 (2016): 4410–4420. DOI:10.1021/acsnano.6b00043.
124. Wang, C.I., Wu, W.C., Periasamy, A.P., and Chang, H.T. "Electrochemical synthesis of photoluminescent carbon nanodots from glycine for highly sensitive detection of hemoglobin". *Green Chemistry* 16, no. 5 (2014): 2509–2514.
125. Zheng, M., Ruan, S., Liu, S., Sun, T., Qu, D., Zhao, H., Xie, Z., Gao, H., Jing, X., and Sun, Z. "Self-targeting fluorescent carbon dots for diagnosis of brain cancer cells". *ACS Nano* 9, no. 11 (2015): 11455–11461. DOI:10.1021/acsnano.5b05575.
126. Mandal, T.K., and Parvin, N. "Rapid detection of bacteria by carbon quantum dots". *Journal of Biomedical Nanotechnology* 7, no. 6 (2011): 846–848.
127. Jhonsi, M.A., Ananth, D.A., Gayathri, N., Sivasudha, T., Yamini, R., Bera, S., and Kathiravan, A. "Antimicrobial activity, cytotoxicity and DNA binding studies of carbon dots". *Spectrochimica Acta Part A: Molecular and Biomolecular Spectroscopy* 196 (2018): 295–302. DOI:10.1016/j.saa.2018.02.030.
128. Kong, W., Liu, R., Li, H., Liu, J., Huang, H., Liu, Y., and Kang, Z. "High-bright fluorescent carbon dots and their application in selective nucleoli staining". *Journal of Materials Chemistry B* 2, no. 31(2014): 5077–5082.
129. Kumar, V.B., Natan, M., Jacobi, G., Porat, Z., Banin, E., and Gedanken, A. "Ga@C-dots as an antibacterial agent for the eradication of Pseudomonas aeruginosa". *International Journal of Nanomedicine* 12 (2017): 725–730.
130. Ma, N., Dooley, C.J., and Kelley, S.O. "RNA-templated semiconductor nanocrystals". *Journal of the American Chemical Society* 128, no. 39 (2006): 12598–12599. DOI:10.1021/ja0638962.

131. Nigar, S., Pottoo, F.H., Tabassum, N., Verma, S., and Javed, M.N. "Molecular insights into the role of inflammation and oxidative stress in epilepsy". *Journal of Advances in Medical and Pharmaceutical Sciences* 10 (2016): 1–9.
132. Hua, X-W., Bao, Y-W., Wang, H-Y., Chen, Z., and Wu, F-G. "Bacteria-derived fluorescent carbon dots for microbial live/dead differentiation". *Nanoscale* 9, no. 6 (2016): 2150–2161. DOI:10.1039/C6NR06558A.
133. Lin, F., Li, C., and Chen, C. "Exopolysaccharide-derived carbon dots for microbial viability assessment". *Frontiers in Microbiology* 9 (2018): 2697–2707. DOI:10.3389/fmicb.2018.02697.
134. Yang, J., Zhang, X., Ma, Y-H., Gao, G., Chen, X., Jia, H-R., Li, Y-H., Chen, Z., and Wu, F-G. "Carbon dot-based platform for simultaneous bacterial distinguishment and antibacterial applications". *ACS Applied Materials and Interfaces* 30, 8, no. 47 (2016): 32170–32181. DOI:10.1021/acsami.6b10398.
135. Song, Y., Lu, F., Li, H., Wang, H., Zhang, M., Liu, Y., and Kang, Z. "Degradable carbon dots from cigarette smoking with broad spectrum antimicrobial activities against drug-resistant bacteria". *ACS Applied Bio Materials* 1, no. 6 (2018): 1871–1879. DOI:10.1021/acsabm.8b00421.
136. Travlou, N.A., Algarra, M., Alcoholado, C., Cifuentes-Rueda, M., Labella, A.M., Lázaro-Martínez, J.M., Rodríguez-Castellón, R., and Bandos, T.J. "CQDs-surface chemistry-dependent Ag release governs the high antibacterial activity of AgMOF composites". *ACS Applied Bio Materials* 1, no. 3 (2018): 693–707. DOI:10.1021/acsabm.8b00166.
137. Ju, B., Nie, H., Zhang, X-G., Chen, Q., Guo, X., Xing, Z., Li, M., and Zhang, S.X-A. "Inorganic salt incorporated solvothermal synthesis of multicolor carbon dots, emission mechanism and anti-bacterial study". *ACS Applied Nano Materials* 1, no. 11 (2018): 6131–6138. DOI:10.1021/acsanm.8b01355.
138. Cao, L., Wang, X., Meziani, M.J., Lu, F., Wang, H., Luo, P.G., Lin, Y., Harruff, B.A., Veca, L.M., Murray, D., and Xie, S.Y. "Carbon dots for multiphoton bio-imaging". *Journal of the American Chemical Society* 129, no. 37 (2007): 11318–11319.
139. Das, P., Bose, M., Ganguly, S., Mondal, S., Das, A.K., Banerjee, S., and Das, N.C. "Green approach to photoluminescent carbon dots for imaging of gram-negative bacteria Escherichia coli". *Nanotechnology* 28, no. 19 (2017): 195501–195513. DOI:10.1088/1361-6528/aa6714.
140. Zhang, J., Liu, X., Wang, X., Mu, L., Yuan, M., Liu, B., and Shi, H. "Carbon dots-decorated Na2W4O13 composite with WO3 for highly efficient photocatalytic antibacterial activity". *Journal of Hazardous Materials* 359 (2018): 1–8. DOI:10.1016/j.jhazmat.2018.06.072.
141. Dong, X., Moyer, M.M., Yang, F., Sun, Y-P., and Yang, L. "Carbon dots' antiviral functions against noroviruses". *Scientific Reports* 7, no. 1 (2017): 519 DOI:10.1038/s41598-017-00675-x.
142. Dong, X., Awak, M.A., Tomlinson, N., Tang, Y., Sun, Y-P., and Yang, L. "Antibacterial effects of carbon dots in combination with other antimicrobial reagents". *PLOS One* 12, no. 9 (2017): 1–16. DOI:10.1371/journal.pone.0185324.
143. Dong, X., Bond, A.E., Pan, N., Coleman, M., Tang, Y., Sun, Y-P., and Yang, L. "Synergistic photoactivated antimicrobial effects of carbon dots combined with dye photosensitizers". *International Journal of Nanomedicine* 13 (2018): 8025–8035. DOI:10.2147/IJN.S183086.
144. Thakur, M., Pandey, S., Mewada, A., Patil, V., Khade, M., Goshi, E., and Sharon, M. "Antibiotic conjugated fluorescent carbon dots as a theranostic agent

for controlled drug release, bio-imaging, and enhanced antimicrobial activity". *Journal of Drug Delivery* (2014): 1–8. DOI:10.1155/2014/282193.
145. Xue, M., Zhan, Z., Zou, M., Zhang, L., and Zhao, S. "Green synthesis of stable and biocompatible fluorescent carbon dots from peanut shells for multicolor living cell imaging". *New Journal of Chemistry* 40, no. 2 (2016): 1698–1703.
146. Hasnain, M.S., Javed, M.N., Alam, M.S., Rishishwar, P., Rishishwar, S., Ali, S., Nayak, A.K., & Beg, S. (2019). Purple heart plant leaves extract-mediated silver nanoparticle synthesis: optimization by Box-Behnken design. *Materials Science and Engineering: C*, 99, 1105–1114.

the controlled drug release, bioimaging, and enhanced antibacterial activity. *Journal of Drug Delivery* (2019): 1–8. DOI:10.1155/[illegible]

14. Xue, M., Zhan, Z., Zou, M., Zhang, L. and Zhao, S.: Green synthesis of stable and biocompatible fluorescent carbon dots from peanut shells for multicolor living cell imaging. *New Journal of Chemistry* 40, no. 2 (2016): 1698–1703.

15. [illegible] Sayed, M. S., Alam, M.S., [illegible], R., [illegible], S. [illegible] & Nguyen, V., & [illegible], S. (2019). Jungle [illegible] plant leaves extract mediated silver nanoparticles synthesis, optimization [illegible] Response surface design. *Materials Science and Engineering: C* 99, [illegible]

16 Nanotechnology
An Emerging and Promising Technology for the Welfare of Human Health

J. Immanuel Suresh and M.S. Sri Janani
PG Department of Microbiology, The American College, Madurai – 625 002, Tamil Nadu, India

16.1 INTRODUCTION

Nanotechnology is a tremendously promising and emerging scientific, systematic technology. This field offers an assortment of advantageous applications to scientific and engineering research with the aid of nanomaterials. Nanotechnology is a term that is derived from a Greek word ("nano", which means "dwarf") and is explained as revisions to domineering, operating and generating methods depending on their atomic specifications [1–6]. Nanotechnology plays a vital role in daily life, and its applications have been quickly used to diagnose human beings for infectious diseases in the clinical field [7]. It has the aptitude to develop innovative scientific devices to functioning at supra-molecular stages. Generally, this field of science deals with nanomaterials and their configurations in the nanometer (nm) range [8–14]. One nanometer (1 nm) corresponds to one billionth of a meter, that is 10^{-9} m. Nano-sized particles and nano-sized materials are habitually used in nanotechnology [15]. Prokaryotic organisms, such as bacterial cells, and eukaryotic organisms, such as mammalian and plant cells, range in size from > 100 nanometer. These prokaryotic and eukaryotic cells can simply engulf the particles of nano-sized viruses ranging from 75 to 100 nanometers. The methods used in the nanotechnology field are based on this fact. Nanoparticles can be further categorized into nanocapsules and nanospheres. Nanocapsules are defined as the reservoir system for drugs [16–18]. Nanospheres are structured in the form of a matrix, where the drug can be encapsulated inside the particles [19]. According to the nanoscale, proteins range from 5 to 50 nanometers; atomic particles are 0.1 nm in size [20]; and nucleic acids, such as DNA and RNA, are 2 nanometers in width. A single distinct human hair diameter is 50 μm in size, which varies to 1 nanometer [21]. So, the single

DOI: 10.1201/9781003220350-16

human hair will be 50,000 times greater than the dimension of 1 nm. The research area of nanotechnology plays a crucial role in the clinical and biomedical fields.

16.2 ATTITUDES OF NANOTECHNOLOGY

16.2.1 Applications in the Medical Field

Nanotechnology has a variety of attitudes and widespread applications in the diagnosis and treatment of human diseases [22]. This branch of science has crucial functions in target-based clinical therapy and reformation of damaged and injured tissues. Nanotechnology is a pioneering and most excellent tool for a non-invasive diagnosing strategy for the internal portions of the individual's body [23]. It is widely useful nowadays to analyze tests and screenings to decide an individual's susceptibility to an assortment of disorders caused by infectious agents [24]. This field of science is developing investigative analysis and diagnostic methodology for infections [25]. The most exclusive physiochemical features of nanomaterials are the performance, sensitivity and stability of nanoparticles, and improved elasticity has been employed in medical healing in the early recognition of illnesses [26–28].

16.2.2 Uniqueness in Diagnostics of Nanotechnology

In the medical field, nanotechnology is an elemental energetic power that has been advanced and upgraded with an assortment of nanoscale therapeutics, such as biosensors, implantable machines, drug delivery methodologies and imaging systems [29]. The entire world has analyzed the glucometer for glucose examination in the blood, which was the outbreak of biosensors [30]. It is enhanced from in-vitro diagnosis to in-vivo analysis of glucose present in the blood. This innovative design was acknowledged and stretched out from the prehistoric enzymatic procedure.

Nano-based devices and biosensors have been revealed from several nanomaterials for screening minute concentrations of biomolecules at the beginning phase. These devices can facilitate detection of contagious disease in the premature stage [31]. These materials and devices can be pioneering and popular devices for tumor cell recognition [32, 33]. Conventional methods are unable to categorize the benign stage of cancer from the malignant stage of cancer. Conventional methods are also powerless to recognize cancer cells in the premature phase [34]. Novel nanomaterials and nanoparticles are constructed with advanced features, such as the capability to discriminate the stages of cancer and provide exact imaging of tainted infectious sections.

16.2.3 Precise Applications in Drug Delivery Using Nanomaterials

The pioneering nanoparticles can distinguish detrimental tissues by serving as competent imaging mediators. Nanomaterials also perform as ultimate transporters to carry salutary therapeutic medicines. More than ever, they carry anticancer drugs with high ability to target the objective location, with little harm to nearby healthy

TABLE 16.1
Nanoparticles and Their Functional Role in Drug Delivery

S.No	Nanoparticles	Functional Role in Drug Delivery
1.	Solid lipid nanoparticles	Used to deliver drugs to the brain
2.	Albumin paclitaxel nanomaterial	Used in cancer therapy
3.	Polyhexylcyanoacrylate	Used to deliver antiviral drugs
4.	Poly-(D,L-lactide-co-glycolide) derived from lisinopril	Used to deliver cardiovascular drugs
5.	Poly-(D,L-lactide-co-glycolide)	Used in gene delivery
6.	Doxorubicin magnetic conjugate nanoparticles	Used in the treatment of lung diseases

tissues [35–37]. Nanoparticles and nanomaterials are used to carry drugs to the brain and deliver antiviral and cardiovascular drugs. Moreover, they are used for cancer therapy, gene delivery, and the treatment of lung diseases [38]. The objectives of nanoparticles are reallocated to the direction of intracellular molecular intensities, for instance, gene-encoding of recombinant DNA and gene silencing proteins in the present scenario [39]. Nanoparticles and their functional role in drug delivery is described in Table. 16.1

Nanomaterials are furthermore used for covering scaffolds in proteins. Modern advancements are upgrading the field of nanotechnology with the help of nanomaterials. Nanoparticles have gained a supplementary curiosity in molecular programming through sensors [40]. They have the aptitude to distinguish between extra- and intra-cellular positions at micro-level intensity.

16.3 ROLE OF REGENERATIVE MEDICINE IN NANOTECHNOLOGY

The ultimate aim of nanotechnology is to deliver growth factors, drugs, and hormones for tissue rejuvenation. Moreover, these goals are the aim of research in regenerative medicine [41]. To tackle those goals, nanomaterials are extensively used to achieve tissue regeneration. Nanomaterials afford the prolonged release of bioactive particles for tissue rejuvenation to sustain continued subsistence, diffusion, and replication of cells [42]. Nanomaterials, such as charged nanoparticles, magnetic nanoparticles, nanosuspensions, mesoporous nanoparticles and micellar nanoparticles, are primarily used in the field of tissue restoration [43]. The estimated consequence of nanotechnology-based therapy is intact tissue replacement with proficient enhancement. Furthermore, nanomaterials can expand and develop the extracellular matrix (ECM) [44, 45].

Biomimetic hydrogels are primarily used to treat bone rejuvenation, [46, 47]. When contrasted with traditional microparticles, nanofill-combined nanomaterials afford superior potentiality with compressibility [48]. Biomimetic nanoparticles are enhanced to support the host and their homeostatic environment. These particles can maintain the host machinery at the nanolevel. Nanomaterials and their role in tissue rejuvenation are summarized in Table 16.2.

TABLE 16.2
Nanomaterials and Their Role in Regenerative Medicine

S.No	Nanomaterials	Role in Regenerative Medicine
1.	Nanofilled crystalline hydroxyapatite	Osteochondral restoration
2.	Nano-based hydrolyzed polyacrylamide	Osteochondral restoration
3.	Nanoparticles with chodritin sulfate	Epidermal skin rejuvanation in abrasions
4.	Polyethylene glycol-based hydrogel	Aid to rejuvenate the heart cells during transplantation
5.	Graphene oxide film	Osteogenic demarcation of stem cells obtained from adipose in human beings
6.	Biomimetic-based hydrogels	Bone rejuvenation
7.	Collagen mutually with chitosan sugar	Used to maintain islet existence post transplantation

16.4 FATAL EFFECTS OF NANOPARTICLES

If there are advantages of nanoparticles, then surely there are disadvantages. Likewise, nanotechnology holds disadvantages, too. One edge of the field possesses eventual healthiness, and the other edge possesses imminent health risks [49]. This branch of science affords bountiful benefits. They are as follows: elevated performance with miniaturized size, which means at the nanolevel, as well as very high power utilization – the most powerful, with enhanced consistency [50–56]. The fatal consequences of nanoparticles must be revised with the intention of determining the advantages [57]. Nanomaterials acquire exclusive biological responses to noxious effects.

16.4.1 Nanomaterials and Their Structure

Nanomaterials are listed as follows: nanotubes, nanospheres, nanorods, silicas, copper, silver and gold. These are all nanoscale in size [58, 59]. These nanomaterials easily interact and penetrate into human organs and cellular tissues. These nanoparticles are bound to the ECM found in the human cell and tissue. They can alter the intracellular function of cells and organs [60, 61]. To be specific, nanorods and nanospheres can release some types of harmful ions. For instance, AgNPss (silver nanoparticles) are nontoxic in small amounts. In greater amounts, they are more hazardous to intracellular human cells and organelles [62]. Therefore, the highly developed methodology has lots of advantages as well as unpleasant consequences.

16.4.2 Soluble Nature of Nanoparticles

Solubility is one of the foremost characteristics of lethal effects in nanomaterials, and it varies for different nanoparticles. Consequently, the interaction that happens between a nanoparticle and a solvent is the ultimate factor for lethality [63, 64].

TABLE 16.3
Compositions of Nanoparticles and Their Toxic Effect

S.No	Nanoparticles	Toxic Effect
1.	Cadmium selenide	Lethal to liver and renal cells
2.	Carbon-based nanoparticles	Formation of cancerous tissue in human lungs
3.	Iron-based nanoparticles	Lethal to nerve cells in humans
4.	Cadmium sulfide	Fatal to renal cells and liver cells
5.	Gold-glutathione	Fatal during dialysis process

When it is soluble in a medium, a particle scattering modifies the surface area of the nanoparticle [65]. For instance, nanofibers such as titanium dioxide are nontoxic in small amounts. In greater amounts, they will be lethal to the host [66–70]. For this reason, larger nanofibers are unable to dissolve in the bloodstream and establish themselves in the alveolar macrophages.

16.4.3 Nanoparticles and Their Surface Charge

The most imperative factor for toxicity is the surface chemistry of nanomaterials [71, 72]. Dejaguin Landau Verwey Overbeek's theory stated that the constancy of elements is characterized by the surface of Vanderwaals forces and electrostatic interactions of the elements. Likewise, silica-dioxide nanoparticles with negative charge had more noxious consequences in contrast to weakly charged silica-dioxide nanoparticles. For instance, polymer-layered AgNPs with greater charge were steadier than AgNPs [73, 74].

16.4.4 Cleanness of Nanoparticles

Another imperative factor for lethal effects is composition and purity of nanomaterials. The lethality ratio varies in all nanomaterials [75]. Some common impurities of nanomaterials that cause dangerous outcomes are listed in Table 16.3.

Antigenicity and aggregation play an indispensable role in nanomaterials [76]. The lethal effects of nanomaterials can be overcome in a nontoxic form, which may lead to accomplishments in medical research.

16.5 NANOPARTICLES AND THEIR USAGE IN THE FIELD OF MEDICINE

1. Target-based magnetic nanomaterials: These particles are designed by a miniaturized device [77]. They examine glioblastoma microvesicles found in the bloodstream.
2. Iron oxide layered with proteases for recognition of cancer: Cathepsins and metallo proteases are used. These nanomaterials can harbor cancerous tissues and cells [78]. They can be recognized by mass spectrometry in a patient's urine.

3. Nanovelcro chip: It is an anti-epithelial cell adhesion molecule layered with silicon nanowires and polydimethylsiloxane [79, 80]. It exploits the rule of laser microdissection. This nanomaterial is used to separate the circulating cancerous cells.
4. Nanoflares: It is the first and foremost hereditary nanomaterial designed for identifying tumors in the bloodstream. It facilitates live cell recognition of messenger RNA [81]. It is based on the working principle of fluorescence.
5. Single-walled carbon nanotubes: It scrutinizes nitric oxide range in inflammatory syndromes. It is based on the working principle of fluorescence.
6. Silver-based DNA probes: It is a nanoparticle aimed at explicit markers in infectious diseases [82, 83]. It depends on the working principle of Raman scattering.
7. Graphene oxide: It identifies the occurrence range of cancerous tissues and cells at the nanolevel.
8. Gold nanomaterials: It is personalized and designed with monoclonal antibodies. It recognizes the existence of Influenza-A virus in the bloodstream [84]. It depends on the working principle of colorimetric immunosensing.
9. Silver nanorod material: It is designed as a chip and used for the recognition of microorganisms like bacteria, viruses in food and biological fluids like urine, blood, and saliva [85, 86]. It depends on the working principle of surface-enhanced Raman spectroscopy.
10. Gold nanoparticles layered with viral antibodies: It is used to identify the specific viral antibodies in biological samples [87]. It depends on the working mechanism of dynamic light scattering.
11. Clump-forming nanoparticles: It is used to detect viral markers. Furthermore, it recognizes the subsistence of tumor biomarkers.
12. Silicon fluorescent nano diamonds: These nanomaterials are biocompatible, constant and nonpoisonous luminescent-based nanoprobes [88–90]. They can be a perfect analytical device for bio imaging. They are also utilized as a nontoxic vector that is used for delivering drugs.
13. Positron emission tomography: It is a group of nanoparticles [^{18}F] FAC. Chemodrugs can be given for cancer as therapeutics [91]. These agents are used in PET scans. Chemotherapeutic drugs come into sight as bright clear images in PET scans.
14. Micro-fabricated Quantum dot-linked immune diagnostic assay: It is an economic and in-vitro diagnostic examination. It is used to identify nanomolar concentration of myeloperoxidase [92, 93]. It is a rapid perceiving immunofluorescence sensor along with the potential of 2 μl of chemical solution as an analyte.
15. Molecular-based gold nanosensor: It recognizes warfarin sensitivity in genetic analysis and is permitted by the FDA. It allows testing for hereditary targets.
16. Iron-oxide magnetic nanoparticles: They are layered with poly dopamine and used to situate tumor clusters with the aid of magnetic resonance imaging [94–98]. They are based on the working principle of infrared laser irradiation.

17. Nanosized MRI agent: It attaches to alpha-v-beta 3 integrin found on the freshly budding exterior of blood vessels.

16.6 NANOMATERIAL-BASED TREATMENT OF INFECTIOUS DISEASES

1. Polycations polymer of polyethylenimine: This nanomaterial is used as a plasmid DNA transporter for the endosomal mechanism.
2. Monodisperse microgels: This nanocapsule encompases the chitosan matrix and recombinantly modified human insulin [99]. It is used to examine the discharge of insulin and the blood sugar level, particularly in diabetes mellitus.
3. Gelatin nanomaterials: These can be used as a carrier for osteopontin. They are administrated intranasally for diagnostic purposes for ischemic stroke.
4. BIND-014 designed by Bind Therapeutics Inc: This targets prostate membrane antigen and docetaxe [100, 101]. This nanomaterial plays an elementary role in chemotherapy.
5. Nano crystalline silver: This nanoparticle is used to treat injuries and acts as an antimicrobial agent.
6. Nanomaterials depend on poly-based polymer: This nanopolymer is utilized as a transporter to deliver insulin exclusively in patients with Type I and II diabetes.
7. siRNA-associated cyclodextrin: This nanoparticle is used to control the enzymatic mechanism in cancerous tissues and cells.
8. Thymosin β4 layered nanoscaffolds: These nanomaterials encourage development and demarcation of cardio myocytes during cardiac attack [102].
9. Bio mimetic nanosponge: To eliminate toxic substances from the human body.
10. Bismuth-based nanomaterials: These types of nanoparticles are used to deliberate emission to treat tumors in radiation therapy.
11. Single-walled nanotube with HER2 antibody: Nanomaterial used for discerning damage of breast cancerous tissues and cells.
12. Small interfering RNA nanoparticles: This fidgetin-like nanomaterial encourages treating wounds and rejuvenation [103]. FL2 is the monitor of cell relocation, and it is aimed at nanoparticle-encapsulated siRNA.
13. Nanofiber scaffold: It depends on carbon nanotubes, and it is utilized in cardiac engineering for heart tissues.
14. Nano-composite film: This is also based on carbon nanotubes and is used for ultrasound treatment. It transfers and produces elevated resonance emission to interrupt tissues and cells [104]. It is otherwise referred to as "Indiscernible knife for non invasive therapy".
15. Polyethylene oxylated nanotubes: This nanomaterial can preserve blood flow in the human brain.
16. Poly-ε-L-lysin with polyethylene glycol: This nanomaterial depends on targeting ligands, such as anions and cations [105–107]. It stimulates the clotting process of blood.
17. Fullerene nanomaterials: This type of nanoparticle is used to decrease allergic responses.

16.7 CONCLUSION

In the present scenario, Nanotherapeutics is an innovative emerging area of research and development in medicine. Nanotechnology plays an incredible role in the field of medicine. It facilitates a large number of products that are extremely trustworthy, authoritative and powerful [108]. It provides indispensable and valuable exploitation in the clinical field [109, 110]. The concepts in nanotechnology are correlated with information technology and biotechnology. The field of nanotechnology has extreme scope worldwide. Nanodevices and nanomedicines have the ability to trounce the drawbacks of conventional methods [111–115], even though difficulties are yet to be overcome and are not convenient in distant areas.

REFERENCES

1. Morrison, D., Dokmeci, M., Demirci, U., and Khademhosseini, A. *Biomedical Nanostructures*. John Wiley & Sons Publications, 2008.
2. Murashov, V., and Howard, J. *The US Must Help Set International Standards for Nanotechnology*, pp. 635–636. John Wiley and Sons, 2008.
3. National Nanotechnology Initiatives, United States. "Nanotechnologyand You". http://www.nano.gov/you#content (accessed 15 Aug 2016).
4. Negri, P., and Dluhy, R.A. *Ag Nanorod Based Surface-Enhanced Raman Spectroscopy Applied to Bioanalytical Sensing- Biophotonics*. 6th edition, pp. 20–35. John Wiley and Sons, 2013.
5. Nireesh, M. "Regulatory frameworks for nanotechnology in foods and medical products: Summary results of a survey activity". *OECD Science, Technology and Industry Policy Papers*, No. 4, OECD Publishing, 24 April 2013. at 11 http://dx.doi.org/10.1787/5k47w4vsb4s4-en (accessed 3 Aug 16).
6. Office of Science and Technology Policy, White House. "OSTP Memo: Policy Principles for the US Decision-Making Concerning Regulation and Oversight of Applications of Nanotechnology and Nanomaterials" (29 Jun 2011). http://www.nano.gov/ node/643 (accessed on 15 Aug 2016).
7. Rocco, M., Williams, S., and Alivisato, P. "Nanotechnology research directions: IWGN, Loyola College in Maryland, "National Nanotechnology Initiative: Leading to the next Industrial Revolution". *A report by the Interagency Working Group on Nanoscience, Engineering and Technology Committee on Technology*. National Science and Technology Council, Washington, DC, 2000.
8. Mansoori, G.A., Soelaiman, T.A.F., and Soelaiman, T.A.F. *Nanotechnology: An Introduction for the Standards Community*, pp. 1–22. Elsevier Publications, 2005.
9. Marchant, G.E., and Abbott, K.W. *International Harmonisation of Nanotechnology Governance through "Soft Law" Approaches*, pp. 393–410. Springer Publications, 2013.
10. Marchant, G.E., and Sylvester, D. *Transnational Models for Regulation of Nanotechnology*, pp. 714–725. Springer Publications, (2006).
11. Marquis, B.J., Love, S.A., Braun, K.L., and Haynes, C.L. Analytical methods to assess nanoparticle toxicity. The Analyst. (2009 Mar) 134(3), 425–439. doi: 10.1039/b818082b.
12. Massachusetts Institute of Technology. "Gold nanoparticles for controlled delivery". *Science News* (Jan. 6, 2009), http://www.sciencedaily.com/releases/2008/12/081231005359.htm (accessed 13 Aug 2016).

13. Matsuura, J.H. *Nanotechnology Regulation and Policy Worldwide*. Elsevier Publications, 2006.
14. Miller, K. (2011). Nanotechnology: how voluntary regulatory programs can both ease public apprehensions and increase innovation in the midst of uncertain federal regulations. *Ind. Health L. Rev.*, 8, 435.
15. Feynman, R. *Infinitesimal Machinery. Lecture reprinted in the Journal of Micro electromechanical Systems*. Springer publications. Elsevier publications, 1993.
16. Verma, A., Kumar, N., Malviya, R., Sharma, P.K., and Verma, A. "Emerging trends in noninvasive insulin delivery, emerging trends in noninvasive insulin delivery". *Journal of Pharmacology* (2014): 123–128. DOI:10.1155/2014/378048
17. Wang, H.N., Fales, A.M., Zaas, A.K., Woods, C.W., and Burke, T. *Surfaceenhanced Raman Scattering Molecular Sentinel Nanoprobes for Viral Infection Diagnostics*, pp 153–158. Springer, 2013.
18. Wu, J., Ding, T., and Sun, J. "Neurotoxic potential of iron oxide nanoparticles in the rat brain striatum and hippocampus". *Neurotoxicology* 34 (2013): 243–253.
19. Kwong, G.A., von Maltzahn, G., Murugappan, G., Abudayyeh, O., and Mo, S. *Mass-Encoded Synthetic Biomarkers for Multiplexed Urinary Monitoring of Disease*, pp. 63–70. Elsevier, 2012.
20. Halo, T.L., McMahon, K.M., Angeloni, N.L., Xu, Y., and Wang, W. "Nanoflares for the detection, isolation, and culture of live tumor cells from human blood". *Proceedings of the National Academy of Sciences of the United States of America* 111 (2014): 17104–17109.
21. Iverson, N.M., Barone, P.W., Shandell, M., Trudel, L.J., and Sen, S. *In Vivo Biosensing via Tissue-localizable Near-infrared-fluorescent Single-walled Carbon Nanotubes*, pp. 873–880. Elsevier publications, 2013.
22. Sarre, R. *Keeping an Eye on Fraud: Proactive and Reactive Options for Statutory Watchdogs*, pp. 283. Springer, 1995.
23. Hamilton, R.F., Wu, N., Porter, D., Buford, M., and Wolfarth, M. "Particle length-dependent titanium dioxide nanomaterials toxicity and bioactivity". *Part Fibre Toxicology* 35 (2009): 6–35.
24. Driskell, J.D., Jones, C.A., Tompkins, S.M., and Tripp, R.A. "One-step assay fordetecting influenza virus using dynamic light scattering and gold nanoparticles". *The Analyst* 136 (2011): 3083–3090.
25. Shvedova, A.A., Kisin, E.R., Mercer, R., Murray, A.R., and Johnson, V.J. "Unusual inflammatory and fibrogenic pulmonary responses to single-walled carbon nanotubes in mice". *American Journal of Physiology and Lung Cell Molecular Physiology* 289 (2005): 698–708.
26. Harper, B., Sinche, F., Wu, R.H., Gowrishankar, M., and Marquart, G. *The Impact of Surface Ligands and Synthesis Method on the Toxicity of Glutathione-Coated Gold Nanoparticles*, pp. 355–371. Springer publications, 2014.
27. Hu, C.M.J., Fang, R.H., Copp, J., Luk, B.T., and Zhang, L. "A biomimetic nanosponge that absorbs pore-forming toxins". *Nature Nanotechnology* 8 (2013): 336–340.
28. Ivask, A., Kurvet, I., Kasemets, K., Blinova, I., and Aruoja, V. *Size-Dependent Toxicity of Silver Nanoparticles to Bacteria, Yeast, Algae, Crustaceans and Mammalian Cells In Vitro*. Elsevier publications, 2014.
29. Shoffstall, A.J., Atkins, K.T., Groynom, R.E., Varley, M.E., and Everhart, L.M. "Intravenous hemostatic nanoparticles increase survival following blunt trauma injury". *Biomacromolecules* 13 (2012): 3850–3857.
30. Albanese, A., and Chan, W.C. *Effect of Gold Nanoparticle Aggregation on Cell Uptake and Toxicity*, pp. 5478–5489. John Wiley and Sons publications, 2011.

31. Thompson, D.K. *Small Size, Big Dilemma: The Challenge of Regulating Nanotechnology*, pp. 621. Elsevier, 2012.
32. Desai, N. *Challenges in Development of Nanoparticle-Based Therapeutics*, pp. 282–295. Springer publications, 2012.
33. Diermeier, D. "Public acceptance and the regulation of emerging technologies: The role of private policies". In *The Nanotechnology Challenge*, Dana D.A., ed., p. 63. Cambridge University Press, 2012.
34. Abrams, M.T., Koser, M.L., Seitzer, J., Williams, S.C., and DiPietro, M.A. "Evaluation of Efficacy, Biodistribution, and Inflammation for a Potent siRNA Nanoparticle: Effect of Dexamethasone Co- treatment". *Molecular Therapy* 18 (2011): 171–180.
35. Lynch, I., Weiss, C., and Valsami-Jones, E. "A strategy for grouping of nanomaterials based on key physico-chemical descriptors as a basis for safer-by-design NMs". *Nanoscience Today* 9 (2014): 266–270.
36. Lyu, C.Q., Lu, J.Y., Cao, C.H., Luo, D., and Fu, Y.X. "Induction of osteogenic differentiation of human adipose-derived stem cells by a novel self-supporting graphene hydrogel film and the possible underlying mechanism". *ACS Applied Materials and Interfaces* 7 (2015): 20245–20254.
37. Macoubrie, J. "Informed public perceptions of nanotechnology and trust in government". PEW. (2005).
38. Strekalova, Y.A. "Informing dissemination research: A content analysis of U.S. Newspaper Coverage of medical nanotechnology news". *Science Community* 37 (2015): 151–172.
39. Stebounova, L.V., Adamcakova-Dodd, A., Kim, J.S., Park, H., and O'Shaughnessy, P.T. "Nanosilver induces minimal lung toxicity or inflammation in a subacute murine inhalation model". *Part Fibre Toxicology* 8 (2011): 8–15.
40. Almodarresiyeh, H.A., Filippovich, L., Shahab, S.N., Ariko, N., and Agabekov, V.E. "Polyvinyl alcohol films modified by organic dyes and zinc oxide nanoparticles". The International Conference Science and Applications of Thin Films, SATF, Izmir, Turkey, (2014).
41. Lupton, M. "Where to draw the line – regulation of therapeutic applications of nanotechnology". *Asian Journal of WTO International Health LawPolicy* 9 (2014): 197–216.
42. Sahoo, S.K. *Nanotechnology in Health Care*. Pan Standard Publishing: Singapore, 2012.
43. Gray, K.A. "Five myths about nanotechnology in the current public policy debate: A science and engineering perspective". In *The Nanotechnology Challenge*, Dana D.A., ed., Cambridge University Press, 2012.
44. Brower, V. "Is nanotechnology ready for primetime?" *Journal of National Cancer Institute* 98 (2006): 9–11.
45. Buzea, C., Pacheco Blandino, I., and Robbie, K. *Nanomaterials and Nanoparticles: Sources and Toxicity*. Springer, 2007.
46. Dresser, R. "First-in-human trial participants: Not a vulnerable population, but vulnerable nonetheless". *Journal of Law and Medicinal Ethics* 37(2009): 40–41.
47. Dresser, R. "Building an ethical foundation for first-in-human nanotrials". *Journal of Medical Ethics* 40 (2012): 802–808.
48. Tibbals, H.F. *Medical Nanotechnology and Nanomedicine*. CRC Press, 2011, pp. 1–528.
49. Paradise, J., Wolf, S.M., Ramachandran, G., Kokkoli, E., and Hall, R. *Developing Oversight Frameworks for Nanobiotechnology*, pp. 9–12. Elsevier, 2008.

50. Perez, O. "Precautionary governance and the limits of scientific knowledge: A democratic framework for regulating nanotechnology". *UCLA Journal of Environmental Law and Policy* (2010): DOI:10.2139/ssrn.1585222
51. Priestly, B., Harford, A., and Sim, M. *Nanotechnology: A Promising New Technology – But How Safe?*, pp. 187–188. Elsevier, 2007.
52. Public Health Association of Australia. *Public Health Association of Australia: Policy-at-a-Glance – Public Health Impacts of Nanotechnology Policy* (Sep 2014).
53. Quinn, J. (2012). EU regulation of nanobiotechnology. Nanotech. L. & Bus, 9, 168.
54. Rajesh, M. "NCI alliance for nanotechnology for cancer". http://nano.cancer.gov/learn/now/safety.asp (accessed on 8 Aug 16).
55. Ramsay, I. "Consumer law, regulatory capitalism and the 'new learning' in regulation". *Sydney Law – a Review* 28 (2006): 19.
56. Riehemann, K., Schneider, S.W., Luger, T.A., Godin, B., Ferrari, M., and Fuchs, H. *Nanomedicine Challenge and Perspectives*, pp. 872–897. John Wiley and Sons Publications, 2009.
57. Rollins, K. *Nanobiotechnology Regulation: A Proposal for Self-regulation with Limited Oversight*, pp. 221–225. Elsevier publications, 2009.
58. Braas, D., Ahler, E., Tam, B., Nathanson, D., and Riedinger, M. "Metabolomics strategy reveals subpopulation of liposarcomas sensitive to gemcitabine treatment". *Cancer Discovery* 2 (2012): 1109–1117.
59. Brazell, L. *Nanotechnology Law: Best Practices*, p. 112. Netherlands, 2012.
60. Lefferts, J.A., Schwab, M.C., Dandamudi, U.B., Lee, H.K., and Lewis, L.D. "Warfarin genotyping using three different platforms". *American Journal of Nanotechnology* 2 (2010): 441–446.
61. Liebert, M.A. "New developments in medical nanotechnology". *Biotechnology Law Report* 27 (2008): 225.
62. Klein, P.M., and Wagner, E. "Bioreducible polycations as shuttles for therapeutic nucleic acid and protein transfection". *Antioxid Redox Signal* 21 (2014): 804–817.
63. Alqathami, M., Blencowe, A., Yeo, U.J., Franich, R., and Doran, S. *Enhancement of Radiation Effects by Bismuth Oxide Nanoparticles for Kilovoltage X-Ray Beams: A Dosimetric Study Using a Novel Multi-compartment 3D Radiochromic Dosimeter*, pp. 434–444. Elsevier, 2013.
64. Australian Government, Department of Health, Australia. *National Industrial Chemicals Notification and Assessment Scheme – Communications.* https://www.nicnas.gov.au/communications/publications/informationsheets/existingc hemical-info-sheets/chemicals-commonly-used-in-cosmeticsfactsheet-1?a=2064 (accessed 4 Aug 16).
65. Baac, H.W., Ok, J.G., Maxwell, A., Lee, K.T., and Chen, Y.C. "Carbon-nanotube optoacoustic lens for focused ultrasound generation and high-precision targeted therapy". Scientific Reports 2 (2012): 989.
66. Cooper, D.R., Bekah, D., and Nadeau, J.L. *Gold Nanoparticles and Their Alternatives for Radiation Therapy Enhancement*, pp. 86. Frontiers Publication, 2014.
67. Cui, H.X., Sun, C.J., Liu, Q., Jiang, J. and Gu, W. "Applications of nanotechnology in agrochemical formulation, perspectives, challenges and strategies". *International conference on Nanoagri* (2010): 20–25.
68. Dana, D.A. "The nanotechnology challenge". In *The Nanotechnology Challenge.*, Dana D.A., ed., pp. 3. Cambridge University Press, 2012.
69. Dana, D.A. "Toward risk-based, adaptive regulatory definitions". In *The Nanotechnology Challenge*, Dana D.A., ed., pp. 105. Cambridge University Press, 2012.
70. Dana, D.A. "Conditional liability relief as an incentive for precautionary study". In *The Nanotechnology Challenge*, Dana D.A., ed., pp. 144–150. Cambridge University Press: New York, 2012.

71. Bobadilla, A.D., Samuel, E.L.G., Tour, J.M., and Seminario, J.M. "Calculating the hydrodynamic volume of poly(ethylene oxylated) single-walled carbon nanotubes and hydrophilic carbon clusters". *Journal of Physical Chemistry B* 117 (2013): 343–354.
72. Bowman, D. M., & Hodge, G. A. (2006). A small matter of regulation: an international review of nanotechnology regulation. Colum. Sci. & Tech. L. Rev, 8, 11–19.
73. Wu, M., Zhang, D., Zeng, Y., Wu, L., and Liu, X. "Nanocluster of superparamagnetic iron oxide nanoparticles coated with poly (dopamine) for magnetic field-targeting, highly sensitive MRI and photothermal cancer therapy". *Nanotechnology* 26 (2015): 115–122.
74. Xiao, Y., Gao, X., Taratula, O., Treado, S., and Urbas, A. "Anti-HER2 IgY antibody- functionalized single-walled carbon nanotubes for detection and selective destruction of breast cancer cells". *BMC Cancer* 9 (2009): 351.
75. Shao, H., Chung, J., Balaj, L., Charest, A., and Bigner, D.D. "Protein typing of circulating microvesicles allows real-time monitoring of glioblastoma therapy". *National Medical Press* 18 (2012): 1835–1840.
76. Charafeddine, R.A., Makdisi, J., Schairer, D., O'Rourke, B.P., and Diaz-Valencia, J.D. *Fidgetin-Like 2: A Microtubule-Based Regulator of Wound Healing*, pp. 2309–2318. Elsevier, 2015.
77. Ryan, J.J., Bateman, H.R., Stover, A., Gomez, G., and Norton, S.K. "Fullerene nanomaterials inhibit the allergic response". *Journal of Immunology* 179(2007): 665–672.
78. Oh, B., and Lee, C.H. *Nanofiber for cardiovascular tissue engineering. Expert on Drug Delivery*, pp. 1565–1582. Elsevier, 2013.
79. Kumar, A., Patel, A., Duvalsaint, L., Desai, M., and Marks, E.D. "Thymosin ß4 coated nanofiber scaffolds for the repair of damaged cardiac tissue". *Journal of Nanobiotechnology* 989 (2014): 12–20.
80. Kurzweil, R. *The Singularity Is Near: When Humans Transcend Biology*. Viking: New York, 2005.
81. Mita, M., Burris, H., LoRusso, P., Hart, L., and Eisenberg, P. "Abstract CT210: A phase 1 study of BIND-014, a PSMA-targeted nanoparticle containing docetaxel, administered to patients with refractory solid tumors on a weekly schedule". *Cancer Res* 74 (19 Supplement) (2014): CT210–CT210.
82. Davis, M.E. "The first targeted delivery of siRNA in humans via a selfassembling, cyclodextrin polymer-based nanoparticle: From concept to clinic". *Molecular Pharmacology* 6 (2009): 659–668.
83. De la Rica, R. and Stevens, M.M. "Plasmonic ELISA for the ultrasensitive detectionof disease biomarkers with the naked eye". *National Nanotechnology Press* 7 (2012): pp. 821–824.
84. Kanasty, R., Dorkin, J.R., Vegas, A., and Anderson, D. "Delivery materials for siRNA therapeutics". *Nature Journal* 12 (2013): 967–977.
85. Joachim, E., Kim, I.D., Jin, Y., Kim, K.K., and Lee, J.K. *Gelatin nanoparticles enhance the neuroprotective effects of intranasally administered osteopontin in rat ischemic stroke model and Drug Delivery*, pp. 395–399. Elsevier, 2014.
86. Jotter, F. and Alexander, A. "Managing the "known unknowns": Theranostic cancer nanomedicine and informed consent". In *Biomedical Nanotechnology: Methods and Protocols*, Hurst S., ed., pp. 413. Humana Press: New York, 2011.
87. Tripathy, N., Hong, T.K., Ha, K.T., Jeong, H.S., and Hahn, Y.B. *Effect of ZnO Nanoparticles Aggregation on the Toxicity in RAW 264.7 Murine Macrophage*, pp. 110–117. Elsevier, 2014.
88. Gu, Z., Dang, T.T., Ma, M., Tang, B.C., and Cheng, H. "Glucose-responsive microgels integrated with enzyme nanocapsules for closed-loop insulin delivery". *ACS Nanotechnology* 7(2013): 6758–6766.

89. Guo, N.L., Wan, Y.W., Denvir, J., Porter, D.W., and Pacurari, M. "Multiwalled carbon nanotube-induced gene signatures in the mouse lung: Potential predictive value for human lung cancer risk and prognosis". *Journal of Toxicology and Environmental Health* 75 (2012): 1129–1153.
90. Gwinn, M., and Vallyathan, V. "Nanoparticles: health effects–pros and cons". *Environmental Health Perspectives* 114 (2006): 1818–1825.
91. Fong, J., and Wood, F. "Nanocrystalline silver dressings in wound management: A review". *International Journal of Nanomedicine* 1 (2006): 441–449.
92. Oupicky, D., and Li, J. "Bioreducible Polycations in Nucleic Acid Delivery: Past, Present, and Future Trends". *Macromolecular Bioscience*, pp. 908–922 (2014). DOI: 10.1002/mabi.201400061
93. Park, Y.H., Bae, H.C., Jang, Y., Jeong, S.H., and Lee, H.N. "Effect of the size and surface charge of silica nanoparticles on cutaneous toxicity". *Molecular Cell Toxicology* 9 (2013): 67–74.
94. Lienemann, P.S., Lutolf, M.P., and Ehrbar, M. "Biomimetic hydrogels for controlled biomolecule delivery to augment bone regeneration". *Advanced Drug Delivery Reviews* 64 (2012): 1078–1089.
95. Liu, L., Sun, M., Li, Q., Zhang, H., and Alvarez, P.J.J. "Genotoxicity and cytotoxicity of cadmium sulfide nanomaterials to mice: Comparison between nanorods and nanodots". *Environmental Engineeering Science*. (2014): 373–380. 10.1089/ees.2013.0417
96. Liu, Y., Zhang, L., Wei, W., Zhao, H., and Zhou, Z. "Colorimetric detection of influenza A virus using antibody-functionalized gold nanoparticles". *Analyst* 140 (2015): 3989–3995.
97. Lu, Y.T., Zhao, L., Shen, Q., Garcia, M.A., and Wu, D. *NanoVelcro Chip for CTC enumeration in prostate cancer patients – Methods*, pp. 144–152. Elsevier, 2013.
98. Ludlow, K., Bowman, D., & Hodge, G. (September 2007). A Review of Possible Impacts of Nanotechnology on Australia's Regulatory Framework (Final Report, Monash University) ('Australian Review').
99. Koushki, N., Katbab, A.A., Tavassoli, H., Jahanbakhsh, A., and Majidi, M. "A new injectable biphasic hydrogel based on partially hydrolyzed polyacrylamide and nanohydroxyapatite as scaffold for osteochondral regeneration". *RSC Advances* 5 (2015): 9089–9096.
100. Deepthi, S., Viha, C.V.S., Thitirat, C., Furuike, T., and Tamura, H. "Fabrication of chitin/poly(butylene succinate)/chondroitin sulfate nanoparticles ternarycomposite hydrogel scaffold for skin tissue engineering". *Polymers* 6 (2014): 2974–2984.
101. Den Hertog, J.A. (1999). General Theories of Regulation. In Encyclopedia of Law and Economics, University of Ghent, 223–270.
102. Grover, G.N., Rao, N., and Christman, K.L. *Myocardial Matrix-Polyethylene Glycol Hybrid Hydrogels for Tissue Engineering*. Elsevier, 2014.
103. Ellis, C.E., Suuronen, E., Yeung, T., Seeberger, K., and Korbutt, G.S. "Bioengineering a highly vascularized matrix for the ectopic transplantation of islets". *Islets* 5 (2013): 216–225.
104. Kennedy, A.J., Melby, N.L., Moser, R.D., Bednar, A.J., and Son, S.F. "Fate and toxicity of CuO nanospheres and nanorods used in Al/CuO nanothermites before and after combustion". *Environmental Science Technology* 47 (2013): 11258–11267.
105. Falkner, R., and Jaspers, N. "Regulating nanotechnologies: Risk, uncertainty and the global governance gap". *Global Environmental Politics* (2012): 30–55. DOI: 10.1162/GLEP_a_00096
106. Faunce, T.A. "Nanotechnology in global medicine and human biosecurity: Private interests, policy dilemmas, and the calibration of public health law". *Journal of Medical Ethics* 35 (2007): 629–642.

107. Faunce, T.A. "Policy challenges of nanomedicine for Australia's PBS". *Australian Health Review* 33 (2009): 258.
108. 108. Bergeson, L.L. "Australia announces adjustments to NICNAS new chemicals processes for industrial nanomaterials". (21 Oct 2010) http://nanotech.lawbc.com/2010/10/australiaannounces-adjustments-to-nicnas-new-chemicals-processesfor-industrial-nanomaterials/ (2016). (accessed 14 Aug 2016).
109. Monica, J.C. *A Nano-Mesothelioma False Alarm*, pp. 319. Elsevier, 2018.
110. Montalti, M., Cantelli, A., and Battistelli, G. "Nanodiamonds and silicon quantum dots: Ultrastable and biocompatible luminescent nanoprobes for long-term bioimaging". *Chemical Society Reviews* 44 (2015): 4853–4921.
111. Yang, X., Gondikas, A.P., Marinakos, S.M., Auffan, M., and Liu, J. "Mechanism of silver nanoparticle toxicity is dependent on dissolved silver and surface coating in Caenorhabditis elegans". *Environmental Science Technology* 46 (2012): 1119–1127.
112. Yeagle, Y. (2007). Nanotechnology and the FDA, Virginia J. Law Technol., 12, 65.
113. Yoon, H.J., Lee, K., Zhang, Z., Pham, T.M., and Nagrath, S. *Nano Assembly of Graphene Oxide for Circulating Tumor Cell Isolation*. John Wiley and Sons publications, 2011.
114. Yu, C., Kim, G.B., Clark, P.M., Zubkov, L., and Papazoglou, E.S. "A microfabricated quantum dot-linked immuno-diagnostic assay (μQLIDA) with an electrohydrodynamic mixing element". *Sensors & Actuators, B: Chemical* 209 (2015): 722–728.
115. Zhang, C. and Liu, Y. "A concise review of magnetic resonance molecular imaging of tumor angiogenesis by targeting integrin αvβ3 with magnetic probes". *International Journal of Nanomedicine* 8 (2013): 1083–1093.

17 Hybrid Perovskite Solar Cells

Principle, Processing and Perspectives

Ananthakumar Soosaimanickam
Crystal Growth Centre, Anna University, Chennai – 600 025, India
Institute of Materials Science (ICMUV), University of Valencia, Spain

Saravanan Krishna Sundaram
Department of Physics, Anna University, Chennai – 600 025, India

Moorthy Babu Sridharan
Crystal Growth Centre, Anna University, Chennai – 600 025, India

17.1 INTRODUCTION

Research on materials with the ability to respond to sunlight has increased in the past decades, thanks to semiconductor nanomaterials for their efficiency in utilizing the sun spectrum to convert sunlight into electrical energy. In this search, new materials for harvesting solar energy have gained attraction due to their promising indications to fulfill the future demand. Materials that belong to II-VI, III-V and I-III-VI_2 were intensively examined in the last decades for photovoltaic applications. After the first-generation and thin-film photovoltaics, the efficacy of third-generation photovoltaics, like dye-sensitized solar cells (DSSCs) and quantum dot sensitized solar cells (QDSSCs), has reached over 12%. Still, crystalline silicon dominates the commercial market. Even though the QDSSCs approach has the ability to cross over the Schockely-Queisser limit theoretically, the drawbacks, such as degradation of quantum dots with electrolytes and recombination of electrons at interface, have placed limits on efficiency. Hence, the search of new, promising materials that overcome all these obstacles has become an inevitable one. Organomethyl ammonium lead halides ($CH_3NH_3PbX_3$), a class of perovskites, has

DOI: 10.1201/9781003220350-17

shown much interest in the last couple of years. A large number of scientific publications with impressive results boost this hope as well. These hybrid perovskite compounds ($CH_3NH_3PbI_3$) possess a narrow bandgap (~1.48 eV) and good thermal stability in bulk form [1]. The outstanding scientific contributions of D. B. Mitzi [2], Miyasaka [3], Snaith [4], N. G. Park [5] and Gratzel [6] in this field have strengthened the expectations of significant achievements in solar cell research in the near future. In an earlier attempt, PbS nanoparticles were covered by a perovskite layer, which made a core-shell arrangement, and enhanced photovoltaic performance was observed [7]. Apart from photovoltaics, application of $CH_3NH_3PbX_3$ is also studied for gas sensor applications to detect gases like NH_3 [8]. Furthermore, single-cell to multi-cell tandem perovskite solar cell is another interesting demonstration for future generation solar cell research [9]. Solution-processed tandem hybrid perovskite solar cells are experimentally demonstrated towards water-splitting applications [10]. Furthermore, synthesizing nanowires of perovskite by means of a template-assisted approach has initiated huge interest in synthesizing such morphologies for photovoltaic applications [11]. Continuous progress in finding new perovskites, surface modifications, different kinds of device configurations, and fabricating a pin-hole free and smooth perovskite active layer through optimizing reaction conditions have stimulated huge interest in the recent years. Both planar and mesoscopic architecture of hybrid perovskite solar cells are actively moving with their own record-breaking results. Besides the upscaling of solid state perovskite, solar cell modules [12] have given big hope on their future dimension. In this scenario, this chapter deals with the present status of hybrid perovskites for advanced photovoltaics and their future trends. Special focus is paid on active layer materials, the hole transport layer and the electron transport layer of perovskite solar cells.

17.2 ORGANOMETAL HALIDE PEROVSKITES ($CH_3NH_3PBX_3$)—INTRODUCTION

The basic unit of hybrid perovskite group compounds is given by the common formula of AMX_3, where A=organic cation ($CH_3NH_3^+$, $CH_3CH_2NH_3^+$, $HC(NH_2)_2^+$ etc.), M = Metallic cation (Pb (or) Sn), and X = halides (Cl, Br, I). Based on corner sharing of MX_6 octahedra, the X atoms occupy the corner, and M atoms occupy the middle of the octahedral. The organic cation (A), located at the interstitial position, is surrounded by eight octahedra in the cuboctahedral group [13] (This is represented in Figure 17.1).

Depending upon the size of the cation (A), one-, two- and three-dimensional networks of perovskites are achieved. The perfectly packed perovskite structure can be identified through the tolerance factor (t) within the limit of unity, which is given by the following equation [14].

$$(R_A + R_X) = t\sqrt{2}(R_B + R_X),$$

Where R_A, R_B, R_X are the ionic radii of the elements A, B and X.

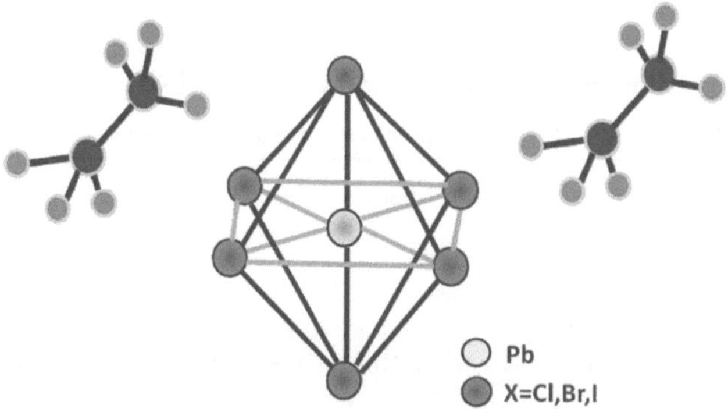

FIGURE 17.1 Structure of the $CH_3NH_3PbX_3$ perovskite (X = Cl, Br, I).

Considering $CH_3NH_3PbI_3$, it belongs to the common group of methyl ammonium trihalogen plumbates. Since most of the high-efficiency solar cells are fabricated using this compound, this chapter discusses the influence of this compound and its physico-chemical characteristics. It is a direct and low bandgap semiconductor possessing the general formula of RNH_3BX_3 [15], ($R = C_nH_{2n+2}$, X = Halogen (I, Br or Cl), B = Pb (or) Sn). The structure of organic-inorganic lead iodide perovskites has been identified as tetragonal at room temperature. The interstitial occupied organic molecules have a very weak interaction with the inorganic lattice through the hydrogen of methyl ammonium ion [16]. Further, this organic cation also undergoes a large orientation motion in cubic phase, which depends upon the dynamics of inorganic cage [17]. This intrinsic motion creates a "giant dielectric constant (GDC)" phenomena under strong illumination in lead perovskites [18]. The tetragonal perovskite, when it undergoes heat treatment, is turned into cubic at higher temperature (300 K) [19], and below 165 K it tends to be orthorhombic [20]. Out of four phases that exists in $CH_3NH_3PbI_3$ (α, ß, γ and δ phase), the notable physical characteristics of two important phases (α, ß) are given in Table 17.1. When perovskite transforms from α (room temperature) to ß (at 333 K) phase, the octrahedra tends to be tilted [21]. The effective absorption co-efficient values of hybrid perovskites are quite high, which is one of the most important characteristic properties. The comparison of effective absorption co-efficient value of perovskites with other semiconductors is given in Figure 17.2. In general, the organometal halide perovskites are obtained by using chemical methods by reacting the precursors of individual components, through grinding, etc. The optical properties of these materials predominantly depend upon the preparation method. Depending upon the processing conditions, the $CH_3NH_3PbI_3$ perovskite behaves as a p- or n-type semiconductor. DFT analysis indicates that the bulky nature of iodide ions in the structure led to its stable structure as pseudocubic [22]. They also possess ferroelectric properties, such as spontaneous electric polarization. This polarization induces a slow rearrangement in the perovskites, which is influenced by grain size and the nature of organic cations [23]. These ferroelectric properties are one of the reasons for the presence of a large hysteresis in methyl ammonium

TABLE 17.1
Important crystallographic data of α and ß phases of $CH_3NH_3PbI_3$

	Crystallographic Data of α and ß Phases of $CH_3NH_3PbI_3$	
	α-$CH_3NH_3PbI_3$	β-$CH_3NH_3PbI_3$
Crystal system	Tetragonal	Tetragonal
Space group	*P4mm*	*I4cm*
Unit cell dimensions	a = 6.3115 (2) Å, α = 90.00°b = 6.3115 (2) Å, β = 90.00°c = 6.3161 (2) Å, γ = 90.00°	a = 8.849(2) Å, α = 90.00°b = 8.849(2) Å, β = 90.00°c = 12.642(2) Å, γ = 90.00°
Volume	251.60 (2) $Å^3$	990.0(4) $Å^3$
Density (calculated)	4.092 g/cm^3	4.159 g/cm^3
Absorption coefficient	25.884 mm^{-1}	26.312 mm^{-1}
Extinction coefficient	0.016(2)	–

Source: Adapted from Stoumpas et al. [21].

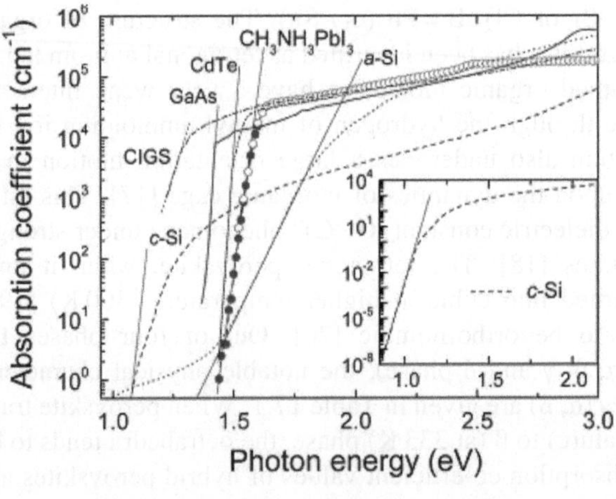

FIGURE 17.2 Effective absorption co-efficient of a $CH_3NH_3PbI_3$ perovskite thin film materials compared with other photovoltaic materials.

Source: Reprinted with permission from De Wolf et al. [30]. Copyright@2014 American Chemical Society.

lead halide perovskites in addition to defect density and excess of ions [4]. Additionally, the re-orientation of permanent dipoles of $CH_3NH_3^+$ ion and resistance of the inorganic {PbI_3} lattice can be correlated as reason for this large hysteresis [24]. This hysteresis mainly depends upon growth condition and type of the organic cation of the perovskite system [23]. This electric dipole formation of trimethyl ammonium ion ($CH_3NH_3^+$) may undergo a photoferroic effect, which is expected to enhance the efficiency [24]. The nature of the halogen combination with the metal ions also influences the structure. When alloying the hybrid lead iodide with chloride, bromide (i.e. $CH_3NH_3PbI_{3-x}Br_x$, $CH_3NH_3PbI_{3-x}Cl_x$) leads to the changes in the optical properties, like absorption and bandgap, and it possesses anomalous alloy properties [25]. Addition of chlorine is leads to enhancement of absorption, whereas addition of bromine shifts the absorption edge towards a lower wavelength [26]. Addition of bromine in excess of a certain percentage leads to phase transition from tetragonal to cubic [27]. The bandgap of the lead perovskite could be tuned from 1.51 to 3.62 by alloying with Cl or Br in proper ratio at the iodide site in the structure. Moreover, the addition of Br into perovskites has led to improvement of V_{oc} and the lifetime of the prepared devices. The lead bromide perovskite ($CH_3NH_3PbBr_3$) supported by alumina produced high V_{oc} up to 1.30 V [28]. In addition to being an effective sensitizer in solar cells, the methyl ammonium iodide perovskites plays a ligand role in the synthesis of semiconductor nanocrystals to passivate them effectively [29]. Mixing of two different cations in the perovskite structure is found to be quite beneficial to improving the efficiency.

17.3 ORGANOMETAL HALIDE PEROVSKITES SOLAR CELLS—IN A NUTSHELL

17.3.1 The Era of Organolead Halide Perovskite Solar Cells

The fascinating material properties of hybrid perovskites, like higher optical absorption than conventional thin film solar cell absorber, small effective masses of electrons and holes due to the strong s-p antibonding coupling, dominant shallow point defects and intrinsically benign grain boundaries, have lead to utilizing them for the fabrication of future generation of photovoltaic devices [31]. The seminal publication of Miyasaka et al. during the year 2009 marked the beginning of the organolead halides as effective sensitizers for the current solid state mesoscopic/planar heterojunction solar cells [3]. The authors achieved about 3.8% efficiency through the traditional iodide redox (I^-/I_3^-) electrolyte. Shortly after that, Im et al. achieved about 6.5% efficiency through the sensitization of quantum dots of perovskites (size: 2–3 nm) with the same electrolyte [32]. The same group also achieved 2.4% efficiency using ethyl ammonium lead iodide ($CH_3CH_2NH_3PbI_3$) [33]. Although there is some computational approach of comparing the interface of DSSCs and perovskite solar cells [34], it is understood clearly that perovskite solar cells function neither like a dye-sensitized solar cells nor the bulk hetero-junction type hybrid solar cells. The breakthrough has come when spiro-OMeTAD polymer is utilized as a hole transport layer to make a solid state perovskite solar cells. In contrast to the organic semiconductors, the diffusion length of the hybrid perovskites crosses over 1 μm [35]. The Wannier-Mott type exciton formed in hybrid perovskites supports the mobility of the electron throughout the layer, which helps to attain high charge collection. Undoubtedly, the internal quantum efficiency (IPCE) of perovskite solar cells has reached nearly 100% [36], which ensures its place for future solar module markets. The active region thickness of the hybrid perovskite solar cells comes around 1 μm and a built-in voltage of 1 V with the

(a)

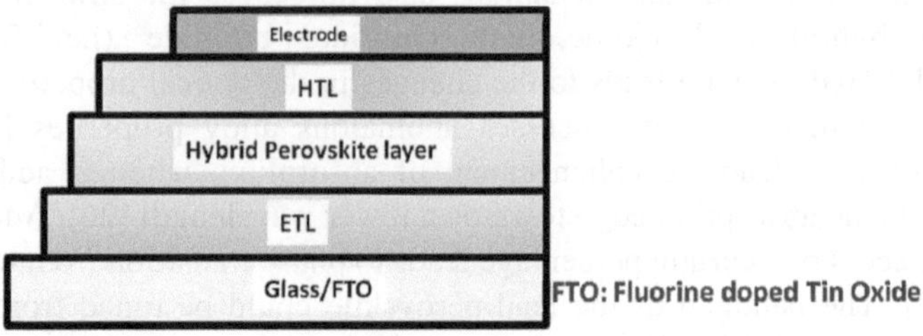

(b)

ETL = Electron transport layer	Hybrid Perovskite Layer	HTL=Hole transport layer
ZrO_2, TiO_2, ZnO, Al_2O_3	RAX_3 (or) RBX_3 (or) $RA_{1-x}B_xX_3$ (or) $RAX_{3-x}Y_x$ (or) $(R)_x\ (R')_{1-x}\ AX_3$ where, R R' = $CH_3NH_3^+$, $NH_2CH{=}CH_2$, Cs^+, $CH_3CH_2NH_3^+$ X,Y = Cl, Br, I A= Pb, B= Sn	P3HT, spiro-OMeTAD, PTTA CuSCN
Electrode : Au, Ag, Al		

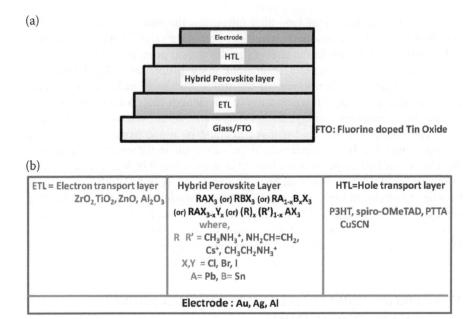

FIGURE 17.3 Schematic diagram of (a) basic configuration of a perovskite solar cell, and (b) commonly employed materials in each layer.

kinetically controlled defect formation [37]. The schematic diagram of basic device construction and materials used for the hybrid perovskite solar cell is given in Figure 17.3.

The most advantageous and also important step in the fabrication of perovskite solar cells is the method of utilizing the precursors. Here, halide salts of metals are dissolved in solvents like N, N' dimethyl formamide (DMF), dimethyl sulphoxide (DMSO) or γ–butyrolactone and spin-coated on the mesoporous TiO_2 or alumina. The lead perovskite is also relatively stable in the nonpolar solvent like ethyl acetate, but unstable in polar solvent like acetonitrile, which induces bleaching [8]. The conductivity of the perovskite layer is around 10^{-3} S cm^{-3}. The development of an active layer of perovskites is carried out by many methods, namely spin coating, dip coating, etc. Advanced methods like aerosol-assisted chemical vapor deposition and hybrid chemical vapor deposition also have given good results for perovskite deposition [38, 39]. Numerous attempts have been made to optimize the device structure, particularly in the hole transporting layer (HTM). The major breakthrough came when the solid state hole transporter 2,2′,7,7′-tetra-kis(N, N′-dimethoxyphenylamine)–9,9′-spirobifuluorene (spiro-MeOTAD) was utilized with the perovskite active layer. Initially, 9.7% efficiency was achieved through this solid hole transporter for the device configuration FTO/TiO_2/$CH_3NH_3PbI_3$/spiro-MeOTAD/Au [40]. Later, through continuous efforts, 16.2% was reached through this solid hole conductor [41]. Through these achievements, the predominant role of the hole transport layer was confirmed. But, the interesting thing is perovskite solar cells could efficiently harvest electrons and conduct the charges, even without the electron transport or hole transport layer. This differentiates the abnormal trends of PSSCs from DSSCs and QSSCs. The important futuristic properties of the $CH_3NH_3PbI_3$ include the large optical absorption co-efficient (4.3×10^5 cm^{-1} at 360 nm), high diffusion length (about 1 μm), high open circuit voltage (~1 V), ambipolar nature (functions as both hole and electron transporter), easily tunable spectrum as well as bandgap by alloying with other halides, and compatible solution processable nature. Further, the photoluminescence quantum efficiency of spin-coated hybrid lead halide perovskite has reached up to 70%, which is unusual in semiconducting systems [42]. The general perovskite solar cell structure can be represented as FTO/m-TiO_2/$CH_3NH_3PbI_3$/HTM/electrode.

The common metal electrodes, like Au, Ag and Al, have been widely reported as top contacts, though some new counter electrodes, such as highly ordered mesoporous carbon with graphite paste, carbon nanotubes (CNT) [43, 44] and metal electrode-free perovskite solar cells [44], were also attempted. The most preferred electrode is Au, since silver (Ag) forms shunt paths with TiO_2 and also there is a possibility of corrosion of silver when contact is made with the perovskite layer [45]. The efficiency of the hybrid perovskite solar cells normally depends on various optimized factors that include type of device configuration, nature of the solvent, rotation speed of the spin-coating process, type of hole conductor, structure of the TiO_2 layer, etc. The overall synthesis and construction steps involved in the hybrid perovskite solar cells are schematically given in Figure 17.4(e, f). The probable charge transport mechanism in perovskite absorbers is under intense investigation. It has been found through DFT analysis that the important factors that play a major role in carrier transport of $CH_3NH_3PbI_3$ are i) small e^- and h^+ mass; ii) large Born effective charges, which results in large static dielectric constant; and iii) iodine interstitial for deep trap and nonradiative recombination centers [46]. When analyzing the excitonic transport properties, the exciton-lifetime and exciton-binding energy of $CH_3NH_3PbI_3$ is 78 ns and 19 ± 3 meV, which is much higher than the values reported for traditional organic semiconductors (5 ns and ≥100 meV) [36]. The reason for the high open-circuit voltage of the hybrid metal perovskites has been investigated by several research groups. Nevertheless, its characteristic properties, which include high charge carrier mobility, high dielectric constant and low-lying valence band, can be ruled out for this cause [28]. Further, the enhanced charge separation and high V_{oc} can be ascribed with the internal electric field present in hybrid perovskites. This clearly promises remarkable efficiency outcome over organic photovoltaics. Results from density functional theory (DFT) analysis show that the intrinsic defects states present in the $CH_3NH_3PbI_3$ layer may play a major role in charge conduction. This study also describes that the Schotky defects are not playing any role in forming the trap states, whereas the Frenkel defects that are formed due to vacancies of elements such as Pb, I and CH_3NH_3 form shallow levels in the band edge [47]. Further, DFT also explains that the presence of dimers and trimers of Pb^{2+} & I^- in intrinsic defects, which form due to the strong covalency between Pb^{2+} and I anions, may act as the recombination center [48]. The origin of high V_{oc} in perovskite solar cells has been less explored so far, but it may owe to the

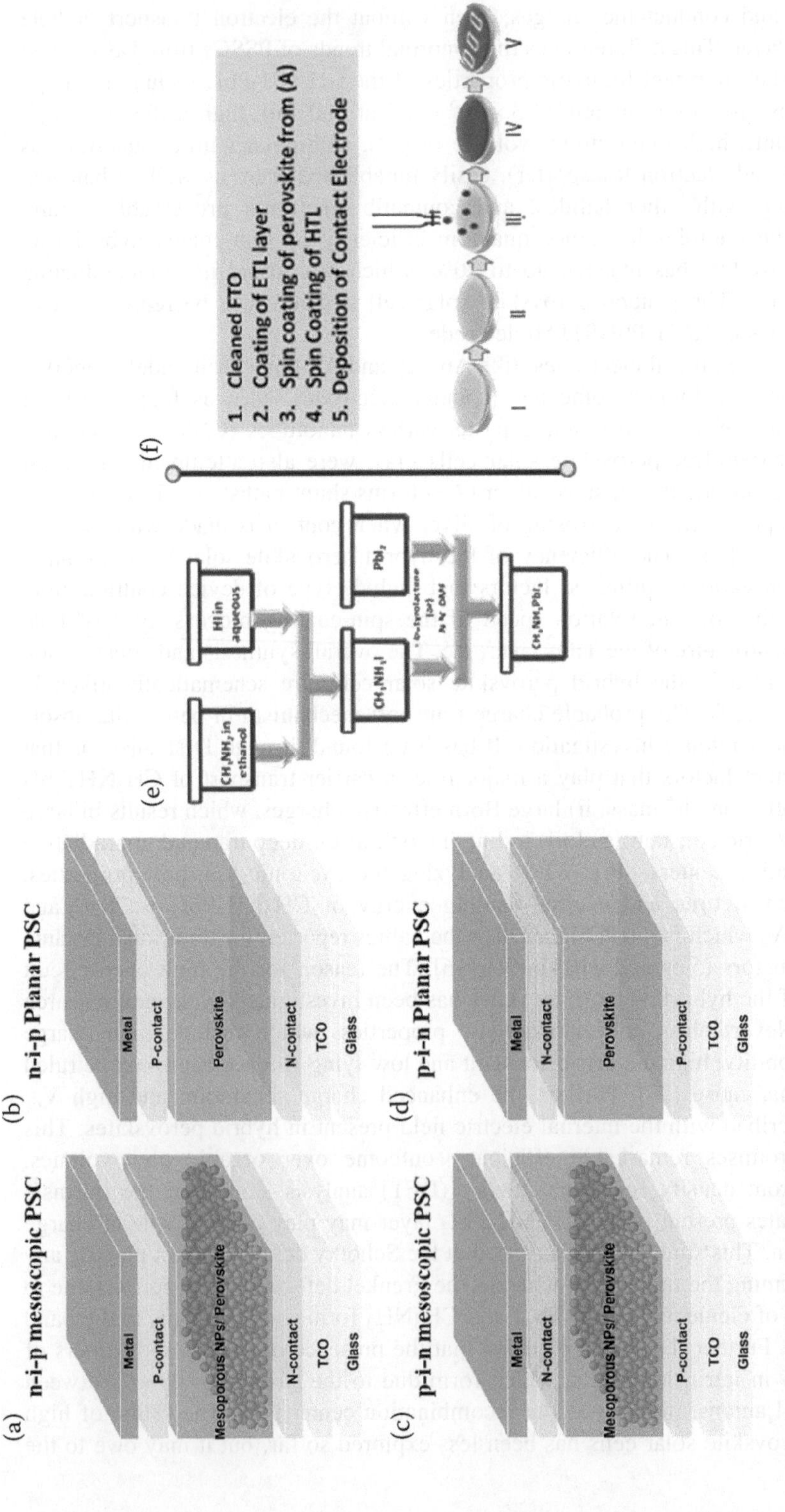

FIGURE 17.4 (a–d) Different kind of perovskite solar cell structures.

Source: Reprinted from Li et al. [31] with permission from Royal Society of Chemistry. (e) Schematic diagram of the synthesis outline of $CH_3NH_3PbI_3$ and (f) typical fabrication steps involved in assembling a mesoscopic perovskite solar cell.

absence of optically detectable deep states, low degree of structural disorder, type of bandgap and absorber thickness, etc. [30]. The excessive thickness of the active perovskite layer may have led to poor efficiency due to recombination of carriers [49, 50]. The photoemission spectroscopy analysis of perovskites with appropriate hole transport layer (i.e. spiro compound) is one of the best experimental evidences, which provides the value of electronic energy levels to find minimum energy loss in order to obtain high efficiency [51].

Uniform deposition or coating of the perovskite layer on the TiO_2 layer with appropriate thickness is very important for high-efficiency solar cells. The improper coating of perovskites creates a "shunt path," which creates a direct contact between the TiO_2 layer and the hole transport layer and ultimately leads to low V_{oc} and efficiency. Another important factor is presence of moisture in the air, which is detrimental to perovskite solar cells. Generally, perovskite solar cells are vulnerable to air in the absence of sealing [52]. The interaction of moisture with perovskite results in decomposition of the compound. This takes place after some time interval and is observed through the formation of yellow-color product on the surface, which is nothing but PbI_2. The overall cycle of this decomposition of the perovskite layer is given in Figure 17.5. Hence, the moisture-free atmosphere (inert) would be an ideal one for the coating of hybrid perovskites. Despite this, about 5.67% efficiency was achieved in the presence of 50% humidity in the air [53]. Since the absorption co-efficient of the perovskites is quite high, the planar hetero-junction approach gives good efficiency results. Several methods are being employed to deposit the

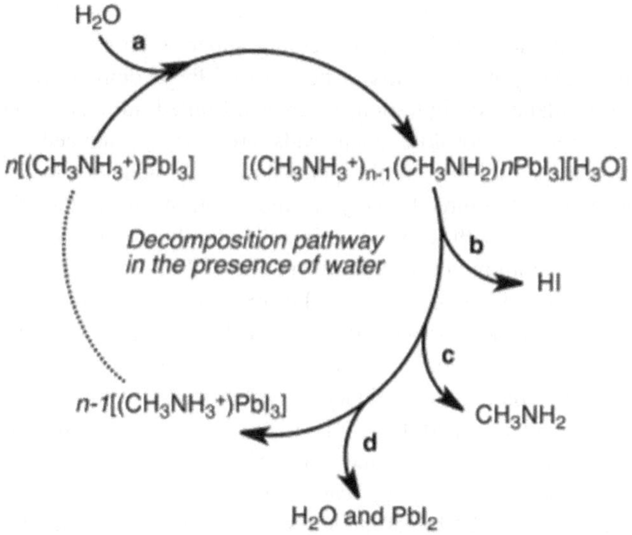

FIGURE 17.5 Decomposition pathway of hybrid perovskites by water (a) the interaction of H_2O with perovskite (b & c) decomposition of perovskite and d) formation of yellow solid of PbI_2.

Source: Reprinted with permission from Forst et al. [54]. Copyright@2014 American Chemical Society.

perovskite layer on the TiO_2 layer. These methods include vapor-assisted deposition, spin coating, vapor-assisted solution process (VASP), sequential deposition method, inter-diffusion method, etc. Out of these, sequential deposition method has emerged as the most successful one in fabrication of a pin-hole free active layer with outstanding efficiencies. Burschka and co-workers have achieved about 15% efficiency for FTO/TiO_2/$MAPbI_3$/spiro-MeOTAD/Au device structure with long-term stability through deposition of perovskite active layers by the two-step assisted sequential deposited method [55]. This sequential deposition method is also used to make a rough interface between perovskite/hole transport material; hence, the enhanced charge transport and PCE was feasible due to the light scattering [56]. The way of depositing precursors using this sequential method plays an important role in order to obtain a good infiltrated perovskite morphology. Ma and co-workers analyzed the sequential deposition of PbI_2 and $PbCl_2$ and then finally dipped into CH_3NH_3I to obtain the mixed hybrid perovskite $CH_3NH_3PbI_{3-x}Cl_x$. It was found that the resultant layer possessed good morphology, and the finally fabricated device was able to produce 11.7% efficiency, which is superior to similar devices fabricated by spin coating [4.8%]. Generally, the efficiency of the vapor phase deposition overcomes the solution phase deposited devices due to high charge carrier mobility [57]. For example, a vapor phase deposited $CH_3NH_3PbI_{3-x}Cl_x$ has shown the mobility value of $33\,cm^{-1}\,V^{-1}\,s^{-1}$, which is much higher than the solution-processed one ($11.6\,cm^{-1}\,V^{-1}\,s^{-1}$) [58]. Another advantage of the vapor phase deposition is the multi-junction devices with large area could be easily achieved. The fabrication of mixed hybrid perovskite solar cell through a dual source vapor phase deposition has led to the efficiency 15.4% [59]. Meso-superstructured perovskite solar cells are possessing superior photo voltage compared with the planar heterojunction. This is because of large density of sub-bandgap states available in planar configuration, whereas a limited number of such states in meso super-structured perovskite solar cells are due to induced doping [60]. Moreover, the carrier mobility of the perovskite-infiltrated mesoporous network dominates the planar structure. Usually, in the solution phase, the trimethyl ammonium iodide is added with lead iodide in solvents like γ-butyrolactone, DMF to obtain the trimethyl ammonium lead iodidie ($CH_3NH_3PbI_3$). These solvents play an important role in perovskites morphology. For example, γ- Valerolactone mediated synthesis results a rod-like morphology of hybrid perovskites [61]. Like organic solar cells, the organic electron acceptor molecule PCBM is utilized in perovskite solar cells for the planar heterojunction solar cells. The active layers of $CH_3NH_3PbI_3$/C60 (30 nm) and $CH_3NH_3PbI_3$/PCBM (25 nm) in planer heterojunction type have produced 1.6% and 2.4% efficiency. Sun et al. used two-step vapor-assisted solution process to fabricate perovskite solar cell with thick layer of $CH_3NH_3PbI_3$ and reached 12.1% efficiency with nearly 100% of IPCE [36]. Other than FTO, the chemically modified ITO electrode also has generated promising results. Cesium carbonate (Cs_2CO_3) modified the ITO surface with a perovskite active layer and produced 15.1% efficiency ($V_{oc} = 1.07\,V$, $J_{sc} = 19.9\,mA/cm^2$, $FF = 0.71$) [62]. Using yitria-doped titania (Y-TiO_2) as the electron transport layer and PEIE-modified ITO as the electrode, 19.3% efficiency was demonstrated through DMF as the solvent [63]. In advance of this, an insitu reaction of PbI_2 and

CH_3NH_3I vapor has resulted a thermodynamically stable $CH_3NH_3PbI_3$ film with good morphology through a low temperature VSAP approach [64]. Through this method, the efficiency achieved for the best cell was 12.1% with $J_{sc} = 19.8\ mA/cm^2$, $V_{oc} = 0.924\ V$, fill factor (FF) = 66.3%, and EQE ~80%. Besides the one-step solution growth method of $CH_3NH_3PbI_3$, the addition of CH_3NH_3Cl into traditional precursors has also resulted good coverage of the perovskite layer [65]. New approaches, such as sequential vapor deposition method (SVD) and low temperature interdiffusion methods, indicate them as good methods to produce excellent morphology of the perovskite layer [66, 67]. Regardless of the method of deposition, one- and two-step deposition of hybrid perovskites has formed almost similar stoichiometric and electronic structure confirmed through photoelectron spectroscopy [68]. The morphology of the perovskites also depends on the concentration of the precursors [69]. To achieve high efficiency in perovskite solar cells, homogenous morphology coverage is an important parameter. Uniform morphology with good perovskite coverage can be obtained by tuning the atmosphere, annealing temperature, film thickness, etc. [70]. To achieve excellent morphology, annealing plays a very important role in obtaining the good crystalline perovskite domains on the mesoporous TiO_2 layer. Annealing perovskite active layer beyond certain limit normally lead to discrete islands like morphology instead of a continuous one.

Though decomposition of methyl ammonium lead halide starts only at >300°C [21], optimized annealing temperature is required to evaporate the solvent, which is used to dissolve the precursors. Generally, the annealing process enhances the large-size formation of grains; hence, the reduction of charge recombination at the interface is largely minimized [71]. Dualeh et al. have investigated the effect of annealing in efficiency of perovskite solar cells. They found that the highest efficiency ($\eta = 11.6\%$ $V_{oc} = 938\ mV$, $J_{sc} = 18.37\ mA\ cm^{-2}$) is reached when the perovskite ($CH_3NH_3PbI_3$) coated layer on TiO_2 mesoporous is annealed at 100 °C [72]. At the optimized temperature and post annealing time, the PbI_2 compound formed from $CH_3NH_3PbI_3$ passivate from recombination at the grain boundaries and improves the device performance [73]. This "self-induced passivation" of $CH_3NH_3PbI_3$ by PbI_2 has to be further explored for halide-mixed systems. Similarly, for the chloride-mixed lead perovskite, $CH_3NH_3PbI_{3-x}Cl_x$, the optimized annealing temperature for high efficiency is around 110°C [74]. The morphological effect of annealing the perovskite layer at different temperatures is given in Figure 17.6. When the temperature exceeds a limit, it deteriorates the perovskite composition and results in formation of PbI_2. Moreover, when the temperature is below 165 K, it reduces the mobility of charge carriers due to the phase transition from tetragonal to orthorhombic (γ-phase) [75]. Hence, systematic annealing of the perovskite layer's deposition is the driving force in obtaining high efficiency. Saliba et al. examined the influence of different type of annealing methods on perovskite films [76]. It is observed that the time dependent-temperature annealing analysis of the hybrid perovskite films provide nanosized domains, whereas the flash annealing method results in micron-sized crystal domains. Moreover, the flash annealed perovskite films result longer PL lifetimes and high steady state PL intensity. In addition, with annealing, solvents and solvent evaporation rate also play an important role in producing perovskite films with good morphology and

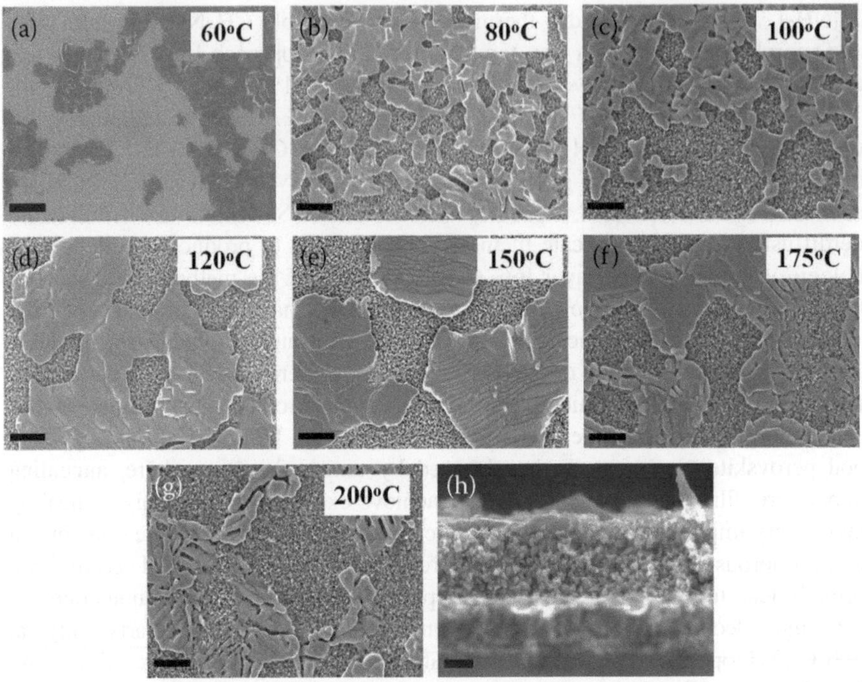

FIGURE 17.6 SEM images of perovskite active layer at different annealing temperatures (a–g) and the cross-section image of sample annealed at 150°C (h).

Source: Reprinted with permission from Dualeh et al. [72]. Copyright@2014 Wiley and Sons.

efficiency [77, 78]. Compared with thermal annealing, the solvent (e.g. DMF) vapor-based annealing is found to be efficient to induce the large grain size with no photocurrent hysteresis [79]. The usage of solvents like dichloromethane (DCM) has been found to produce more-uniform, good morphology perovskite films, which resulted in improved efficiency [80]. Additives, such as 1,8-diiodooctane (DIO), have played a vital role in increasing crystallization and reaching the smooth surface of hybrid perovskites [81]. The theoretical model predicted by Burlakov et al. explains elegantly about the surface coverage and effect of annealing of the perovskite layer [82]. Solvents are predominantly helping to achieve the smooth surface of the perovskite active layer. The mixture of N, N'-dimethylformamide and γ-butyrolactone gives a very smoothy layer with improved efficiency in planer heterojunction solar cells compared with the individual [83]. Because of the reduced interfacial area at the heterojunction and high charge carrier mobility (over $20\,cm^{-2}\,V^{-1}s^{-1}$), the planar heterojunction approach has been highly researched in recent years. Even though the bulk heterojunction type of configuration is followed in the planar structure of perovskite solar cells, the method of functioning of active layer is different. The mixed perovskite system, such as addition of other organic metal cation and alloying of other halogens with iodine in the native $MPbX_3$

structure, would lead to drastic changes in their optical properties and device efficiency. Also, the organic cation replacement leads to considerable expansion/ shrinkage of volume of the cell, depending on their size. Most of the high-efficiency perovskite solar cells reported so far belong to the mixed lead halide system only. The reason for this behavior cannot be ascertained as such. Meanwhile, analysis by Even et al. describes that the exciton screening phenomena, which generates the free carriers in the mixed halide perovskite system, is absent in the pure one ($CH_3NH_3PbX_3$) [84]. This effect also enhances the absorption of mixed systems in comparison with pure. The addition of chloride and bromide ions with the hybrid lead halide perovskite forms the reduction of bandgap and thus increases the absorption, which results in high V_{oc}. For example, the addition of bromide $CH_3NH_3PbI_2Br$ causes an increase of conduction band edge energies that are higher than the parent compound $CH_3NH_3PbI_3$ [85]. It has been found that lack of iodine (I) in hybrid perovskite film formation is also a reason for high photovoltaic performance [86]. Not only halide ions, but the organic cation replacement in $CH_3NH_3PbX_3$ is also drastically changed by the structural and optical properties of hybrid perovskites. When the trimethyl ammonium ion is replaced by formamidimium ion ($HC(NH_2)_2^+$) (space group = P3 *m*I), which has smaller radii than tri-methyl ammonium ion, the bandgap is decreased (about 1.43 eV) in addition to change in symmetry. Improved photovoltaic performance is observed when trimethyl ammonium ion is replaced by the formamidimium ion [87]. This "nearly cubic" structure of formamidimium ion ($FAPbI_3$) exhibits two polymorphs in nature, a trigonal black perovskite and a hexagonal yellow non-perovskite. The initial efforts on replacing trimethyl ammonium ion by formamidimium ion in solid state perovskite solar cells resulted 4.3% efficiency with low EQE [88]. In this case, for the device structure FTO/TiO_2/$FAPbI_3$(spin-coat)/P3HT/Au, 3.7% efficiency is achieved, which is approximately equal with the device with $MAPbI_{3-x}Cl_x$ (3.8%) as active layer. However, a similar kind of structure with $FAPbI_3$ formed by the in-situ dipping reaction of FTO/TiO_2/PbI_2 into FAI resulted in efficiency of 7.5% [89]. Pellet and his co-workers succeeded in enhancing the efficiency of perovskite solar cell by putting the mixed system of methyl ammonium and formamidinium ion as the perovskite active layer through the sequential deposition method [87]. In their study, for the mixed system of $MA_{0.6}FA_{0.4}PbI_3$ composition, the efficiency increased impressively and reached 14.9% ($V_{oc} = 1.003$, $J_{sc} = 21.2\ mA\ cm^{-2}$). This efficiency is higher than the solar cell achieved through respective individual compositions (i.e. $MAPbI_3$ and $FAPbI_3$). All these efforts have paved a new way in the direction of deposition of perovskite active layers. When inorganic elements like cesium are doped with mixed hybrid halide perovskites (e.g. $Cs_xCH_{3-x}NH_3PbI_3$), considerable improvement in efficiency is observed with the planar heterojunction architecture. For 10% doping of cesium, 7.68% efficiency is achieved (η=5.51% for the undoped one) for the device structure FTO/PEDOT:PSS/$Cs_xCH_{3-x}NH_3PbI_3$/ PCBM/Al [90]. The remarkable enhancement in short circuit current density from 8.55 to 10.22 mA cm^{-2} after doping reveals the potential use of dopants in hybrid perovskite compounds. Furthermore, DFT study reveals that the iodine (p-type) doping could be achieved by doping elements like Na, K, Rb, Cu and O in iodine-rich growth conditions, whereas the non-equilibrium conditions promote n-type perovskite formation [91]. The sandwich-type $ITO/TiO_2/CH_3NH_3PbI_2Cl/P3HT/Au$ structure has resulted in the highest short circuit current density ($J_{sc} = 21.3\ mA\ cm^{-2}$ with $\eta = 10.8\%$). The reason for this high J_{sc} may be due to the n-type behavior of perovskites [92]. Eperon et al. investigated through the planar heterojunction approach that the complete replacement of trimethyl ammonium ion with formamidimium ions in perovskites results in 14.2% efficiency, one of the highest in planar heterojunction architecture [93]. Besides, the study has also explored the replacement of iodide with bromide in organolead perovskites of $FAPbI_yBr_{3-y}$. The increase of bromide content in the lead iodide causes an increase in the V_{oc}. Moreover, the $MA)_x(FA)_{1-x}PbI_3$ (x = 0–1) 3-D perovskites have showed red-shifted absorption edge in the spectrum and also enhanced the lifetime of the fabricated solar cell. Using one-dimensional TiO_2 nanowire array, 4.87% efficiency is observed through $CH_3NH_3PbI_2Br$, which is higher than the device fabricated using $CH_3NH_3PbI_3$ (4.29%). The bromide content in $CH_3NH_3PbI_nBr_{3-n}$ at any percentage level could act as the best hole conductor with change in lattice parameter values and improved efficiency (from 7.2% to 8.54%) [94]. Moreover, the addition of bromide ions modifies the crystal structure from tetragonal to cubic. The high photovoltage and V_{oc} of $MPbBr_3$ also depend on the selection of suitable hole transport material. Experiments using a $MAPbBr_3$-layered device displayed $V_{oc} = 1.4\ V$, FF = 79% with a remarkable efficiency 6.7%. This effect is due to the deeper HOMO level of HTM near the valence band edge of $MAPbBr_3$ [95]. The bromide addition with trimethyl lead chloride further shifts the absorption spectrum towards blue region [96, 97]. The influence of bromide ions in absorption of the active layer and also in the bandgap is represented in Figure 17.7.

The incorporation of a lower concentration of chloride in the place of iodide (i.e. $MAPbI_{3-x}Cl_x$) has been found to improve the charge transport in perovskite solar cells [99]. The specific reason for this improved charge transport due to the inclusion of chloride ion is under investigation. However, DFT analysis shows that the inclusion of chlorine mainly presents at close with the TiO_2/perovskite interface [100]. The experimental investigations on hybrid perovskites explore that the presence of chlorine in perovskite strongly affects the crystallization dynamics through preferred orientation of the crystalline grains [101,102]. Through optical analysis, the exciton binding energy of the $CH_3NH_3PbI_{3-x}Cl_x$ is found to be about 55 ± 20 meV, which clearly ensured the presence of free charge carriers like typical inorganic semiconductors [103]. Moreover, the chloride-mixed perovskite system $CH_3NH_3PbI_{3-x}Cl_x$ has been found to have an order of magnitude lower bi-molecular recombination rate than $CH_3NH_3PbI_3$, which can be described as a plausible reason for its larger diffusion length (>1 μm) in order to apply for the planer heterojunction solar cells effectively [104]. This large diffusion length (L_D) has been confirmed through the PL quenching experiments [105, 106] and also through the impedance spectroscopy analysis [107]. Moreover, it is commonly observed that the iodide chloride-mixed perovskites are more stable in air than the pure lead iodide. In examining the origin of this change in properties, it has been found that the insertion of chloride with iodide reduces the lattice constant of iodide. It also reduces the formation of interstitial defects and density of trapping levels of iodide ions [108]. Interestingly, functioning of

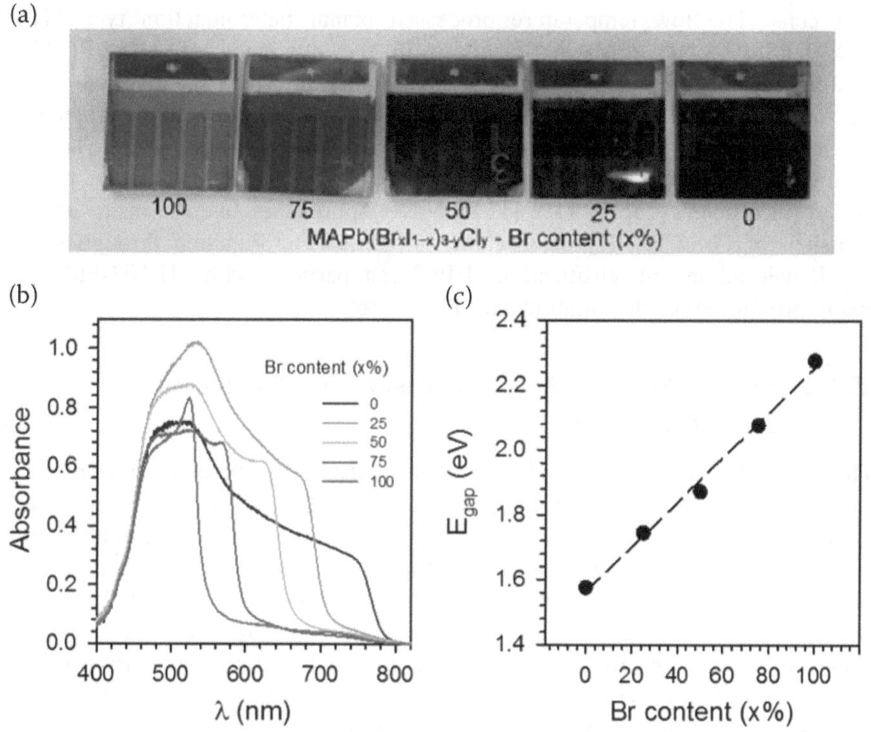

FIGURE 17.7 (a) Picture of $MAPb(Br_xI_{1-x})_{3-y}Cl_y$, ($0 \leq x \leq 1$) devices with different Br/I molar ratios (b) absorption spectra of the prepared samples with different bromide content and c) energy bandgap plot of the samples.

Source: Reprinted with permission from Suarez et al. [98]. Copyright@2014 American Chemical Society.

$CH_3NH_3I_{3-x}Cl_x$ in DMF as a liquid electrolyte in dye-sensitized solar cells has triggered more intriguing research in this compound [109]. Here, the complex ions of $[PbX_6]^{4-}$ in DMF provide the stability in the device, which withstands the efficiency of 8.19% up to 100 hrs. When methyl ammonium ions are replaced by formamidimium ions, the structural changes occur from tetragonal to quasi cubic structure with variable electronic properties. This owes to the spin-orbit coupling and octahedra tilting behavior, which induces this change [110]. In fact, the spin-orbit coupling effect has a very strong influence on the conduction band levels of the perovskites [84]. Thus, the planar hetero junction type of perovskite solar cells emerged as the promising approach with high efficiency. Interestingly, the effect of "neutral-colored semi transparency" in micro-structured planar heterojunction perovskite solar cells are studied through partially de-wetting the solution-casted films [111]. Adaptation of materials and concepts from dye-sensitized and organic solar cells has shown fascinating results in hybrid perovskite solar cells. Due to the favorable energy levels, the acceptors used in organic solar cells like PCBM, PC61BM are also applied for the perovskite solar cells. The low temperature processed planar heterojunction type ITO/PEDOT:PSS/$CH_3NH_3PbI_{3-x}Cl_x$, resulting in the efficiency of 11.5% [112]. Docampo et al. demonstrated a planar heterojunction with the "inverted" type structure FTO/PEDOT:PSS/$CH_3NH_3PbI_{3-x}Cl_x$/PCBM/TiO_x/Al with the efficiency of nearly 10% [113]. The formation of the perovskite layer through the vacuum sublimation method in between two organic electron and hole transport layers resulted in efficiency up to 15% [114]. Novel approaches like the insitu grown core-shell of PbS/$CH_3NH_3PbI_3$ organic-inorganic heterojunction through spin-SILAR method and co-sensitization of PbS nanoparticles with $CH_3NH_3PbI_3$ are also improving efficiency significantly [115, 116].

17.3.2 Role of Alumina (Al_2O_3) in Meso Super Structured Solar Cells (MSSCs)

When the TiO_2 layer is replaced by alumina (Al_2O_3) for the electron transport, enhancement in the efficiency is observed. This Al_2O_3 is also found to be an efficient encapsulant to protect the perovskite layer from degradation by moisture [117]. Further, the typical recombination effect of electrons, which are ejected from the perovskites into the electron transport layer, is avoided when TiO_2 is replaced by Al_2O_3. The transient studies also reveal the absence of signal, which ensures the absence of the charge transfer from perovskite to alumina. Simply, the wide bandgap ($E_g = 7$ to 9 eV) of insulating Al_2O_3 acts as the "scaffold" for the perovskites. The mode of charge transport through Al_2O_3 in perovskite solar cells is represented in Figure 17.8. Such solar cells with alumina instead of TiO_2 are generally called meso super structured solar cells (MSSCs). The first attempt to make alumina-based MSSCs was achieved by Lee et al. with 10.9% efficiency [118]. Furthermore, the transient studies also support the faster charge collection rate of alumina over TiO_2. These phenomena can also be correlated with fewer numbers of trap sites in alumina compared with TiO_2. This observation put the platform for the consequent analysis in

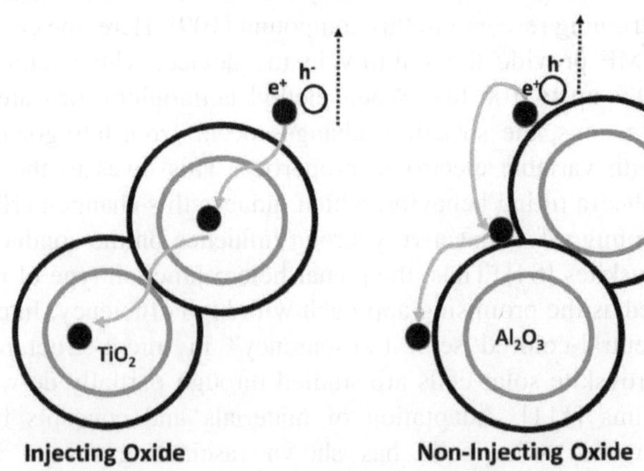

FIGURE 17.8 Charge transport through TiO_2 and Al_2O_3. *Source*: Adopted from Le et al. [118].

MSSCs. For low temperature (less than 150°C), processed meso structured solar cells have the geometry of FTO/Al_2O_3 (400 nm)/$CH_3NH_3PbI_{3-x}Cl_x$/Spiro-OMeTAD/Electrode; about 12.3% efficiency was achieved [119]. Similarly, the co-deposited alumina-perovskite layer through spin coating at low temperature (less than 110°C) gave the efficiency of about 7.16% (V_{oc} = 0.925, J_{sc} (mA cm^{-2}) = 12.78, FF = 0.61) [120]. The Al_2O_3 deposited through the atomic layer deposition (ALD) method to fabricate the metal-insulator-semiconductor structure enhanced the efficiency of the solar cell from 3.30% to 5.07% [121]. Here, it was found that the introduction of the alumina layer reduced the e-injection from perovskite to electrode, which decreased the depleted width of the perovskite in the back contact region. This effect enhanced the carrier transport as well as efficiency. Interestingly, the insertion of Au@SiO_2 core-shell nanostructures on the alumina was found to increase the short-circuit photocurrent in the efficiency of the perovskite solar cell [122]. The growth of perovskite by the insertion of amino acid on alumina's surface also affected the resultant efficiency [123]. These results clearly indicate the role of additives, which influences the carrier transport in alumina layer. Still, this meso structured solar cells shows their potential over TiO_2-based solar cells. Future developments in this area will solve many setbacks to attain high efficiency.

17.4 ORGANOMETAL HALIDE PEROVSKITE IN FLEXIBLE SUBSTRATES

The specific interest on fabrication of the perovskite solar cell has turned out to make a flexible substrate like poly-ethylene-terephthalate (PET) because of its low temperature processing, high resistance against water and oxygen, and high mechanical properties [124]. Flexible PET substrates have shown successive outcomes for the organic electronic devices. In this aspect, the low temperature processed planar heterojunction type of flexible perovskite solar cells have attracted considerable interest. The first report in this category was demonstrated by Docampo et al., with the device structure PET/ITO/$CH_3NH_3Pb_{1-x}X_{3-x}$, and the authors were able to produce 6.4% efficiency [125]. Consequently, a sandwich model of PET/AZO/Ag/AZO/PEDOT:PSS/polyTPD/$CH_3NH_3PbI_3$/PCBM/Au produced 7% efficiency [126]. The substrate used was sequential layers of a polymeric conductor (PET), aluminium-doped zinc oxide (AZO), silver and AZO. You et al. reported a perovskite solar cell with structure PET/ITO/PEDOT:PSS/$CH_3NH_3PbI_{3-x}Cl_x$/PCBM/Al, which resulted in 9.2% (V_{oc} = 0.86 V, J_{sc} = 16.5 mA/cm^2, FF = 64%) via a low temperature process (<120°C) [127]. Through a similar kind of structure with $CH_3NH_3PbI_3$ as an active layer and BCP (bathocuproine) as the hole transporting layer, 4.54% efficiency was reported with moderate open-circuit voltage (V_{oc} = 0.92 V) [128]. It was concluded that high work function of PET/ITO substrate could reduce the energy loss due to the hole transfer. Carbon nanotubes (CNTs) deposited on the perovskite active layer showed their potential to function as a flexible transparent perovskite solar cell. In spite of being a good conductor, a CNT electrode-based sequentially deposited perovskite solar cell achieved about 6.7% efficiency [129]. Flexible solar cells based on a ZnO layer have also been analyzed with considerable interest. As an alternative to TiO_2, ZnO nanorods coated with a PET substrate through a traditional chemical bath deposition (CBD) method resulted in 2.62% efficiency [130]. These results indicate the promising avenues of perovskite solar cells for roll-to-roll production. The rapid progress in this flexible approach would increase the current efficiency level in the near future.

17.5 THE EMERGENCE OF LEAD-FREE HYBRID PEROVSKITE SOLAR CELLS

The organometal lead halides provide outstanding efficiency in solar cells, but the toxic element Pb^{2+} is hazardous to health as well as polluting to the atmosphere. Hence, there are many attempts to replace Pb^{2+} with suitable elements that could solve these problems. Tin (Sn), the element that is located in the same group of Pb^{2+}, possesses similar kinds of electronic and structural properties and has been identified as a suitable replacement. The spin-orbit splitting of conduction band of the tin halide materials have found to be about three times smaller than the lead halide compounds [131]. Tin-based distorted structure of perovskites like p-type $CsSnCl_3$ is studied elaborately in the place of liquid electrolyte as hole conductor for solid state DSSCs [132–134]. The smaller bandgap of $CsSnI_3$ (1.3 eV) is comparable to hybrid lead perovskite $CH_3NH_3PbI_3$ (1.5 eV), which could produce high photocurrent (22 mA/cm^2) with very low V_{oc} (0.24 V) [135]. Replacing cesium with methyl ammonium or formamidimium ion ($CH_3NH_3^+$, $NH_2CHNH_2^+$) could increase the bandgap and lattice constant further [136]. The structural, electrical and transport properties of organic-inorganic hybrid tin perovskites (e.g. CH_3NH_3 $SnCl_3$) were thoroughly examined by Mitzi et al. in past decades [137]. The electronic properties of these hybrid perovskites mainly depend upon their inorganic cage and the displacement of inorganic cation. Distortion and tilting of octahedral network strongly influences their bandgap [138]. The high Hall mobility value of holes (200 cm^{-2}V^{-1}S^{-1}) in hybrid tin perovskites, which is closer to the value of silicon (500 cm^2 V^{-1} s^{-1}), has forced researchers to explore them for solar cell devices. Furthermore, the conductivity of tin perovskites could also be enhanced through artificial hole doping [139]. It is observed that low bandgap $CH_3NH_3SnI_3$ (E_g = 1.3 eV) hybrid perovskites deliver high J_{sc} values and also function as a better electron transporter compared with $CH_3NH_3PbI_3$ (E_g = 1.55 eV) [140]. The first report on the fabrication and analysis of hybrid tin perovskite solar cells is concurrently examined by two groups [141, 142]. Kanatzidis et al. have fabricated the solar cells with spiro compound as hole transporter, which showed efficiency for $CH_3NH_3SnI_3$ as 5.23% and for $CH_3NH_3SnI_{3-x}Br_x$ as 5.73%. Similar successful attempt by Snaith and his colleagues has demonstrated 6.4% efficiency (V_{oc} = 0.88 V, J_{sc} = 16.8 mA cm^{-2}, FF = 0.42) for the device fabricated under an inert atmosphere. Furthermore, 4.18% efficiency is achieved for the alloyed Sn/Pb in $CH_3NH_3Sn_{0.5}Pb_{0.5}I_3$ at 1060 nm (NIR region), with large IPCE in FTO/TiO_2/m-TiO_2/$CH_3NH_3Sn_xPb_{(1-x)}I_3$/rr-P3HT structure [143] (Figure 17.9a, b). These results may be due to the shallow level of valence band of mixed perovskites than spiro-OMeTAD and P3HT. Tuning pin-hole free path of morphology and altering the diffusion length of tin perovskites (~30 nm against 1 μm of lead perovskites) by

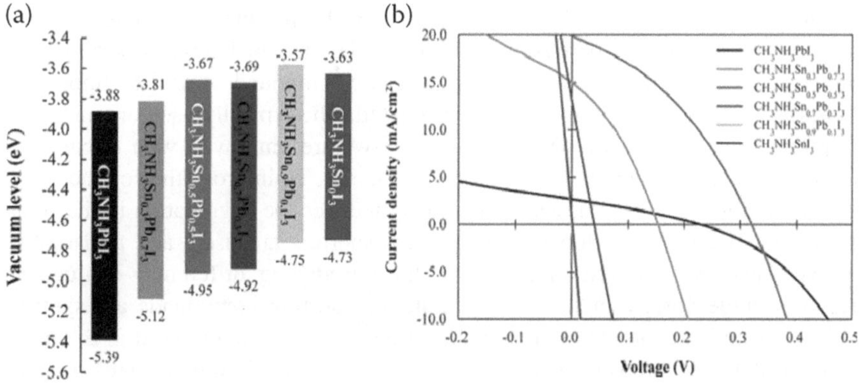

FIGURE 17.9 (a) Energy diagram for $CH_3NH_3Sn_xPb_{(1-x)}I_3$ perovskite and (b) I–V curves for $CH_3NH_3Sn_xPb_{(1-x)}I_3$ perovskite solar cells.

Source: Reprinted with permission from Ogomi et al. [143]. Copyright@2014 The American Chemical Society.

minimizing the hole concentration would improve further the efficiency of tin-based perovskite solar cells in the near future.

17.6 RECENT DEVELOPMENTS IN HYBRID PEROVSKITE SOLAR CELLS

The past few years have evidently seen that performance of perovskite solar cells has greatly progressed through massive development in materials, methods, electrodes and device configuration. In the case of materials, use of mixed cations of hybrid perovskites improves the efficiency over 20%. In this case, usually a triple cation system (e.g. $Cs_xMA_yFA_{1\text{-}x\text{-}y}PbI_3$) is examined with different random compositions, and tunable efficiency performance is achieved [144–146]. The structural and optical properties of these triple cation systems are intensively analyzed to improve performance. Achieving smooth and defect-free perovskite film is important for the efficient charge transport of carriers. The solvent chemistry used for the fabrication of perovskite films is well organized in the recent years, and defect and pin-hole free films are achieved through suitable solvents with carefully tuned parameters. In terms of passivation, since halide perovskite surface is quite vigorous with the additives, several potential compounds are efficient in improving the carrier transport. Upscaling perovskite solar cells has also reached to significant milestone, and there are several potential approaches, such as screen printing, ink jet printing, slot-die coating and blade coating, that are performed to fabricate highly efficient perovskite modules on glass and flexible substrates [147–150]. Recently, a Polish company Saule Technologies has launched the industrial production of perovskite solar cells, and this shows that many innovative commercial products may emerge in the near future based on halide perovskite compounds.

There is considerable development achieved in finding efficient hole-transporting materials for perovskite solar cells. The use of hole-transporting materials is analyzed with new organic and inorganic compounds, and their important role in the efficiency improvement is investigated. The typically used organic-hole transporting materials Spiro-oMeTAD and PTAA are employed with dopants in order to improve their conductivity and hole-mobility. To improve the conductivity, hydrophilic dopants are used, which obviously induces the degradation in the solar cell. To avoid this issue, dopant-free hole-transporters are used, and remarkable improvements are achieved in this area. By introducing different π-conjugated systems, a wide variety of dopant-free hole-transporting compounds are synthesized. A collective example of such dopant-free hole-transporters used in the p-i-n type perovskite solar cells is presented in Figure 17.10 [151]. Similar to the organic hole-transporting materials, there is considerable progress also achieved in developing potential inorganic hole-transporting materials.

Tandem perovskite solar cells have achieved a new dimension in the recent years. Through the tandem approach, over 25% efficiency is achieved by few research groups. Combining a wide-bandgap semiconductor with a narrow-bandgap semiconductor is indeed useful for several applications, including photocatalytic

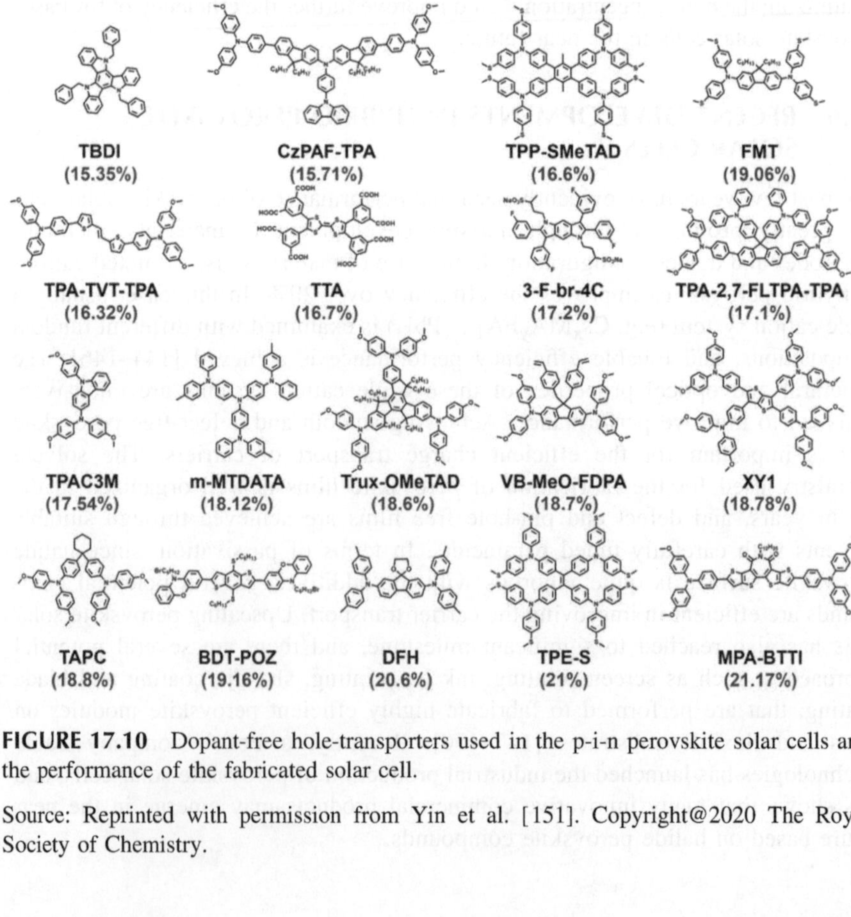

FIGURE 17.10 Dopant-free hole-transporters used in the p-i-n perovskite solar cells and the performance of the fabricated solar cell.

Source: Reprinted with permission from Yin et al. [151]. Copyright@2020 The Royal Society of Chemistry.

water splitting. In tandem solar cells, the perovskite solar cell serves as the top cell, and the silicon solar cell serves as the bottom cell. Specifically, silicon/perovskite tandem solar cells are getting interesting due to their high photon conversion ability. It has been proved that a perovskite-perovskite tandem perovskite solar cell can reach 80% of the theoretical photovoltage limit [152]. Also, such two-terminal perovskite solar cells have the capability of producing high V_{oc}, for example, up to 1.85 eV [153]. Since tandem solar cells are based on multiple layers, morphology of each layer imparts critical output in delivering final efficiency. Furthermore, photon management in perovskite solar cells is an important issue, and even a small texture of the surface influences a lot in the light coupling phenomena [154]. In tandem architecture, formation of large grains in the perovskite layer usually suppresses performance; however, this could be resolved by incorporating suitable additives [155]. Using a four-terminal tandem approach, Yang et al. obtained 28.3% efficiency through an ultra-thin transparent electrode, which is one of the best efficiency records in this category [156]. Other than silicon, tandem perovskite solar cells are fabricated with copper indium diselenide ($CuInSe_2$) or copper indium gallium sulphide (CIGS), and significant results have been achieved in this architecture [157–159]. Impressively, this tandem approach also was successfully demonstrated on flexible substrates, which shows commercializing such things for a wide variety of applications. These flexible perovskite solar cells are predicted to overcome the theoretical limit of an ideal solar cell, which is more beneficial for commercial use. Other than these two, perovskite-perovskite tandem architecture shows promising results in improving the V_{oc} of the resultant device. For example, it shows that the V_{oc} of the $CH_3NH_3PbBr_3$-$CH_3NH_3PbI_3$ tandem solar cell exceeds 2.2 eV, which is quite impressive for further improvement [160]. The comparative graph that represents the current trend of single junction, 2 T and 4 T tandem perovskite solar cells is given in Figure 17.11 [161].

Incorporating dopants with the perovskite active layer is one of the good approaches to improving the performance of perovskite solar cells. There are several kinds of metal ions, such as transition metals, alkali metals, alkaline-earth metals [162, 163] and rare-earth metals [164, 165]. Metal-ion doping is significantly improving the performance of the solar cell and altering the structural, optical and morphological characteristics of the perovskite layer [166]. Besides, the charge carrier dynamics of the metal-ion-doped perovskite layer are quite different from the pristine layer, which is more interesting for further investigation. Among the metals, alkali-metal-doped perovskite shows improved carrier transport due to the reduced interfacial recombination effect. This is due to the fact that doping of Na^+ and K^+ into the perovskite layer improves the crystallinity and crystallite size [167]. Also, there are different kinds of studies reported regarding the doping effect on the TiO_2 hole-blocking layer [168–170]. All these analyses imply that incorporation of foreign metal ions into the perovskite solar cells will result in superior performance in future generations of perovskite solar cells.

Interesting concepts, such as bio-inspired perovskite solar cells [171], recycled perovskite solar cells [172,173], ambient processed perovskite solar cells [174], additive engineered perovskite solar cells [175] and semi-transparent perovskite solar cells [176] appear to be quite promising in fabricating high-performance

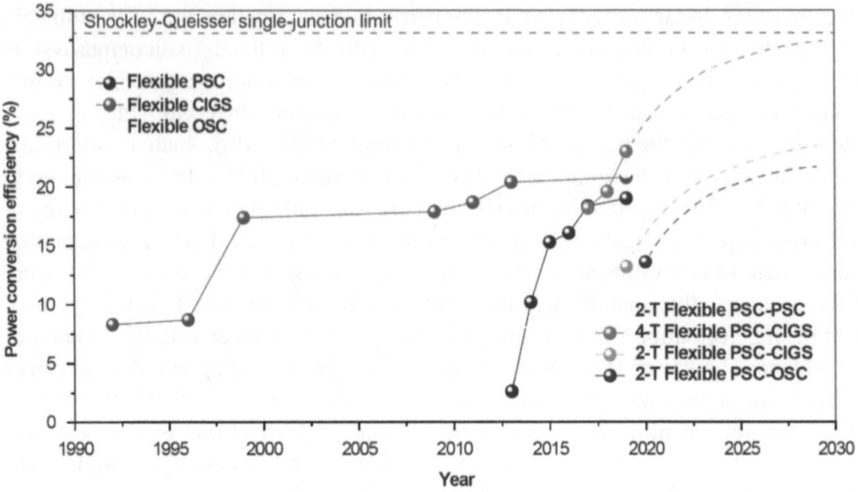

FIGURE 17.11 Efficiency of single-junction and flexible 2 T and 4 T tandem perovskite solar cells with respect to year.

Source: Reprinted with permission from Jiang and Qi [161]. Copyright@2021 The Royal Society of Chemistry.

perovskite solar cells. Specifically, semi-transparent perovskite solar cells are the growing area of research, and progress in the transparent conducting electrode research has developed new dimensions in this area. Intensifying research in all these directions may result in new findings that may be useful for other applications of halide perovskites.

In the case of lead-free perovskite solar cells, progress has been made in the passivation of the active layer to avoid defects-related issues [177]. Considerable research has also focused on the development of mixed Pb-Sn perovskites, and efficiency of such combinations has been measured. In the beginning of the lead-free perovskites, tin (Sn) was considered to be the only alternative element for Pb^{2+}. Later, different elements such as bismuth (Bi), stibium (Sb) and germanium (Ge) were used, and their different compositions with other halides were investigated. Although all these lead-free hybrid perovskites were more stable than Pb^{2+}-based perovskite solar cells, they still had a large number of intrinsic defects. Also, the poor stability of Sn^{2+}-based perovskite solar cells was a major problem in moving forward with this element. However, a mixed system of Sn/Ge appears to be promising for the stable solar cells; perhaps the efficiency is within 10%. These materials, however, can be used for other potential applications, such as photodetectors, light-emitting diodes (LEDs) and thermoelectric devices.

17.7 CONCLUSION

The remarkable achievement of high efficiency using hybrid perovskites has given a strong hope of surpassing the shockley-queisser limit in these solar cells, provided some critical issues in the operation principle and materials handling are resolved. The tandem architecture construction of perovskite solar cells is expected to be a major breakthrough for crossing the present efficiency level. Besides, the exact composition of halides has to be identified for the mixed perovskite halide system. Further, a hybrid of quantum dots and carbon-based materials (like functionalized fullerene) in the interlayer of perovskites also enhances the charge transport to a considerable level. The instability of perovskites under moisture and obtaining good active layer morphology in the large-scale production are the major detrimental exits in the journey of high-efficiency perovskite solar cells. However, recently these issues are somewhat resolved through different solvent and synthesis strategies. It is believed that future results with good surface tuning of hybrid perovskite and its performance would eliminate many obstacles. It is expected that complete unraveling of the mechanism of charge transfer at interface of perovskite solar cells would grab the researchers' attention. The issue of alternative metal electrode contacts to replace the precious elements like silver (Ag) and gold (Au) has to be dealt with severely for the large-scale low-cost production. Semi-transparent perovskite solar cells appear to be a promising approach for building integrated photovoltaic applications. The lead-free hybrid perovskites have shown promising avenues as toxic-free solar converting materials in order to attain efficiency, which is near or equal to lead containing hybrid perovskites. Double perovskites are identified as one of the stable materials for the long-term applications. Perhaps the defects-related issues still exist in this category. The mixed cation and mixed halide perovskite compounds also should further be scrutinized in order to obtain high efficiency. Since the emission spectra of lead perovskites compounds come near-IR region, their application for the luminescent solar concentrator (LSC) are also an interesting direction. With proper encapsulation of the fabricated devices and functionalization of hole transporter to enhance the charge collection, improved lifetime of devices would be a new milestone in this research. The inverted type of solar cell architecture of lead and tin perovskites would also be a new paradigm for the high-efficiency devices in future applications. It is expected that large-scale production of highly stable hybrid perovskite-based solar panels through cost-effective approaches for building integrated applications will come true in the near future.

ACKNOWLEDGMENTS

The authors sincerely thank DRDO (ERIP/EP/201808007/M01/1740) for funding the research work. S. Ananthakumar sincerely thanks Ministry of New and Renewable Energy (MNRE), Govt. of India, for providing fellowship under National Renewable Energy Fellowship (NREF) scheme for his doctoral studies.

REFERENCES

1. Dang, Y., Liu, Y., Sun Y., Yuan D., Liu, X., Lu, W., Liu, G., Xia, H., and Tao, X. "Bulk crystal growth of hybrid perovskite material $CH_3NH_3PbI_3$." *CrystEngComm.* 17 (2015): 665–670.

2. Mitzi, D.B., Field, C.A., Harrison, W.T.A., and Guloy, A.M. "Conducting tin halides with a layered organic-based perovskite structure." *Nature* 369 (1994): 467–469.
3. Kojima, A., Teshima, K., Shirai, Y., and Miyasaka, T. "Organometal halide perovskites as visible-light sensitizers for photovoltaic cells." *J. Am. Chem. Soc.* 131 (2009): 6050–6051.
4. Snaith, H.J. "Perovskites: The Emergence of a new era for low-cost, high efficiency solar cells." *J. Phys. Chem. Lett.* 4 (2013): 3623–3630.
5. Kim H.S., Im S.H., and Park N.G. "Organolead halide perovskites: New horizons in solar cell research." *J. Phys. Chem. C* 118 (2014): 5615–5625.
6. Marchioro, A., Teuscher, J., Friedrich, D., Kunst, M., Krol, R.V., Moehl, T., Gratzel, M., and Moser, J.E. "Unravelling the mechanism of photo induced charge transfer processes in lead iodide perovskite solar cells." *Nature Photonics* 8 (2014): 250–255.
7. Seo, G., Seo, J., Ryu, S., Yin, W., Ahn, T.K., and Seok, S. "Enhancing the performance of sensitized solar cells with PbS/$CH_3NH_3PbI_3$ core/shell quantum dots." *J. Phys. Chem. Lett.* 5 (2014): 2015–2020.
8. Zhao, Y., and Zhu, K. "Optical bleaching of perovskite $(CH_3NH_3)PbI_3$ through room-temperature phase transformation induced by ammonia." *Chem. Comm.* 50 (2014): 1605–1607.
9. Sobus, J., and Zeolek, M. "Optimization of absorption bands of dye-sensitized and perovskite tandem solar cells based on loss-in-potential values." *Phys. Chem. Chem. Phys.* 16 (2014): 14116–14126.
10. Luo, J., Im, J.H., Mayer, M.T., Schreier, M., Nazeeruddin, M.K., Park, N.G., Tilley, S.D., Fan, H.J., and Gratzel, M. "Water photolysis at 12.3% efficiency via perovskite photovoltaics and earth-abundant catalysts." *Science* 26 (2014): 1593–1596.
11. Rauda, E., Senter, R., and Tolbert, S.H. "Directing anisotropic charge transport of layered organic-inorganic hybrid perovskite semiconductors in porous templates." *J. Mater. Chem. C* 1 (2013): 1423–1427.
12. Matteocci, F., Razza, S., Di Giacomo, F., Casaluci, S., Mincuzzi, G., Brown, T.M., D'Epifanio, A., Licoccia, S., and Di Carlo, A. "Solid-state solar modules based on mesoscopic organometal halide perovskite: a route towards the up-scaling process." *Phys. Chem. Chem. Phys.* 16 (2014): 3918–3923.
13. Bretschneider, S.A., Weickert, J., Dorman, J.A., and Schmidt-Mende, L. "Research update: Physical and electrical characteristics of lead halide perovskites for solar cell applications." *APL Mat.* 2 (2014): 040701.
14. Galasso, F.S. *Structure, properties and preparation of perovskite type compounds*, ed. R. Smoluchowski and N. Kurti, Pergamon Press, 41–45, 1969.
15. Cheng, Z., and Lin. J. "Layered organic-inorganic hybrid perovskites: structure, optical properties, film preparation, patterning and templating engineering." *CrystEngComm* 12 (2010): 2646–2662.
16. Umari, P., Mosconi, E., and Angelis, F.D. "Relativistic GW calculations on $CH_3NH_3PbI_3$ and $CH_3NH_3SnI_3$ perovskites for solar cell applications." *Scientific Reports* 4 (2014): 1–7.
17. Mosconi, E., Amat, A., Nazeeruddin, Md.K., Gratzel, M., and De Angelis, F. "First-principle modeling of mixed halide organometal perovskites for photovoltaic applications." *J. Phys. Chem. C* 117 (2013): 13902–13913.
18. Juarez-Perez, E.J., Sanchez, R.S., Badia, L., Garcia-Belmonte, G., Kang, Y.S., Moro-Sero, I., and Bisquert, J. "Photoinduced giant dielectric constant in lead halide." *J. Phys. Chem. Lett.* 5 (2014): 2390–23914.
19. Baikie, T., Fang, Y., Kadro, J.M., Schreyer, M., Mhaisalkar, S.G., Gratzel, M., and White, T.J. "Synthesis and crystal chemistry of the hybrid perovskite $(CH_3NH_3)PbI_3$ for solid-state sensitized solar cell applications." *J Mater Chem A* 1 (2013): 5628–5641.

20. Savenije, T.J., Ponseca, Jr C.S., Kunneman, L., Abdellah, M., Zheng, K., Tian, Y., Zhu, Q., Canton, S.E., Scheblykin, I.G., Pullerits, T., Yarsev, A., and Sundstrom, V. "Thermally activated exciton dissociation and recombination control the carrier dynamics in organometal halide perovskite." *J. Phys. Chem. Let.* 5 (2014): 2189–2194.
21. Stoumpos, C.C., Mallikakas, C.D., and Kanatzidis, M.G. "Semiconducting Tin and lead iodide perovskites with organic cations: Phase transitions, high mobilites, and near-infrared photoluminescent properties." *Inorg. Chem.* 52 (2013): 9019–9038.
22. Giorgi, G., Fujisawa, J.I., Segawa, H., and Yamashita, K. "Small photocarrier effective masses featuring ambipolar transport in methylammonium lead iodide perovskite: A density functional analysis." *J. Phys. Chem. Lett.* 4 (2013): 4213–4216.
23. Sanchez, R.S., Gonzalez-Pedro, V., Lee, J.W., Park, N.G., Kang, Y.S., Mora-Sero, I., and Bisquert, J. "Slow dynamic processes in lead halide perovskite solar cells." *Characteristic times and hysteresis. J. Phys. Chem. Lett.* 5 (2014): 2357–2363.
24. Brivio, F., Walker, A.B., and Walsh, A. "Structural and electronic properties of hybrid perovskites for high-efficiency thin-film photovoltaics from first-principles." *APL Materials* 1 (2013): 042111.
25. Yin, W.J., Yan, Y., and Wei, S.H. "Anomalous alloy properties in mixed halide perovskites." *J. Phys. Chem. Lett.* 5 (2014): 3625–3631.
26. Suarez, B., Gonzalez-Pedro, V., Ripolles, T.S., Sanchez, R.S., Otero, L., and Mora-Sero, I. "Recombination study of combined halides (Cl, Br, I) Perovskite solar cells." *J. Phys. Chem. Lett.* 5 (2014): 1628–1635.
27. Bretschneider, S.A., Weickert, J., Dorman, J.A., and Schmidt-Mende, L. "Research update: Physical and electrical characteristics of lead halide perovskites for solar cell applications." *APL Maeterials.* 2 (2014): 040701.
28. Edri, E., Kirmayer, S., Cahen, D., and Hodes, G. "High open-circuit voltage solar cells based on organic-inorganic lead bromide perovskite." *J. Phys. Chem. Lett.* 4 (2013): 897–902.
29. Dirin, D.N., Dreyfuss, S., Bodnarchuk, M.I., Nedelcu, G., Papagiorgis, P., Itskos, G., and Kovalenko, M.V. "Lead halide perovskites and other metal halide complexes as inorganic capping ligands for colloidal nanocrystals." *J. Am. Chem. Soc.* 18 (2014): 6550–6553.
30. De Wolf, S., Holovsky, J., Moon, S.J., Loper, P., Niesen, B., Ledinsky, M., Haug, F.J., Yum, J.H., and Ballif, C. "Organometallic halide perovskites: Sharp optical absorption edge and its relation to photovoltaic performance." *J. Phys. Chem. Let.* 5 (2014): 1035–1039.
31. Li, M-H., Shen, P-S., Wang, K-C., Guo, T-F., Guo, and Chen. P. "Inorganic p-type contact materials for perovskite-based solar cells." *J. Mater. Chem. A* 3 (2015): 9011–9019.
32. Im, J.H., Lee, C.R., Lee, J.W., Park, S.W., and Park, N.G. "6.5% efficient perovskite quantum-dot-sensitized solar cell." *Nanoscale* 3 (2011): 4088–4093.
33. Im J.H., Chung, J., Kim, S.J., and Park, N.G. "Synthesis, structure and photovoltaic property of a nanocrystalline 2H perovskite type novel sensitizer ($CH_3CH_2NH_3$) PbI_3." *Nanoscale Res. Lett.* 7 (2012): 1–7.
34. Angelis, F.D. "Modeling materials and processes in hybrid/organic photovoltaics: from dye sensitized to perovskite solar cells." *Acc. Chem. Res.* 47 (2014): 3349–3360.
35. Hodes, G., "Perovskite-based solar cells." *Science* 342 (2013): 317–318.
36. Sun, S., Salim, T., Mathews, N., Duchamp, M., Boothroyd, C., Xing, G., Sum, T.C., and Lam, Y.M. "The origin of high efficiency in low-temperature solution-processable bilayer organometal halide hybrid solar cells." *Energy Environ. Sci.* 7 (2014): 399–407.

37. Forst, J.M., Butler, K.T., Brivio, F., Hendon, C.H., Schilfgaarde, M.V., and Wash. A. "Atomisitc origins of high-performance in hybrid halide perovskite solar cells." *Nano Lett.* 14 (2014): 2584–2590.
38. Lewis, D.J., and O'Brien, P. "Ambient pressure aerosol-assisted chemical vapour deposition of $(CH_3NH_3)PbBr_3$ an inorganic-organic perovskite important in photovoltaics." *Chem. Comm.* 50 (2014): 6319–6321.
39. Leydan, M.R., Ono, L.K., Raga, S.R., Kato, Y., Wang, S., and Qi, Y. "High performance perovskite solar cells by hybrid chemical vapor deposition." *J. Mater. Chem. A* 2 (2014): 18742–18745.
40. Kim, H. S., Lee, C. R., Im, J. H., Moehl, T., Marchioro, A., Moon, S. J., Baker, R. H., Im, J. H., Lee, K. B., Moehl, T., Marchioro, A., Moon, S. J., Baker, R. H., Yum, J. H., Moser, J. E., Gratzel, M., and Park, N. G. "Lead iodide perovskite sensitized all solid state submicron thin film mesoscopic solar cell with efficiency exceeding 9%." *Sci. Rep.* 2 (2012): 591(1-7).
41. Green, M. A., Baillie, A. H., and Snaith, H. J. "The emergence of perovskite soar cells." *Nature Photonics* 8 (2014): 506–514.
42. Deschler, F., Price, M., Pathak, S., Klintberg, L. E., Jarausch, D. D., Ruben Higler, Huttner, S., Leijetens, T., Stranks, S. D., Snaith, H. J., Atature, M., Philips, R. T., and Friend, R. H. "High photoluminescence efficiency and optically pumped lasing in solution processed mixed halide perovskite semiconductors." *J. Phys. Chem. Lett.* 5 (2014): 1421–1426.
43. Xu, M., Rong, Y., Ku, Z., Mei, A., Liu, T., Zhang, L., Li, X., and Han, H. "Highly ordered mesoporous carbon for mesoscopic $CH_3NH_3PbI_3/TiO_2$ heterojunction solar cell." *J. Mater. Chem. A* 2 (2014): 8607–8611.
44. Li, Z., Kulkarni, S. A., Boix, P. P., Shi, E., Cao, A., Fu, K., Batabyal, S. K., Zhang, J., Xiong, Q., Wong, L. H., Mathews, N., and Mhaisalkar, S. G. "Laminated carbon nanotubes networks for metal electrode-free efficient perovskite solar cells." *ACS Nano.* 8 (2014): 6797–6804.
45. Leijtens, T., Lauber, B., Eperon, G.E., Stranks, S. D., and Snaith, H.J. "The importance of perovskite pore filling in organometal mixed halide sensitized TiO_2 based solar cells." *J. Phys. Chem. Lett.* 5 (2014): 1096–1102.
46. Du, M.H. "Efficient carrier transport in halide perovskites: Theoretical perspectives." *J. Mater. Chem. A.* 2 (2014): 9091–9098.
47. Kim, J., Lee, S. H., Lee, J.H., and Hong, K.H. "The role of intrinsic defects in methylammonium lead iodide perovskite." *J. Phys. Chem. Lett.* 5 (2014): 1312–1317.
48. Agiorgousis, M.L., Sun, Y.Y., Zeng, H., and Zhang. S. "Strong covalency-induced recombination centres in perovskite solar cell material $CH_3NH_3PbI_3$." *J. Am. Chem. Soc.* 136 (2014): 14570–14575.
49. Heo, J. H., Im, S.H., Noh, J.H., Mandal, T.N., Lim, C.S., Chang, J.A., Lee, Y.H., Kim, H.J., Sarkar, A., Nazeeruddin, M.K., Gratzel, M., and Seok. S. II. "Efficient inorganic-organic hybrid heterojunction solar cells containing perovskite compound and polymeric hole conductors." *Nature photonics* 7 (2013): 486–491.
50. Liu, D., Gangishetty, M.K., and Kelly, T.L. "Effect of $CH_3NH_3PbI_3$ thickness on device efficiency in planar heterojunction perovskite solar cells." *J. Mater. Chem. A* 2 (2014): 19873–19881.
51. Schulz, P., Edri, E., Kirmayer, S., Hodes, G., Cahen, D., and Kahn, A. "Interface energetic in organo-metal halide perovskite based photovoltaic cells." *Energy Environ. Sci.* 7 (2014): 1377–1381.
52. Law, C., Miseikis, L., Dimitrov, S., Shakya-Tuladhar, P., Li, X., Barnes, P.R.F., Durrant, J., and O'Regan, B.C. "Performance and stability of lead perovskite/TiO_2, Polymer/PCBM." *Adv. Mater.* 26 (2014): 6268–6273.

53. Seetharaman, S.M., Nagarjuna, P., Kumar, P.N., Singh, S.P., Deepa, M., Namboothiriy, M.A.G. "Efficient organic inorganic hybrid perovskite solar cells processed in air." *Phys. Chem. Chem. Phys.* 16 (2014): 24691–24696.
54. Forst, J. M., Butler, K.T., Brivio, F., Hendon, C.H., Schilfgaarde, M.V., and Walsh. A. "Atomistic origins of high-performance in hybrid halide perovskite solar cells." *Nano. Lett.* 14 (2014): 2584–2590.
55. Burschka, J., Pellet, N., Moon, S.J., Humphry-Baker, R., Gao, P., Nazeeruddin, M.K., and Gratzel, M. "Sequential deposition as a route to high-performance perovskite-sensitized solar cells." *Nature* 499 (2013): 316–319.
56. Zheng, L., Ma, Y., Chu, S., Wang, S., Qu, B., Xiao, L., Chen, Z., Gong, Q., Wu, Z., and Hou, X. "Improved light absorption and charge transport for perovskite solar cells with rough interfaces by sequential deposition." *Nanoscale* 6 (2014): 8171–8176.
57. Ma, Y., Zheng, L., Chung, Y.H., Chu, S., Xiao, L., Chen, Z., Wang, S., Qu, B., Gong, Q., Wu, Z., and Hou, X. "A highly efficient mesoscopic solar cell based on $CH_3NH_3PbI_{3-x}Cl_x$ fabricated via sequential solution deposition." *Chem. Comm.* 50 (2014): 12458–12461
58. Wehrenfennig, C., Eperon, G.E., Jhonston, M.B., Snaith, H.J., and Hertz, L.M. "High charge carrier mobilities and lifetimes in organolead trihalide perovskites." *Adv. Mater.* 26 (2014): 1584–1589
59. Liu, M., Johnston, M.B., and Snaith, H.J. "Efficient planar heterojunction perovskite solar cells by vapour deposition." *Nature* 501 (2013): 395–398.
60. Leijtens, T., Lauber, B., Eperon, G.E., Stranks, S.D., and Snaith, H.J., "The importance of perovskite pore filling in organometal mixed halide sensitized TiO_2 based solar cells." *J. Phy. Chem. Lett* 5 (2014): 1096–1102.
61. Chen, Z., Li, H., Tang, Y., Huang, X., Ho, D., and Lee, C.S. "Shape controlled synthesis of organolead halide perovskite nanocrystals and their tunable optical absorption." *Mater. Res. Express* 1 (2014): 015034.
62. Hu, Q., Wu, J., Jiang, C., Liu, T., Que, X., Zhu, R., and Gong, Q. "Engineering of electron-selective contact for perovskite solar cells with efficiency exceeding 15%." *ACS Nano* 8 (2014): 10161–10167.
63. Zhou, H., Chen, Q., Li, G., Luo, S., Song, T.B., Duan, H.S., Hong, Z., You, J., Liu, Y., and Yang, Y. "Interface engineering of highly efficient perovskite solar cells." *Science* 345 (2014): 542–546.
64. Chen, Q., Zhou, H., Hong, Z., Luo, S., Duan, H.S., Wang, H.H., Li, Y., Li, G., and Yang, Y. "Planar heterojunction perovskite solar cells via vapor-assisted solution process." *J. Am. Chem. Soc.* 136:622–625.
65. Zhao, Y., and Zhu, K. "CH_3NH_3Cl assisted one-step solution growth of $CH_3NH_3PbI_3$: structure, charge-carrier dynamics, and photovoltaic properties of perovskite solar cells." *J. Phy. Chem. C.* 118 (2014): 9412–9418.
66. Hu, H., Wang, D., Zhou, Y., Zhang, J., Lv, S., Pang, S., Chen, X., Liu, Z., Padture, N.P., and Cui, G. "Vapour-based processing of hole-conductor-free $CH_3NH_3PbI_3$ perovskite/C60 fullerene planar solar cells." *RSC Adv.* 4 (2014): 28964–28967.
67. Xiao, Z., Bi, C., Shao, Y., Dong, Q., Wang, Q., Yuan, Y., Wang, C., Gao, Y., and Huang, J. "Efficient, high yield perovskite photovoltaic devices grown by interdiffusion of solution-processed precursor stacking layers." *Energy Environ. Sci.* 7 (2014): 2619–2623.
68. Lindblad, R., Bi, D., Park, B.W., Oscarsson, J., Gorgoi, M., Siegbahn, H., Odelius, M., Johansson, E.M.J., and Rensmo, H. "Electronic structure of TiO_2/$CH_3NH_3PbI_3$ perovskite solar cell interfaces." *J. Phys. Chem. Lett.* 5 (2014): 648–653.

69. Conings, B., Baeten, L., De Dobbelaere, C., D'Haen, J., Manca, J., and Boyen, H.G. "Perovskite based hybrid solar cells exceeding 10% efficiency with high reproducibility using a thin film sandwich approach." *Adv. Mater.* 26 (2014): 2041–2046.
70. Eperon, G.E., Burlakov, V.M., Docampo, P., Goriely, A., and Snaith, H.J. "Morphological control for high performance, solution-processed planar heterojunction perovskite solar cells." *Adv. Fun. Mater.* 24 (2014): 151–157.
71. Bi, C., Shao, Y., Yuan, Y., Xiao, Z., Wang, C., Gao, Y., and Huang, J. "Understanding the formation and evolution of interdiffusion grown organoloead halide perovskite thin films by thermal annealing." *J. Mater. Chem. A* 2 (2014): 18508–18514.
72. Dualeh, Tetreault, N., Moehl, T., Gao, P., Nazeeruddin, M.K., and Gratzel, M. "Effect of annealing temperature on film morphology of organic-inorganic hybrid perovskite solid-state solar cells." *Adv. Fun. Mater.* 24 (2014): 3250–3258.
73. Chen, Q., Zhou, H., Hong, Z., Luo, S., Duan, H.S., Wang, H.H., Liu, Y., Li, G., and Yang, Y. "Planar heterojunction perovskite solar cells via vapor assisted solution process." *J. Am. Chem. Soc.* 136 (2014): 622–625.
74. Chavhan, S., Miguel, O., Grande, H.J., Pdero, V.G., Sanchez, R.S., Barea, E.M., Mora-Sero, I., and Tena-Zaera, R. "Organo-metal halide perovskite based solar cells with CuSCN as the inorganic hole selective contact." *J. Mater. Chem. A* 2 (2014): 12754.
75. Savenije, T.J., Ponseca, C.S., Kunneman, L., Abdellah, M., Zheng, K., Tian, Y., Zhu. Q., Canton, S.E., Scheblykin, I.G., Pullerits, T., Yartsev, A., and Sundstrom, V. "Thermally activated exciton dissociation and recombination control the carrier dynamics in organometal halide perovskite." *J. Phys. Chem. Lett.* 5 (2014): 2189–2294.
76. Saliba, M., Tan, K.W., Sai, H., Moore, D.T., Scott, T., Zhang, W., Estroff, L.A., Wiesner, U., and Snaith, H.J. "Influence of thermal processing protocol upon the crystallization and photovoltaic performance of organic-inorganic lead trihalide perovskites." *J. Phys. Chem. C* 118 (2014): 17171–17177.
77. Jeng, J.Y., Chiang, Y.F., Lee, M.H., Peng, S.R., Guo, T.F., Chen, P., and Wen, T.C. "$CH_3NH_3PbI_3$ perovskite/fullerene planar-heterojunction hybrid solar cells." *Adv. Mater.* 25 (2013): 3727–3732.
78. Kang, R., Kim, J.E., Lee, S., Jeon, Y.J., and Kim, D.Y. "Optimized horganometal halide perovskite planar hybrid solar cells via control of solvent evaporation rate." *J. Phys. Chem. C* 118 (2014): 26513–26520.
79. Xiao, Z., Dong, Q., Bi, C., Shao, Y., Yuan, Y., and Huang, J. "Solvent annealing of perovskite-induced crystal growth for photovoltaic-device efficiency enhancement." *Adv. Mater.* 26 (2014): 6503–6509.
80. Bi, D., El-Zohry, A.M., Hagfeldt, A., and Boschloo, G. "Improved morphology control using a modified two-step method for efficient perovskite solar cells." *ACS Appl. Mater. Interfaces* 6 (2014): 18751–18757.
81. Liang, P.W., Liao, C.Y., Cheuh, C.C., Zuo, F., Williams, S.T., Xin, X.K., Lin, J., and Jen, A.K.Y. "Additive enhanced crystallization of solution-processed perovskite for highly efficient planar-heterojunction solar cells." *Adv. Mater.* 26 (2014): 3748–3754.
82. Burlakov, V.M., Eperon, G.E., Snaith, H.J., Chapman, S.J., and Goriely, A. "Controlling coverage of solution cast materials with unfavourable surface interactions." *Appl. Phys. Lett.* 104 (2014): 091602.
83. Kim, H.B., Choi, H., Jeong, J., Kim, S., Walker, B., Song, S., and Kim, J.Y. "Mixed solvents for the optimization of morphology in solution-processed, inverted-type perovskite/fullerene hybrid solar cells." *Nanoscale* 6 (2014): 6679–6683.

84. Even, J., Pedesseau, L., Jancu, J.M., and Katan, C. "Importance of spin-orbit coupling in hybrid organic/inorganic perovskites for photovoltaic applications." *J. Phys. Chem. Lett.* 4 (2013): 2999–3005.
85. Qiu, J., Qiu, Y., Yan, K., Zhong, M., Mu, C., and Yang, S. "All-solid-state hybrid solar cells based on a new organometal halide perovskite sensitizer and one-dimensional TiO_2 nanowire arrays." *Nanoscale* 5 (2013): 3245–3248.
86. Buin, Pietsch, P., Xu, J., Voznyy, O., Ip, A.H., Comin, R., and Sargent, E.H. "Materials processing routes to trap-free halide perovskites." *Nano Lett* 14 (2014): 6281–6286.
87. Pellet, N., Gao, P., Gregori, G., Yang, T.Y., Nazeeruddin, M.K., Maier, J., and Gratzel, M. "Mixed-organic-cation perovskite photovoltaics for enhanced solar-light harvesting." *Angew. Chem. Int. Ed.* 53 (2014): 3151–3157.
88. Koh, T.M., Fu, K., Fang, Y., Chen, S., Sum, T.C., Mathews, N., Mhaisalkar, S.G., Boix, P.P., and Baikie, T. "Formamidinium-containing metal-halide: An alternative material for near-IR absorption perovskite solar cells." *J. Phys. Chem. C* 118 (2014): 16458–16462.
89. Pang, S., Hu, H., Zhang, J., Lv, S., Yu, Y., Wei, F., Qin, T., Xu, H., Liu, Z., and Cui, G., "$NH_2CH{=}NH_2PbI_3$ An alternative organolead iodide perovskite sensitizer for mesoscopic solar cells." *Chem. Mater.* 26 (2014): 1485–1491.
90. Choi, H., Jeong, J., Kim, H.B., Kim, S., Walker, B., Kim, G.H., and Young, J. "Cesium-doped methylammonium lead iodide perovskite light absorber for hybrid solar cells." *Nano Energy* 7 (2014): 80–85.
91. Shi, T., Yin, W.J., and Yan, Y. "Predictions for p-type $CH_3NH_3PbI_3$ perovskites." *J. Phys. Chem. C.* 118 (2014): 25350–25354.
92. Conings, B., Baeten, L., Dobbelaere, C.D., D'Haen, J., Manca, J., and Boyen, H.G. "Perovskite based hybrid solar cells exceeding 10% efficiency with high reproducibility using a thin film sandwich approach." *Adv. Mater.* 26 (2014): 2041–2046.
93. Eperon, G.E., Stranks, S.D., Menelaou, C., Johnston, M.B., Herz, L.M., and Snaith, H.J., "Formamidinium lead trihalide: a broadly tunable perovskite for efficient planar heterojunction solar cells." *Energy Environ. Sci.* 7 (2014): 982–988.
94. Aharon, S., Cohen, B.E., and Etgar, L. "Hybrid lead halide iodide and lead halide bromide in efficient hole conductor free perovskite solar cell." *J. Phys. Chem. C.* 118 (2014): 17160–17165.
95. Ryu, S., Noh, J.H., Jeon, N.J., Kim, Y.C., Yang, W.S., Seo, J., and Seok, S. II. "Voltage output of efficient perovskite solar cells with high open-circuit voltage and fill factor." *Energy Environ. Sci.* 7 (2014): 2614–2618.
96. Mosconi, E., Amat, A., Nazeeruddin, M.K., and Angelis, F.D. "First-principles modeling of mixed halide organometal perovskites for photovoltaic applications." *J. Phys. Chem. C* 117 (2013): 13902–13913.
97. Kulkarni, S.A., Baikie, T., Boix, P.P., Yantara, N., Mathews, N., and Mhaisalkar. S. "Band-gap tuning of lead halide perovskites using a sequential deposition process." *J. Mater. Chem. A* 2 (2014): 9221–9225.
98. Suarez, B., Pedro, V.G., Ripolles, T.S., Sanchez, R.S., Otero, L., and Sero, I.M. "Recombination study of combined halides (Cl, Br, I) perovskite solar cells." *J. Phys. Chem. Lett.* 5 (2014): 1628–1635.
99. Colella, S., Mosconi, E., Fedeli, P., Listorti, A., Gazza, F., Orlandi, F., Ferro, P., Besagni, T., Rizzo, A., Calestani, G., Gigli, G., Angelis, F.D., and Mosca. R. "$MAPbI_{3-x}Cl_x$: The role of chloride as dopant on the transport and structural properties." *Chem. Mater.* 25 (2013): 4613–4618.
100. Collela, S., Mosconi, E., Pellegrino, G., Alberti, A., Guerra, V.L.P., Masi, S., Listorti, A., Rizzo, A., Condorelli, G.G., Angelis, F.D., and Gigli, G. "Elusive

presence of chlorine in mixed halide perovskite solar cells." *J. Phys. Chem. Lett.* 5 (2014): 3532–3538.

101. Grancini, G., Marras, S., Prato, M., Giannini, C., Quarti, C., Angelis, F.D., Bastiani, M.D., Eperon, G.E., Snaith, H.J., Manna, L., and Petrozza, A. "The impact of the crystallization process on the structural and optical properties of hybrid perovskite films for photovoltaics." *J. Phys. Chem. Lett.* 5 (2014): 3836–3842.
102. Williams, S.T., Zuo, F., Chueh, C.C., Liao, C.Y., Liang, P.W., and Jen, A.K.Y. "Role of chloride in the evolution of organo-lead halide perovskite thin films." *ACS Nano* 8 (2014): 10640–10654.
103. D'Innocenzo, V., Grancini, G., Alcocer, M.J.P., Kandada, A.R.S., Stranks, S.D., Lee, M.M., Lanzani, G., Snaith, H.J., and Petrozza, A. "Exitons versus free charges in organo-lead tri-halide perovskites." *Nature Communications* 5 (2014): 1–6.
104. Wehrenfennig, C., Liu, M., Snaith, H.J., Jhonston, M.B., and Herz, L.M. "Homogeneous emission line broadening in the organo lead halide perovskite $CH_3NH_3PbI_{3-x}Cl_x$." *J. Phys. Chem. Lett.* 5 (2014): 1300–1306.
105. Stranks, S.D., Eperon, G.E., Grancini, G., Menelaou, C., Alcocer, M.J.P., Leijtens, L., Herz, L.M., Petrozza, A., and Snaith, H.J. "Electron-hole diffusion lengths exceeding 1 micrometer in an organometal perovskite absorber." *Science* 342 (2013): 341–344.
106. Xing, G., Mathews, N., Lim, S.S., Yantara, N., Liu, X., Sabba, D., Gratzel, M., Mhaisalkar, S., and Sum, T.C. "Low temperature solution processed wavelength-tunable perovskites for lasing." *Nature Materials* 13 (2014): 476–480.
107. Gonzlaez-Pedro, V., Juarez-Perz, E.J., Arsyad, W.S., Barea, E.M., Febregat-Santiago, F., Sero, I.M., and Bisquert, J. "General working principles of CH_3NH_3 PbX_3 perovskite solar cells." *Nano Lett.* 14 (2014): 888–893.
108. Du, M.H. "Efficient carrier transport in halide perovskite: theoretical perspectives." *J. Mater. Chem. A* 2 (2014): 9091–9098.
109. Wang, Q., Yun, J.H., Zhang, M., Chen, H., Chen, Z.G., and Wang, L. "Insight into the liquid state of organo-lead halide perovskites and their new roles in dye-sensitized solar cells." *J. Mater. Chem. A* 2 (2014): 10355–10358.
110. Amat, Mosconi, E., Ronca, E., Quarti, C., Umari, P., Nazeeruddin, M.K., Gratzel, M., and De Angelis, F. "Cation-induced band-gap tuning in organohalide perovskites." *Nano Lett.* 14 (2014): 3608–3616.
111. Eperon, G.E., Burlakov, V.M., Goriely, A., and Snaith, H.J. "Neutral color semitransparent microstructured perovskite solar cells." *ACS Nano* 8 (2013): 591–598.
112. You, J., Hong, Z., Yang, Y., Chen, Q., Cai, M., Song, T.B., Chen, C.C., Lu, S., Liu, Y., Zhou, H., and Yang, Y. "Low-temperature solution processed perovskite solar cells with high efficiency and flexibility." *ACS Nano* 8 (2014): 1674–1680.
113. Docampo, P., Hanusch, F.C., Stranks, S.D., Doblinger, M., Fecki, J.M., Ehrensperger, M., Minar, N.K., Jhonston, M.B., Snaith, H.J., and Bein. T. "Solution deposition-conversion for planar heterojunction mixed halide perovskite solar cells." *Adv. Energy Mater.* 4 (2014): 140035.
114. Malinkiewicz, O., Yella, A., Lee, Y.H., Espallargas, G.M., Gratzel, M., Nazeeruddin, M.K., and Bolink, H.J. "Perovskite solar cells employing organic-inorganic charge transport layers." *Nature Photonics* 8 (2014): 128–132.
115. Seo, G., Seo, J., Ryu, S., Yin, W., Ahn, T.K., and Seok, S. II. "Enhancing the performance of sensitized solar cells with $PbS/CH_3NH_3PbI_3$ core/shell quantum dots." *J. Phy. Chem. Lett.* 5 (2014): 2015–2020.
116. Etgar, L., Gao, P., Qin, P., Greatzel, M., and Nazeeruddin, M.K. "A hybrid lead iodide perovskite and lead sulfide QD heterojunction solar cell to obtain a panchromatic response." *J. Mater. Chem. A* 2 (2014): 11586–11590.

117. Niu, G., Li, W., Meng, F., Wang, L., Dong, H., and Qiu, Y. "Study on the stability of $CH_3NH_3PbI_3$ films and the effect of post-modification by aluminum oxide in all-solid-state hybrid solar cells." *J. Mater. Chem. A* 2 (2014): 705–710.
118. Lee, M.M., Teuscher, J., Miyasaka, T., Murakami, T.N., and Snaith, H.J. "Efficient hybrid solar cells based on Meso-superstructured organometal halide perovskites." *Science* 2 (2012): 643–647.
119. Ball, J.M., Lee, M.M., Hey, A., and Snaith, H.J. "Low-temperature processed meso-superstructured to thin-film perovskite solar cells." *Energy Environ. Sci.* 6 (2013): 1739–1743.
120. Carnie, M.J., Charbonneau, C., Barnes, P.R.F., Davies, M.L., Mabbett, I., Watson, T.M., O'Regan, B.C., and Worseley, D.A. "Ultra-fast sintered TiO_2 films in dye-sensitized solar cells: phase variation, electron transport and recombination." *J. Mater. Chem. A* 1 (2013): 2225–2230.
121. Jiang-Jian, S., Wan, D., Yu-Zhuan, X., Chun-Hui, L., Song-Tao, L., Li-Feng, Z., Juan, D., Yan-Hong, L., Qing-Bo, M., and Qiang, C. "Enhanced performance in perovskite lead iodide heterojunction solar cells with metal-insulator-semiconductor back contact." *Chinese Phys. Lett.* 30 (2013): 128402.
122. Zhang, W., Saliba, M., Stranks, S.D., Sun, Y., Shi, X., Wiesner, U., and Snaith, H.J. "Enhance of perovskite-based solar cells employing core-shell metal nanoparticles." *Nano Lett.* 13 (2013): 4505–4510.
123. Ogomi, Y., Morita, A., Tsukamoto, S., Saitho, T., Shen, Q., Toyoda, T., Yoshino, K., Pandey, S.S., Ma, T., and Hayase, S. "All-solid state perovskite solar cells with HOCO-R-NH_3+I- anchor group inserted between porous titania and perovskite." *J. Phys. Chem. C* 118 (2014): 16651–16659.
124. Logothetidis, S. "Flexible organic electronic devices: materials, process and applications." *Materials Science and Engineering: B* 152 (2008): 96–104.
125. Docampo, P., Ball, J.M., Darwich, M., Eperon, G.E., and Snaith, H.J. "Efficient organometal halide perovskite planar heterojunction solar cells on flexible polymer substrate." *Nature Communications* 4 (2013): 1–6.
126. Carmona, C.R., Malinkiewicz, O., Soriano, A., Espallargas, G.M., Garcia, A., Reinecke, P., Kroyer, T., Dar, M.I., Nazeeruddin, M.K., and Bolink, H.J. "Flexible high efficiency perovskite solar cells." *Energy Environ. Sci.* 7 (2014): 994–997.
127. You, J., Hong, Z., Yang, Y., Chen, Q., Cai, M., Song, T.B., Chen, C.C., Lu, S., Liu, Y., Zhou, H., and Yang, Y. "Low-temperature solution-processed perovskite solar cells with high efficiency and flexibility." *ACS Nano* 8 (2014): 1674–1680.
128. Chiang, Y.F., Jeng, J.Y., Lee, M.H., Peng, S.R., Chen, P., Guo, T.F., Wen, T.C., Hsu, Y.J., and Hsu, C.M. "High voltage and efficient bilayer heterojunction solar cells based on an organic-inorganic hybrid perovskite absorber with a low-cost flexible substrate." *Phys. Chem. Chem. Phys.* 16 (2014): 6033–6040.
129. Li, Z., Kulkarni, S.A., Boix, P.P., Shi, E., Cao, A., Fu, K., Batabyal, S.K., Zhang, J., Xiong, Q., Wong, L.H., Mathews, N., and Mhaisalkar, S.G. "Lamiated carbon nanotubes networks for metal electrode-free efficient perovskite solar cells." *ACS Nano* 8 (2014): 6797–6804.
130. Kumar, M.H., Yantara, N., Dharani, S., Graetzel, M., Mhaisalkar, S., Boix, P.P., and Mathews, N. "Flexible, low-temperature solution processed ZnO based perovskite solid state solar cells." *Chem. Comm.* 49 (2013): 11089–11091.
131. Even, J., Pedesseau, L., Jancu, J.M., and Katan, C. "DFT and k. p modeling of the phase transitions of lead and tin halide perovskites for photovoltaic cells." *Phys. Status Solidi RRL* 8 (2014): 31–35.

132. Chung, Lee, B., He, J., Chang, R.P.H., and Kanatzidis, M.G. "All solid state dye sensitized solar cells with high efficiency." *Nature* 485 (2012): 486–489.
133. Bach, U., Tachibana, Y., Moser, J.E., Haque, S.A., Durrant, J.R., Gratzel, M., and Klug, D.R. "Charge separation in solid-state dye-sensitized heterojunction solar cells." *J. Am. Chem. Soc.* 121 (1999): 7445–7446.
134. Zhang, Q., and Liu, X. "Dye-sensitized solar cell goes solid." *Small* 8 (2012): 3711–3713.
135. Hemant Kumar, M.H., Dharani, S., Leong, W.L., Boix, P.P., Prabhakar, R.R., Baike, T., Shi, C., Ding, H., Ramesh, R., Asta, M., Graetzel, M., Mhaisalkar, S.G., and Mathews, N., "Lead free halide perovskite solar cells with high photocurrents realized through vacancy modulation." *Adv. Mater.* 26 (2014): 7122–7127.
136. Lang, L., Yang, J.H., Liu, H.R., Xiang, H.J., and Gong, X.G. "First principles study on the electronic and optical properties of cubic ABX_3 halide perovskites." *Phys. Lett. A* 378 (2014): 290–293.
137. Mitzi, D.B. "Synthesis structure and properties of organic-inorganic perovskites and related materials." *Progress in Inorganic Chem.* 48 (1999): 1–121.
138. Borriello, Cantele, G., and Ninno, D. "Ab inito investigation of hybrid organic-inorganic perovskites based on tin halides." *Phys. Rev. B* 77 (2008): 235214.
139. Takahashi, Y., Obara, R., Nagakawa, K., Nakano, M., Tokita, J.Y., and Inabe, T. "Tunable charge transport in soluble organic-inorganic hybrid semiconductors." *Chem. Mater.* 19 (2007): 6312–6316.
140. Umari, P., Mosconi, E., and Angelis, F.D. "Relativistic GW calculations on $CH_3NH_3PbI_3$ and $CH_3NH_3SnI_3$ perovskites for solar cell applications." *Sci. Rep.* 4 (2014): 4467(1–7).
141. Stoumpos, C.C., Malliakas, C.D., and Kanatzidis, M.G. "Semiconducting Tin and lead iodide perovskites with organic cations: phase transitions, high mobilities, and near- infrared photoluminescent properties." *Inorg. Chem.* 52 (2013): 9019–9038.
142. Noel, N.K., Stranks, S.D., Abate, A., Wehrenfenning, C., Guarnera, S., Haghighirad, A.A., Sadhanala, A., Eperon, G.E., Pathak, S.K., Johnston, M.B., Petrozza, A., Herz, L.M., and Snaith, H.J. "Lead-free organic-inorganic tin halide perovskites for photovoltaic applications." *Energy Environ. Sci.* 7 (2014): 3061–3068.
143. Ogomi, Y., Morita, A., Tsukamoto, S., Saitho, T., Fujikawa, N., Shen, Q., Toyoda, T., Yoshino, K., Pandey, S.S., Ma, T., and Hayase, S. "$CH_3NH_3Sn_xPb(1\text{-}x)I_3$ perovskite solar cells covering upto 1060 nm." *J. Phys. Chem. Lett.* 5 (2014): 1004–1011.
144. Bi, D., Tress. W., Dar, M.I., Gao, P., Luo, J., Renevier, C., Schenk, K., Abate, A., Giordano, F., Baena, J-P.C., Decoppet. J-D., Zakeeruddin, S.M., Nazeeruddin, M.K., Gratzel. M., and Hagfeldt, A. "Efficient luminescent solar cells based on tailored mixed-cation perovskites." *Sci. Adv.* 2 (2016): e1501170.
145. Alsalloum, A.Y., Turedi, B., Almasabi, K., Zheng, X., Naphade, R., Stranks, S.D., Mohammad, O.F., and Bakr, O.M. "22.8%-efficient single-crystal mixed-cation inverted perovskite solar cells with a near-optimal bandgap." *Energy Environ. Sci.* 14 (2021): 2263–2268.
146. Wang, L., Li, Y., Gao, P., Xu, B., Lin, H., Li, X., and Miyasaka, T. "Artemisinin-passivated mixed-cation perovskite films for durable flexible perovskite solar cells with over 21% efficiency." *J. Mater. Chem. A* 9 (2021): 1574–1582.
147. Wu, R., Wang, C., Jiang, M., Liu, C., Liu, D., Li, S., Kong, Q., He, W., Zhan, C., Zhang, F., Liu, X., Yang, B., and Hu, W. "Progress in blade-coating method for perovskite solar cells toward commercialization." *J. Renew. Sust. Energy* 13 (2021): 012701.
148. Li, J., Dager, J., Shargaieva, O., Flatken, M. A., Kobler, H., Fenske, M., Schultz, C., Stegemann, B., Just, J., Tobbens, D.M., Abate, A., Munir, R., and Unger, E. "20.8%

slot-die coated $MAPbI_3$ perovskite solar cells by optimal DMSO-content and age of 2-ME based precursor Inks." *Adv. Energy Mater.* 11 (2021): 2003460.

149. Burkitt, D., Patidar, R., Greenwood, P., Hooper, K., McGettrick, J., Dimitrov, S., Colombo, M., Stoichkov, V., Richards, D., Deynon, D., Davies, M., and Watson, T. "Roll-to-roll slot-die coated P-I-N perovskite solar cells using acetonitrile based single step perovskite solvent system." *Sustain. Energy Fuels* 4 (2020): 3340–3351.
150. Zuo, C., and Ding, L. "Drop-casting to make efficient perovskite solar cells under high humidity." *Angew. Chem.* 60 (2021): 11242–11246.
151. Yin, X., Zong, Z., Li, Z., and Tang, W. "Toward ideal hole transport materials: a review on recent progress in dopant-free hole transport materials for fabricating efficient and stable perovskite solar cells." *Energy Environ. Sci.* 13 (2020): 4057–4086.
152. Rajagopal, A., Yang, Z., Jo, S.B., Braly, I.L., Liang, P-W., Hillhouse, H.W., and Jen, A.K.-Y. "Highly efficient perovskite-perovskite tandem solar cells reaching 80% of the theoretical limit in photovoltage." *Adv. Mater.* 29 (2017):1702140.
153. Bett, A.J., Schulze, P.S.C., Winkler, K.M., Kabakli, O.S., Ketterer, I., Mundt, L.E., Reichmuth, K.S., Siefer, G., Cojocaru, L., Tutsch, L., Bivor, M., Hermle, M., Glunz, S.W., and Goldschmidt, J.C. "Two-terminal perovskite silicon tandem solar cells with a high-bandgap perovskite absorber enabling voltages over 1.8%." *Progress in Photovoltaics* 28 (2020): 99–110.
154. Qarony, W., Hossain, M.I., Jovanov, V., Salleo, A., Knipp, D., and Tsang, Y.H. "Influence of perovskite interface morphology on the photon management in perovskite/silicon tandem solar cells." *ACS Appl. Mater. Interfaces* 12 (2020): 15080–15086.
155. Chen, B., Rudd, P.N., Yang, S., Yuan, Y., and Huang, J. "Imperfections and their passivation in halide perovskite solar cells." *Chem. Soc. Rev.* 48 (2019): 3842–3867.
156. Yang, D., Zhang, X., Hou, Y., Wang, K., Ye, T., Yoon, J., Wu, C., Sanghadasa, M., Liu, S., and Pirya, S. "28.3%-efficiency perovskite/silicon tandem solar cell by transparent electrode for high efficient semitransparent top cell." *Nano Energy* 84 (2021): 105934.
157. Shen, H., Duong, T., Peng, J., Jacobs, D., Wu, N., Gong, J., Wu, Y., Karuturi, S. K., Fu, X., Weber, K., Xiao, X., White, T.P., and Catchpole, K. "Mechanically-stacked perovskite/CIGS tandem solar cells with efficiency of 23.9% and reduced oxygen sensitivity." *Energy Environ. Sci.* 11 (2018): 394–406.
158. Lee, J-W., Hsieh, Y-T., Marco, N., Bae, S-H., Han, Q., and Yang, Y. "Halide perovskites for tandem solar cells." *J. Phys. Chem. Lett.* 8 (2017): 1999–2011.
159. Li, H., and Zhang, W. "Perovskite tandem solar cells: From fundamentals to commercial deployment." *Chem. Rev.* 120 (2020): 9835–9950.
160. Heo, J.H., and Im, S.H. "$CH_3NH_3PbBr_3$-$CH_3NH_3PbI_3$ perovskite-perovskite tandem solar cells with exceeding 2.2 V open circuit voltage." *Adv. Mater.* 28 (2015): 5121–5125.
161. Jiang, Y., and Qi, Y. "Metal halide perovskite-based flexible tandem solar cells: next-generation flexible photovoltaic technology." *Mater. Chem. Front.* (2021). DOI: 10.1039/D1QM00279A.
162. Wu, M-C., Lin, T-H., Chan, S-H., Liao, Y-H., and Chang, Y-H. "Enhanced photovoltaic performance of perovskite solar cells by tuning Alkaline earth metal-doped perovskite-structured absorber and metal-doped TiO_2 hole blocking layer." *ACS Appl. Energy Mater.* 1 (2018): 4849–4859.
163. Chan, S-H., Wu, M-C., Lee, K-M., Chen, W-C., Lin, T-H., and Su, W-F. "Enhancing perovskite solar cell performance and stability by doping barium in methylammonium lead halide." *J. Mater. Chem. A*, 5 (2017): 18044–18052.

164. Kakavelakis, G., Petridis, K., and Kymakis, E. "Recent advances in plasmonic metal and rare-earth element upconversion nanoparticle doped perovskite solar cells." *J. Mater. Chem. A* 5 (2017): 21604–21624.
165. Karunakaran, S.K., Arumugam, G.M., Yang, W., Ge, S., Khan, S.N., Lin, X., and Yang, G. "Research progress on the application of lanthanide-ion-doped phosphor materials in perovskite solar cells." *ACS Sustainable Chem. Eng.* 9 (2021): 1035–1060.
166. Rudd, P.N., and Huang, J. "Metal ions in halide perovskite materials and devices." *Trends in Chemistry* 1 (2019): 394–409.
167. Chang, J., Lin, Z., Zhu, H., Isikgor, F.H., Xu, Q-H., Zhang, C., Hao, Y., and Ouyang, J. "Enhancing the photovoltaic performance of planar heterojunction perovskite solar cells by doping the perovskite layer with alkali metal ions." *J. Mater. Chem. A* 4 (2016): 16546–16552.
168. Nwankwo, U., Ngqoloda, S., Nkele, A.C., Arendse, C.J., Ozoemena, K.I., Ekwealor, A.B.C., Jose, R., Maaza, M., and Ezema, F.I. "Effects of alkali and transition metal-doped TiO2 hole blocking layers on the perovskite solar cells obtained by a two-step sequential deposition method in air and under vacuum." *RSC Adv* 10 (2020): 13139–13148.
169. Wu, T., Zhen, C., Zhu, H., Wu, J., Jia, C., Wang, L., Liu, G., Park, N-G., and Chen, H-M. "Gradient Sn-doped heteroepitaxial film of faceted rutile TiO_2 as an electron selective layer for efficient perovskite solar cells." *ACS Appl. Mater. Interfaces* 11 (2019): 19638–19646.
170. Kim, J.Y., Rhee, S., Lee, H., An, K., Biswas, S., Lee, Y., Shim, J.W., Lee, C., and Kim, H. "Universal elaboration of Al-doped TiO_2 as an electron extraction layer in Inorganic-organic hybrid perovskite and organic solar cells." *Adv. Mater. Interfaces* 7 (2020): 1902003.
171. Meng, X., Cai, Z., Zhang, Y., Hu, X., Xing, Z., Huang, Z., Huang, Z., Cui, Y., Hu, T., Su, M., Liao, X., Zhang, L., Wang, F., Song, Y., and Chen, Y. "Bio-inspired vertebral design for scalable and flexible perovskite solar cells." *Nat. Comm.* 11 (2020): 1–10.
172. Augustine, B., Remes, K., Lorite, G.S., Varghese, J., and Fabritius, T. "Recycling perovskite solar cells through inexpensive quality recovery and reuse of patterned indium tin oxide and substrates from expired devices by single solvent treatment." *Sol. Energy Mater Sol. Cell.* 194 (2019): 74–82.
173. Chhillar, P., Dhamaniya, B.P., Dutta, V., and Pathak, S.K. "Recycling of perovskite films: Route toward cost-efficient and environment-friendly perovskite technology." *ACS Omega* 4 (2019): 11880–11887.
174. Fong, P.W-K., Hu, H., Ren, Z., Liu, K., Cui, L., Bi, T., Liang, Q., Wu, Z., Hao, J., and Li, G. "Printing high-efficiency perovskites solar cells in high-humidity ambient environment-An In Situ guided Investigation." *Adv. Sci.* 8 (2021): 2003359.
175. Mahapatra, A., Prochowicz, D., Tavakoli, M.M., Trivedi, S., Kumar, P., and Yadav, P. "A review of aspects of additive engineering in perovskite solar cells." *J. Mater. Chem. A* 8 (2020): 27–54.
176. Rahmany, S., and Etgar, L. "Semitransparent perovskite solar cells." *ACS Energy Lett.* 5 (2020): 1519–1531.
177. Li, B., Chang, B., Pan, L., Li, Z., Fu, L., He, Z., and Yin, L. "Tin-based defects and passivation strategies in Tin-related perovskite solar cells." *ACS Energy Lett.* 5 (2020): 3752–3772.

18 Energy Storage Systems in View of Nanotechnology towards Wind Energy Penetration in Distribution Generation Environment

Dimpy Sood
College of Technology and Engineering, Udaipur, Rajasthan, India

Ritesh Tirole
Department of Electrical Engineering, Sir Padampat Singhania University, Udaipur, Rajasthan, India

Sujit Kumar
Department of Electrical and Electronics Engineering, Jain (Deemed-To-Be-University), Bengaluru, Karnataka, India

18.1 INTRODUCTION

Energy consumption has increased significantly over the last three decades with the surge in industries and population growth. India's power demand is increasing at the rate of 6% per year. The Government's Policy Wing, Niti Aayog, has speculated that the energy demand of India is expected to be magnified 3.5 to 4.5 times between 2020 and 2040. In recent years, renewable power generation capacity of India has grown from 85 GW to 560 GW at a growth rate of about 600%, and globally it grew from 1000 GW to 2195 GW. Globally, the capacity of total renewable energy generation has reached 2351 GW, which is one third of total installed electricity capacity. India has been working on the largest Global Green Energy Programme [1]. This is one of the largest renewable energy-producing countries, using various renewable

DOI: 10.1201/9781003220350-18

sources. India has committed that by the end of 2026–2027, 40% of its installed power-generating capacity will be non-fossil fuel sources, and it is projected to get 56% from clean energy sources by 2030, which is currently at 20% in the present fiscal year. Figure 18.1 depicts India's total installed power capacity.

Among various renewable energy sources, solar and wind has a major contribution in fulfilling the demand of the world's renewable electricity consumption. Out of total renewable energy generation capacity of about 25.1%, contribution of wind energy is at 4.8%, which is the second highest after hydropower energy, as shown in Figure 18.2.

Wind energy, generated from wind power plants, is the most efficient and viable solution in places where adequate wind potential is available to harness wind energy. Moreover, wind power plants do not produce greenhouse gases. Thus, energy from renewable energy sources is also termed clean energy. Unlike other renewable sources like solar and tidal, however, wind is unpredictable as

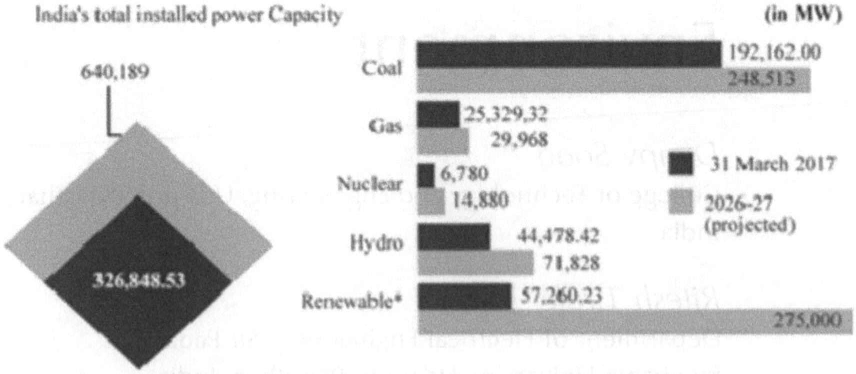

FIGURE 18.1 Installed power capacity of India.

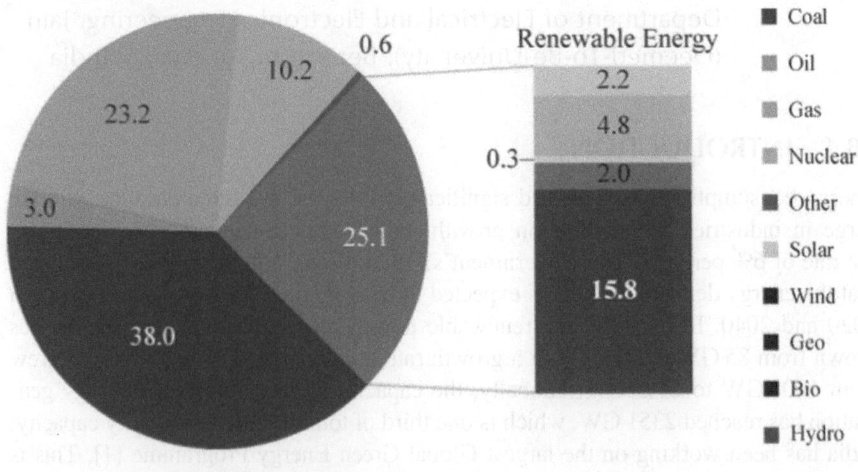

FIGURE 18.2 Wind energy capacity.

well as uncontrollable. This irregular nature of wind increases the technical difficulty in ensuring good quality power, threatens the stability of the power supply and challenges the reliability of the power grid [2–6].

The main challenge in increasing penetration of wind generation into the existing power system network is to manage the uncertain amount of power generated at high wind speeds than at lower ones. The prescribed range of wind speeds for a turbine is called productive wind speed, which normally ranges between 6 m/sec and 25 m/sec. The optimal speed prescribed for a large-scale wind farm is about 9 m/s. The output power of the turbine is cubically proportional to the wind speed. So, with every small variation in the wind speed, the output power from the wind turbine changes significantly and directly affects the power quality of the wind generation system.

Power quality is defined as the power that enables proper working of the equipment. If the power quality is inadequate, it may experience voltage swings, sag, or swell. Various causes and effects of having poor power quality are listed in Table 18.1.

An electrical power system can be utilized to its full potential by using an energy storage system (ESS). The ESS reduces the quantity of main fuel consumed. The ESS also works as an alternative buffer source and provides better security along with enhanced quality of power supplied. By using an ESS, the excess amount of electricity generated due to the invariable wind speed can be stored and released as an additional power support to the grid at the time of contingencies. An ideal ESS is capable of handling exigency situations like wind gusts and varying load demands arising for a few seconds to minutes or even longer. The ESS also helps to enhance the efficiency of the power system by reducing the amount of primary fuel used. This application can strengthen power networks, maintain load levels even during critical service hours and avoid stability issues [7–9].

TABLE 18.1
Causes and Effect of Various Power Quality Issues

Power Quality Issues	Causes	Effects
Voltage Fluctuations	Switching load variations	Over-voltages, flickers and lighting issues
Voltage Sag	System fault, high starting load, excessive load on the network, variation in the source voltage, high inrush current	Overloading issues, erratic lock-up
Voltage Swell	Sudden load variations, variations in the voltage source, high inrush current, wiring issues	Loss of data, damaged equipment, irregular lock-up, grabbling in data
Long-Term Voltage Interruption	Failure of protecting devices, insulation failure and malfunctioning of control system	Malfunctioning in data processing of the equipment

The International Energy Association (IEA) estimated that the total storage capacity requirement of the world, which is currently at 178 GW, will be around 267 GW by the end of 2018. Various energy storage techniques available these days to fulfil the rising energy demand and to provide a smooth power output for each time interval are as follows: batteries, super capacitors, flywheel energy storage (FES), pumped hydropower storage (PHS) and compressed air energy storage (CAES) systems.

The ESS on the basis of form of energy storage can be categorized into mechanical type (PHS, CAES and FES), electrochemical type (flow and rechargeable batteries), electrical type (super capacitor and superconducting magnetic energy storage [SMES]), thermochemical type (solar fuels), chemical type (hydrogen storage along with fuel cells) and thermal type energy storage, as shown in Table 18.2.

The working parameters, such as energy, power, life span, response time and cost of various ESSs, are shown in Table 18.3.

PHS (also known as mechanical storage) is now used in the storage sector for its capacity of approximately 127,000 MW. The utilization rate is nearly 99% of the total storage deployment to date, which is followed by CAES, which has a capacity of 440 MW; sodium sulphur, which has a capacity of 316 MW; lead acid, which has a capacity of 35 MW; nickel–cadmium, which has a capacity of 27 MW; and FES, lithium-ion and flow batteries, all of which have 25 MW, 20 MW, and 3 MW storage capacity, respectively.

Enormous forms of storing energy are available. The detailed description and small comparison based on the strengths and weaknesses of these storage systems are described further.

TABLE 18.2
ESS with Different forms of Stored Energy

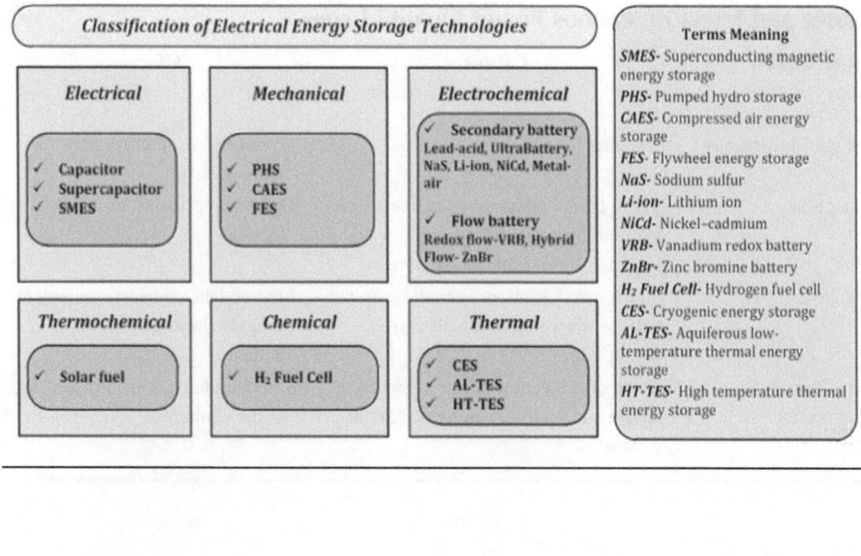

TABLE 18.3
Energy Storage Systems Based on Working Parameter

Type of ESS	Energy Density	Power Density	Life Span	Time of Response	Estimated Cost
Chemical Battery	High	Low	Short	Medium	Low
Sodium-Sulphur Battery	Medium	Low	Short	Slow	Medium
Flywheel	Low	High	Long	Fast	High
Super Capacitor	Low	High	Long	Fast	Medium
SMES	Medium	High	Long	Fast	High

18.2 PUMPED HYDROPOWER STORAGE

The world's leading type of energy storage technology is PHS, which can store a tremendous amount of energy for extended periods. These storage plants use gravitational force for electricity generation. At the time of high renewable energy generation, water is stored at a higher elevation and released to fall with the help of earth's gravity to generate electricity at the time of additional energy requirements. Recent technological advancements have made the PSH capable of facilitating adjustable speeds according to the grid demand and operating in a closed-loop system. A closed-loop PSH does not require continuously flowing water supply, so it can be installed in all types of locations. The installation of PSH can be above-ground, which is called conventional, or it can be underground, per the requirements of the specific location [10].

If the quality of the PHS is very good, it may be used for regulating power and stabilizing the frequency. Cheaper and more efficient than the other energy storage methods, PSH systems offer substantial capacity. This storage device provides 10 hours of energy, which is approximately 6 hours less than the amount of electricity from lithium-ion batteries (LIBs) but lasts for an extremely long time and retains 80% of its original capacity. The highest installed capacities of PSH are found in the United States, where 21.8 GW are installed, as well as in Japan, where 24.6 GW are in use. Countries such as Spain, with a much lower installation capacity of 5.3 MW, round out the list. Over 300 PHS systems are in place in many locations throughout the globe (Figure 18.3).

The main drawbacks of PHS are its high capital cost, long development time, long payback periods and uncertain profitability.

18.3 COMPRESSED AIR ENERGY STORAGE

The second largest and greatest rival to PHS is CAES. The CAES system utilizes compression, storage and expansion as three main processing steps. With CAES, air is pumped into an underground hole, like a salt cavern, during off-peak hours. During peak load demand, the air from the underground cave is released back up

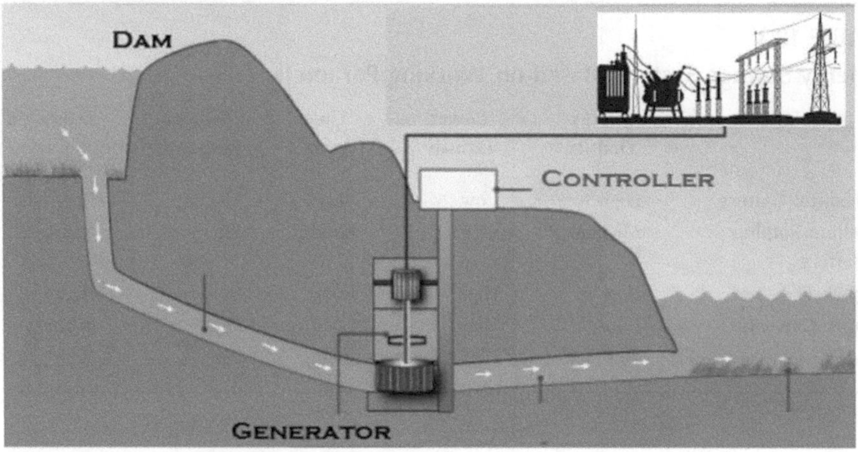

FIGURE 18.3 Pumped storage plant.

into the system, where this compressed air is pre-heated, mixed with natural gas, and ignited through a gas turbine, thus turning on the electric generator to produce electrical energy. This heating process using natural gas releases carbon, which increases the output efficiency of CAES by almost three times [11].

CAES also has many advantages similar to PHS like high reliability, flexibility and longer life but at moderately small operation and preservation costs and has low self-discharge rates. It can be used for long as well as for short durations to provide stored energy at many levels.

The are two operating CAES plants. One is in McIntosh, Alabama, built in 1991 with an installed capacity of 110 MW, and the other one is in Huntorf, Germany. An Anderson County, Texas, plant with a capacity of 317 MW is under construction.

18.4 FLYWHEEL ENERGY STORAGE

Storage systems in this category hold energy in a spinning mass. The basic parts of an FES system are a flywheel, a bearing, an electrical machine, a power converter and a containment chamber. These storage devices are highly efficient for short-term energy needs and have very high discharge capability independent of temperature. The storage systems need little monitoring of the battery state of charge and are highly resistant to environmental factors. High power density and excellent energy density are attributes of FES. The systems provide a fast response operation with very low degradation. Thus, they have a long life, have high scalability and need no periodic maintenance. The two main drawbacks of FES systems are high self-discharge rate and safety issues.

The lower cost of materials and a lower speed flywheel drive system price allows lower speed FES systems to be about five times cheaper than higher speed versions. The operating range of a high-speed FES system falls within the 8000 to 9000 rpm regime. The higher speed levels of the flywheel exhibit quadratic improvements and better energy-generating capacity of the system; there are seldom times when flywheel speeds are reduced to below 50% of the maximum speed levels of the flywheel [12, 13].

Load-balancing and load-shifting applications are the ideal uses for FES systems, although they are not appropriate for long-term storage. When it comes to extended life duration, high energy density, cheap maintenance costs and fast reaction rates, flywheels are renowned for their use. Motors rotate the flywheel at a very high speed of about 50,000 rpm and store energy. This stored energy is later used to generate electricity by rotating the flywheel in the opposite direction. The rotation of the flywheel is reduced by placing the wheel in a vacuum to lessen air resistance, which would make the wheel sluggish.

18.5 CHEMICAL ENERGY STORAGE

Chemical energy is basically stored in bonds that connect atoms and molecules with other atoms and molecules. It is stored in the form of potential energy and is released when a chemical reaction takes place. Chemical energy storage has the greatest range of research so far. Devices that possess chemical energy storage include traditional batteries, metal or metal-air batteries, fuel cells and flow batteries.

18.5.1 Chemical Battery

The most commonly utilized forms of chemical energy storage are lead acid, alkaline zinc manganese dioxide, lithium-ion and nickel-zinc batteries. The most important factor when it comes to battery performance is the material of the electrodes; other factors that influence battery performance include the condition of the electrolyte–electrode interface and the steadiness of the electrode. Because of this, most of these batteries are susceptible to changes in temperature and have a capacity that decreases according to both ambient and charge-discharge conditions.

18.5.1.1 Usage of Nanotechnology in Liquefied Electrolytes

Battery performance is affected by a number of factors, many of which are interrelated. The required chemical stability of the electrolyte must be maintained, but the battery must also be able to produce the SEI layer. This should not cause more adverse effects, and should just provide the protecting layer. All extra deposited layers on the electrodes increase the total cell resistance. Additionally, no additives, nanomaterials included, are allowed to introduce any hazardous or destructive activities while the battery is running. New cell chemistries should be evaluated under various circumstances for months or years to see whether they have had any benefit. It often occurs that hopeful cell chemistry succeeds for a brief period of time, only to fail spectacularly once expectations for commercialization are met [14].

The quest for better battery chemistries, for example, includes the hunt for:

1. Huge electrochemical firmness
2. Great thermal steadiness
3. Extensive functioning voltage assortment
4. Extensive functioning temperature variety

5. Short vapor compression
6. Great conductivity
7. Great capability
8. Extended storage life cycle
9. Extended sequence life span
10. General security and misuse lenience
11. Cheap cost

This is because of the wide range of electrolytes and chemistries available. The above items may be addressed to the greatest extent by using ionic liquids, with the exception of price and, in many instances, low temperature performance. It is sometimes possible to enhance conductivity by adding nanoparticulate solid powders of Al2O3, TiO2 and ZrO2, particularly in electrolyte solutions. This is because free ions and ion pairs have changed their balance, as a result of differing physicochemical interactions. The conductivity doubled within the volume fraction range of 0.2% to 0.5% of the solution [15]. The addition of nanomaterials to the electrolyte–electrode interface and in the electrode enhances charge transport, both within the electrolyte and in the electrode. This provides increased characteristics for both the electrolyte and the electrode. Electrolytes feverously affect just the liquid phase; nevertheless, there are a multitude of advantages in the whole cell system, particularly in the electrodes [15].

18.5.2 Fuel Cells

Fuel cells are very smilar to metal-air batteries, but there are widespread technological developments when they are used as an alternative generating source. Fuel cells or electrochemical cells use an external supply of fuel and an oxidizing agent to transform chemical energy of the fuel into electrical energy. Fuel cells use natural gas, ammonia or hydrocarbon gases as the fuel to produce electricity either directly or by converting these gases to hydrogen-rich gases first and then to electrical output. The life span and cost of the fuel cell on a commercial scale are its major drawbacks.

18.5.3 Flow Batteries

An alternative to LIBs are flow batteries. Flow batteries have a small market share (representing only 10% of the market) and are utilized in numerous long-term energy storage projects. Flow battery storage is used at the Avista Utilities facility in Washington State. A flow battery in Dalian, China, with a capacity of 200 MW (800 MWh) is under construction. As well as being the world's largest flow battery installation, this system will also be the first large-scale battery (with a capacity more than 100 MW) constructed out of flow batteries.

18.5.3.1 Nanostructured Scheme and Mixture of Cathode Constituents for Li Batteries

There are many cathode-limited LIBs; therefore, in order to improve them, researchers must develop cathode materials to increase performance. New and varied methods to develop and create cathode materials may be obtained through the use of nanotechnology. We describe several nanotechnology-enabled LIB cathode-manufacturing techniques in this section. Although most methods used for nano-synthesis of cathode materials include a bottom-up approach, which usually ends with one-dimensional or zero-dimensional cathode nanostructures, a few strategies may be categorized as a top-down approach since they use polyhydric nanostructures. Generally, liquid-phase techniques are used, but solid-state approaches are attempted from time to time. The following techniques have shown to be helpful for the creation of cathode materials: template-based approaches, solvo-thermal/hydrothermal approaches, co-precipitation approaches and solid-state approaches. We therefore address these methods by considering discharge capacity, retention capacity and the pace at which nanostructured cathode materials are able to conduct electrons.

Use of nanotechnology is recommended when you need a small, controllable, porous or hierarchical structure that has a nanostructure in it. An article authored by Vu et al. used the colloidal crystal technique to manufacture a 3D, hierarchically porous LiFePO4–carbon (LFP/C) composite cathode material [16]. To get both meso- and macropores in the structure, they utilized a dual templating process, poly (ethylene oxide)-poly (propylene oxide)-poly (ethylene oxide) triblock copolymer from BASF (PEO106PPO70PEO106). They have utilized poly (methyl methacrylate) (PMMA) for macropores, as well as colloidal crystals. LFP and carbon as well as phenol-formaldehyde sol and a nonionic surfactant are all mixed together using a multiconstituent synthesis technique, which uses a colloidal crystal template. An LFP/C monolithic composite is obtained when several heat treatments are performed at low ramp rates, followed by a final pyrolysis at high temperature (600, 700 or 800°C) that provides a three-dimensional LFP/C macroporous and meso-/microporous (3DOM/m) structure.

18.6 THERMAL STORAGE SYSTEM

Thermal energy storage devices use rocks, salts, water or other materials to store energy. When additional energy is available, they are maintained heated, kept in an insulated area and cooled down as needed. A small volume of cold water is poured over warm rocks, salts or hot water to create steam, which is then utilized to spin steam turbines in order to generate electricity. In addition to serving as a renewable energy source, thermal energy storage may also be utilized to heat and cool buildings rather than to generate electricity. Ice may be made overnight and then utilized to cool a structure during the day by using thermal storage. The efficiency range for thermal energy may be anywhere between 50% and 90% depending on the kind of energy utilized.

18.7 HYBRID ENERGY STORAGE SYSTEM

In recent times, the most relevant topic of research is to combine ESS with a wind energy generation system to enable smooth and high-quality power output. To achieve the distinctive characteristics of the storage system, a combination of two

TABLE 18.4
Parameter comparison between Battery and SC

Function	Super Capacitor	Lithium-Ion
Charging time	5-20 seconds	20-80 minutes
Voltage of each cell	2.6 to 2.8 V	3.5-3.8 V
Specific energy specific power	6 Wh/kg	110-220
	Up to 10000 W/hr	1000 to 3000 W/hr
Life span	12 to 18 years	6 to 12 years
Cost	Less	More than SC

different energy storage devices has been proposed. In this combination, one of the devices possesses high power density, and the other one possesses large energy density, to enhance power stability of the system irrespective of the wind speed variations.

Performance comparisons between super capacitors and LIBs are shown in Table 18.4.

Due to the intermittent nature of wind, the output power of the system gets flickers and sometimes long-term voltage fluctuations. This disturbed and poor quality of power can distort many of the electrical as well as electronic equipment that requires continous and stable power to operate. This requiremenrt demands a storage system with both high-power response as well as large energy storing capacity. However, it is very difficult for a single ESS to possess both of these qualities. So, a hybrid ESS with a combination of batteries and a super capacitor has been suggested. A battery ESS with high energy density provides stability for long-term voltage fluctuations, whereas a supercapacitor with large power density provides the bulk of the power for deep and short duration power fluctuations or flickers. There are many options to submerge short-term voltage fluctuations like SMES, FES and supercapacitor storage system (SCSS). Among all these options, SCSS seems to be the most effective perspective because of its lower cost and lower environmental impact. This combination helps to provide a firm power dispatch.

The installation of large power batteries to store energy increases the per-unit cost of energy stored, which is not feasible from an economic point of view. In an ideal hybrid ESS, the super capacitor compensates for the fast-occurring short-term voltage fluctuations or deep voltage sags, whereas the battery handles only slowly occurring and long-term voltage fluctuations to dispatch a stable power into the grid. So, from an economic as well as an operational point of view, hybrid ESSs are more efficient compared to single ESSs.

18.7.1 Life Span Valuations and Battery-Operated Nanotechnology

Using nanomaterials in the area of traction LIBs is seen as favorable because they provide both mechanical and electrical characteristics in a compact package. Currently, most of the research is focused on developing improved battery anodes utilizing nanomaterials, working with new materials to enhance the battery's total energy density [17, 18]. Until now, assessing the benefit of Energy Based Nanomaterials in batteries has been challenging. Nanomaterials offer significant technological benefits in batteries, such as increased charge density, higher current ability and large surface area per mass, rapid charge/discharge time cycles, and the like. Recent Life Cycle Assessment (LCA) studies have shown that the use of battery nanomaterials in the manufacturing process has negative consequences. Material and energy intensity, as well as the use of hazardous compounds, are major environmental issues when it comes to nanoscale processes used in various applications [19].

To evaluate the energy trade-offs involved in using nano-enabled LIBs, Wender and Seager [20] used LCA. The evaluation is based on laboratory-scale data from a laser vaporization production technique that has been used to produce single wall carbon nanotubes (SWCNT). The authors discussed the diverse materials and energy inputs as well as SWCNT manufacturing waste output while providing a breakdown of the per-kilowatt hour of battery energy storage capacity. It was found that two orders of magnitude more energy is required to make freestanding SWCNT anodes than is used by all other LIB production methods. It is anticipated that additional nano-manufacturing methods would provide similar results. Accordingly, in the use of SWCNTs in LIBs for cars, the author [21] completed an environmental study. An SWCNT electrode uses considerably more energy than a typical graphite electrode, as generated in a laboratory setting. Both studies alluded to the opportunities, significance and issues with future commercial SWCNT manufacturing.

More recently, Li et al. [17] conducted research on SiNW (silicon nanowire) that is used in LIB anodes. This included investigation of the entire life cycle of the novel battery, from the manufacturing of silicon nanowires and nanoparticle emissions during SiNW synthesis through characterization and reporting of nanoparticles and nanowastes emitted from the final product. Over 50% of global warming, human toxicity (HT) and photochemical oxidation potential are due to battery usage. In the case of the life cycle effects, 15, 10 and 17% of the manufacture of the SiNW anode impact contributes to 15, 10 and 17% of overall life cycle impacts, respectively. In the majority of cases, the decrease in overall life expectancy is due to the huge quantities of embedded energy and hazardous substances. Using ordinary graphite anode-based batteries as a reference, SiNW anode batteries exhibit a 6% and a 43% greater environmental effect in the categories studied. Most importantly, when the SiNW materials are made using hazardous chemicals (e.g. HF and HNO3), the different in the HT category is established. Based on lab-scale inventory data uncertainties and the scale-up in production levels, the authors conclude that the difference in effect produced by the two battery types is modest [18].

18.8 CONCLUSION

The hybrid ESS ascertains to be the best in comparison to the other storage systems. The combination of two storage devices completely fulfills the storage demand of the system for both long and short durations. The SCSS, with high power density and low energy density, compensates for fast-occurring short-term voltage fluctuations or deep voltage sags, whereas the battery ESS with high energy density and low power density handles only slowly occurring and long-term voltage fluctuations to dispatch a stable power into the grid.

REFERENCES

1. A. Abrantes. "Overview of power quality aspects in wind generation," in: North American Power Symposium (NAPS), Sept. 2012, pp.1–6.
2. Katsuhisa Yoshimoto, Toshiya Nanahara, Gentaro Koshimizu, Yoshihsa Uchida. "New control method for regulating state-of-charge of a battery in hybrid wind power/battery energy storage system", in: 2006 IEEE PES Power Systems Conference and Exposition, 2006, pp. 1244–1251.
3. S. Teleke, M.E. Baran, S. Bhattacharya, and A.Q. Huang. "Rule-based control of battery energy storage for dispatching intermittent renewable sources," *IEEE Trans. Sustain. Energy*, 1, no. 3 (Oct. 2010): 117–124.
4. Chen Z., Hu Y., Blaabjerg F. "Stability improvement of induction generator-based wind turbine systems." *IET. Renew. Power. Gener.* 1, no. 1 (2007): 81–93
5. P.F. Ribeiro, B.K. Johnson, M.L. Crow, A. Arsoy, and Y. Liu. "Energy storage systems for advanced power applications." *Proc. of the IEEE*, 89 (Dec. 2001): 1744–1756
6. M. Beaudin, H. Zareipour, A. Schellenberglabe, and W. Rosehart. "Energy storage for mitigating the variability of renewable electricity sources: an updated review." *Energy. Sustain. Dev.* 14 (2010): 302–314.
7. C. Abbey and G. Joos. "Supercapacitor energy storage for wind energy applications." *IEEE Trans. Ind. Appln.*, 49, no. 4 (2013): 1649–1657.
8. L. Qu and W. Qiao. "Constant power control of DFIG wind turbines with supercapacitor energy storage." *IEEE Trans. Ind. Appl.*, 47 (Jan.-Feb. 2011): 359–367.
9. Q. Li, S.S. Choi, Y. Yuan, and D.L. Yao. "On the Determination of Battery Energy Storage Capacity and Short-Term Power Dispatch of a Wind Farm." *IEEE Trans Sustain Energy*, 2 (Aprl. 2011): 148–158.
10. A. Abedini and H. Nikkhajoei. "Dynamic Model and Control of a Wind-turbine Generator with Energy Storage." *IET Renew. Power Gen.*, 5 (Jan. 2011): 67–78.
11. D.J. Swider. "Compressed air energy storage in an electricity system with significant wind power generation." *IEEE Trans. Energy Convers.*, 22, no. 1 (Mar. 2011): 95–102.
12. M.H. Ali, J. Tamura, and B. Wu. "SMES strategy to minimize frequency fluctuations of wind generator system." in Proc. IECON, 2008, pp. 3382–3387.
13. J. Kondoh, I. Ishii, H. Yamaguchi, A. Murata, K. Otani, K. Sakuta, N. Higuchi, S. Sekine, and M. Kamimoto. "Electrical energy storage systems for energy networks." *Energy Convers. Manag.*, 41, no. 17 (Nov. 2000): 1863–1874.
14. J. Salminen, T. Kallio, N. Omar, P.V.D. Bossche, J.V. Mierlo, and H. Gualous. "Transport energy—lithium ion batteries." L.M., Rodriguez-Martinez and N., Omar *Future Energy: Improved, Sustainable and Clean Options for our Planet*, second ed., pp. 292–307. Elsevier, London, UK, 2013.
15. P. Bruce, B. Scrosati, and J.M. Tarascon. "Nanomaterials for rechargeable lithium batteries." *Angew. Chem. Int. Ed.* 47 (2008): 2930–2946.
16. A. Vu, and A. Stein. "Multiconstituent synthesis of LiFePO4/C composites with hierarchical porosity as cathode materials for lithium-ion batteries." *Chem. Mater.* 23 (2011): 3237–3245.

17. B. Li, X. Gao, J. Li, and C. Yuan. "Life cycle environmental impact of high-capacity lithium ion battery with silicon nanowires anode for electric vehicles." *Environ. Sci. Technol.* 48 (2014): 3047–3055.
18. G. Majeau-Bettez, T.R. Hawkins, and A.H. Stromman. "Life cycle environmental assessment of lithium-ion and nickel metal hydride batteries for plug-in hybrid and battery electric vehicles." *Env. Sci. Technol.* 45 (2011): 4548–4554.
19. T.P. Seager, R.P. Raffaelle, and B.J. Landi. "Sources of variability and uncertainty in LCA of single wall carbon nanotubes for Li-ion batteries in electric vehicles." in: 2008 IEEE International Symposium on Electronics and the Environment, IEEE, San Francisco, CA, USA, May 19–22, 2008, pp. 1–5.
20. B.A. Wender, and T.P. Seager, "Towards prospective life cycle assessment: single-wall carbon nanotubes for lithium-ion batteries." in: 2011 IEEE International Symposium on Sustainable Systems and Technology, IEEE, Chicago, IL, USA, May 16–18, 2011, pp. 1–4.
21. S. Amarakoon, J. Smith, and B. Segal. *Application of Life-Cycle Assessment to Nanoscale Technology: Lithium Ion Batteries for Electric Vehicles*, pp. 1–119. U.S. Environmental Protection Agency, 2013.

Index